The Earth: Our Physical Environment

7:30 A.M.

10:30 A.M.

NOON

3:30 P.M.

7:30 P.M.

WILLIAM L. DONN

Professor of Earth and Planetary Sciences
The City College of New York

Senior Research Scientist
Lamont-Doherty Geological Observatory
of Columbia University

The Earth: Our Physical Environment

JOHN WILEY & SONS, INC.   New York · London · Sydney · Toronto

Library of Congress Catalogue Card Number: 79-37431

ISBN 0-471-21785-9

Printed in the United States of America.

10 9 8 7 6 5 4 3 2 1

or speak to the earth, and it shall teach thee

Book of Job, chapter 12

For Renée, Matthew and Tara

Preface

Writing a general book about the earth and its properties is immensely difficult and is a task full of compromises. Many major areas of science are involved and must be reasonably integrated. If the textbook is to be covered in one course, or at least covered with the possible omission of certain sections, it must not be encyclopedic or too long. If it emphasizes either the "discovery" or historical aspects, there will hardly be space for setting the basic foundation of the subject. Also, no author can be a scholar in all of the areas covered, but he must be reasonably knowledgeable and proficient. He certainly should convey the spirit of science in at least the areas in which he has a particular expertise.

Courses in earth science have become very diverse in approach and content. Different instructors and different schools focus their courses on various aspects of the subject. But a textbook must provide a sound organization of principles, ideas, and facts. It cannot sway with trends and styles and still meet the needs of a broad audience. With a sound textbook to draw from, individual instructors can expand or modify particular areas.

In this book I have emphasized processes and principles instead of facts, realizing that facts provide the framework for the development of processes and principles. Where appropriate to the subject matter, the "discovery" approach has been used, as in Chapter 8 on continental drift and sea floor spreading. Parts of Chapter 10 (earthquakes), and others, are treated similarly.

Basically, I have avoided summary types of chapters. I believe that an adequate understanding on even the elementary level requires a certain depth rather than a spread of knowledge. As a result, some topics included in other texts have necessarily been omitted. A major departure from previous works on earth science and geology has been a decrease of emphasis on land forms and erosional processes. Although this subject still forms the largest chapter in the book, the study of the earth has progressed far from the time when our knowledge was limited to surface geology. Other areas have therefore been given approximate parity with surface geology.

I have tried to weave a thread of continuity from chapter to chapter so that the entire subject is properly integrated, and have avoided setting up compartmented groups of chapters. At the same time, particular chapters or groups of chapters can stand alone. Depending on the nature of a particular course, certain chapters or groups of them can be omitted or rearranged without affecting the logic. The organization of chapters, although logical, is not necessarily sacrosanct. For example, some instructors may prefer to place Chapter 10 (earthquakes) before Chapters 8 and 9 on sea floor spreading and crustal deformation, respectively; or the chapters on the atmosphere and oceans can follow the first two chapters on space science, a scheme common in many texts. In a lighter course where the atmosphere is introduced into the course simply as the gaseous phase of the earth, Chapter 11 will suffice. But if a more complete picture, to include winds, pressure, and other "weather elements" is desired, Chapter 12 can be included. Chapter 13 provides a thorough introduction to the nature of storms and weather forecasting.

In the case of other subjects treated in a single chapter, the latter portion of a chapter may be omitted without affecting the logic of the presentation. Chapter 10 on earthquakes, for example, presents a rather traditional, mostly descriptive, approach in the first portion. The latter part is somewhat more advanced; it includes the discovery method by which seismology has revealed the internal structure of the earth.

Although the book has been developed as a one-term course in earth science, it is likely that a single course cannot cover the entire contents. The sequence and number of chapters that are utilized must depend on the course purpose and the hours available. I have tried to make the book sufficiently general to give adequate preparation for teachers of earth science in elementary and high schools.

The last chapter is really more of an essay on the relationship of man to his environment than a chapter of pure exposition. It is an attempt to point out that man developed on the earth from a natural process of evolution involving a balanced ecology and that we must do something to reverse the disruption of our ecology or we will "self-destruct."

Finally, I thank once again all who have given generously of information and illustrations so important to the preparation of the book. I sincerely regret that in some cases ideas or modified illustrations may have been used out of my notes or memory, both being inadequate to recall the source.

William L. Donn

Grand View-on-Hudson, New York
July, 1971

Contents

Front Endpaper: The earth from 22,300 miles in space. Photographed by the National Aeronautics and Space Administrations's ATS-III satellite. (Photograph courtesy of NASA.)

Back Endpaper: Technological Satellite (ATS-III) photograph of North and South America. (Photograph courtesy of NASA.)

The Earth: Our Physical Environment

Introduction

"The earth is a spinning globe. Vast though it seems to us, it is a mere speck of matter in the greater vastness of space." With these eloquent words, H. G. Wells opened his famous *Outline of History*. All of the efforts of modern research have done nothing to change this view; they have only refined it.

To each person concerned with the natural sciences, the earth with its atmosphere and oceans has its own significance and interest. Geologists, for example, are concerned primarily with the earth's crust and how it got to be the way it is. Geophysicists delve deeper to satisfy their interests, trying to understand the nature and structure of the earth beneath the crust. Meteorologists worry about the atmosphere, while astronomers worry about it in a different way—they wish it weren't present to obscure the weak starlight from which so much about the universe is deduced. Oceanographers investigate the properties of sea water, its currents, waves, and tides, while marine geologists, interested in the sediments and rocks beneath the sea, regard the oceans as a murky mess that obscures the bottom.

The era beginning with artificial earth satellites has brought a new look to earth science. Previously writers and teachers tried to present imagined views of the earth as a whole, as seen by an observer in space. Now we have visual astronaut reports and high-quality pictures. Such results may dull our imagination, but they have improved our knowledge of the earth and its environment.

Earth science as conceived today involves more than the study of the solid earth but includes its astronomical background as well as the nature of the surrounding atmospheric and oceanic envelopes. The major emphasis is, of course, focused on the solid sphere. The goal of research in earth science is twofold: (1) to describe and understand the earth as it is now, and (2) to apply this knowledge in unraveling its natural history. In the language of modern computer science, the earth is a vast data bank. A major task of the geologist is to seek out these data through observation and then synthesize them into a body of knowledge that describes conditions during the earth's five-billion-year past and that describes the processes involved in the earth's evolution.

One often hears the earth referred to as a cold, inert mass. This is far from true; it is torn by frequent catastrophic earthquakes and violent volcanic eruptions; it is distorted further by scores of microearthquakes that take place in the crust every day. If the earth's surface had been photographed by a time-lapse camera over the last several million years and the film played back at normal motion-picture speed, the view would be that of an undulating, almost-living surface as mountains rise out of watercovered regions and are then deeply eroded, only to rise again; as volcanoes and lava flows pour forth streaming luminous masses of molten rock; as huge continental-sized sheets of miles-thick glacial ice advance over land, recede and advance again as the sea level rises and falls hundreds of feet, washing over and then laying bare huge coastal areas. Then, too, if the camera could view the large central mass of the earth we would see molten rock material at a temperature of about 10,000°F. The earth is far from a quiet, cold, dead object.

The earth rising above the lunar horizon as the Lunar Ascent Module of the Apollo 11 mission rises to rendezvous with the Command Module from which the photograph was taken. The large dark area on the moon is Smyth's Sea. (NASA Photo.)

chapter 1

The solar corona photographed at the time of the total eclipse of the sun on March 7, 1970. Although always present, the delicately illuminated corona is normally obscured by the brightness of the daytime sky.

The Solar System

A GLIMPSE OF THE UNIVERSE

We who live on the earth appear to be at the center of the universe. The sun, moon, planets, and stars seem to rotate endlessly about us. This illusion led to the age old *geocentric* concept of the Greek, Roman, and Egyptian scholars according to which the earth was considered to be fixed at the center of the universe and to be orbited daily by the known heavenly bodies.

The distances of the planets, stars, and stellar systems, which are now known to be so great as to stagger the imagination, preclude the possibility of these objects circling the earth in a day. But without the aid of observational techniques developed in the past few hundred years, it was quite impossible to perceive these huge distances. So far are the stars and so great is the separation among them that we do not easily perceive any change in their positions during hundreds to thousands of years despite their nearly fantastic rates of speed. Night after night the "fixed" stars drift across the sky, always maintaining their same relative position.

Ancient astronomers soon noticed that a few starlike objects appeared to move among the fixed stars, changing their positions from night to night. The moving objects were called *planets* by Greek astronomers (meaning "wanderers"). Adjustments were then made in the geocentric theory to explain the special motions of the planets. The most complete early explanation of the organization of the universe was given in the second century A.D. by the Greco-Egyptian astronomer Claudius Ptolemy in his compendium of astronomy, the *Almagest,* which stands high among the great works of man.

In the Ptolemaic theory, the behavior of the planets was explained as in Fig. 1-1. In addition to moving about the earth in a primary circle, the planets were also imagined to travel in small circles along their orbits, thus permitting them to move relative to the stars. This seemed to explain nicely the difference between the behavior of planets and stars as they circled the earth. So reasonable was this version of the geocentric theory that it controlled and constrained man's view of the universe for nearly 1500 years.

The Ptolemaic view persisted so long for three important reasons: it satisfied man's ego, placing him at the center of things; it satisfied man's eternal quest to find order in the natural world; and it satisfied a basic philosophic precept of science that the simplest hypothesis that can explain all of the observations is the most acceptable.

Unfortunately, the grand, simple explanations of initial observations usually become more complicated as observations and knowledge increase.

We now know that each twinkling star in the night sky is a giant incandescent sphere of gas much like our sun, which is a rather ordinary

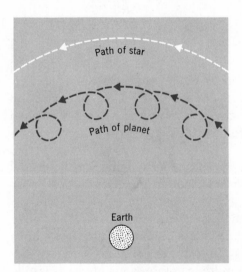

FIG. 1-1. The geocentric view of the universe according to Ptolemy. The stars orbited the earth at the same rate in uniform circular paths, but planets described smaller circles within their orbits causing their positions to shift relative to the stars.

star. All of these stars plus the myriads more that are detected with telescopes are parts of a huge *galactic system.* More than two billion other galaxies, each consisting of a billion or more individual stars, constitute the known universe (see Figs. 1-2 and 1-3). Great as the distances are among individual stars within a galactic system, far greater are the distances among the galaxies. We can see the most distant ones by light (traveling at a speed of 186,000 miles per second), which began its journey through space billions of years ago when the earth was young.

Real understanding of the true relationships of the earth to its celestial environment was the result of the work and imagination of such scientific titans of the Renaissance as Nicolaus Copernicus (1473–1543), Tycho Brahe (1546–1601), Galileo Galilei (1564–1642), Johannes Kepler (1571–1630) and Isaac Newton (1642–1727), whose individual contributions are described later in this chapter. We owe to these men our understanding of the heliocentric (sun-centered) nature of the solar system, which consists of the central sun and its retinue of nine planets, their satellites, and a belt of asteroids. Asteroids are a great swarm of objects thought to be the remains of a disrupted planet and lie mostly between the orbits of Mars and Jupiter.

THE SUN

In this era of accelerated space study and of the beginning of space travel, we have a new and urgent reason for examining the nature of the solar system. This study arose initially from man's great quest to establish order in nature. It continued as our pursuit of knowledge relating to the origin of the earth led to the realization that the earth and solar system had essentially a common beginning. An understanding of the origin of the earth required an understanding of the nature of both the earth and the solar system.

It is most appropriate to begin with the sun. Because it contains 96 percent of the matter (mass) of the solar system, the sun is dominant in the gravitational control of the planetary bodies of the system.

As large and as bright as it appears to us, our sun is but an average sort of star. "Old Sol" would appear as merely another twinkling stellar object if it were displaced to the distance of even the nearest of the stars visible in the night sky. According to the most authoritative modern theories of the origin of the earth and solar system (to be reviewed later in the chapter), it is quite likely that many of the stars are the "suns" of other planetary families.

The sun is not only the largest of the solar-system inhabitants, it is also the chief source of energy and has been so for all of geologic time. For us on earth, the sun provides the energy that drives the atmosphere, and all of its weather processes, that disturbs the sea with its restless waves and circling currents and that provides all of the energy for life. When we burn fuel—coal, oil, or gas—we are releasing the radiant energy that was stored during the life of the organisms whose decay provides the fossil fuels of the world.

Temperature and Energy

The sun, a tremendous mass of incandescent gas, is powered by thermo-nuclear processes that produce central temperatures estimated to be in

NGC 488

NGC 628 M74

NGC 2859

NGC 2523

NGC 175

NGC 1073

FIG. 1-2. Examples of distant stellar galaxies. The spiral pattern, which is so common, has led to the term *spiral nebulas* for such galaxies. Spiral forms have either circular or barred central regions. NGC refers to the "New General Catalogue" of stars and galaxies. Because the galaxies are so distant and faint, the long exposure necessary to photograph them causes overexposure of stellar images which appear as small bright circles. (Photographs from the Hale Observatories.)

excess of 20 million degrees centigrade (36 million degrees Fahrenheit). We do not see this hot interior. The part of the sun visible to us by direct sunlight is the surface layer or *photosphere,* which has a temperature determined from direct radiation to be 5500°C (10,000°F). The remarkably high internal temperatures are obtained by calculating how much energy must be radiated from the interior to maintain the observed temperature of the photosphere.

The temperature of the sun's surface, despite its remoteness, can be estimated quite closely from direct observations of solar radiation. From these observations we have learned that the sun radiates 1500 calories* per second from each square centimeter of its surface. Now, according to Stephen's Law (a fundamental law of physics that relates radiation to temperature) the energy (E) radiated from an object varies directly with the fourth power of the absolute temperature (the temperature scale that begins at absolute zero, −273°C, or −459°F), or

$$E \: \alpha \: T^4$$

The energy and temperature are related numerically by the equation:

$$E = KT^4$$

where K is the radiation constant (0.0000572),

$$E = 0.0000572 \: T^4$$

or

$$T = \sqrt[4]{\frac{E}{0.0000572}}$$

The surface temperature of the sun given above is found by making the appropriate substitution of energy in this equation.

How does the sun develop and maintain temperatures that range from thousands of degrees at the surface to tens of millions of degrees in its interior? The answer has been found in solar-energy processes.

With the exception of a small number of atomic-energy plants, most energy used on earth comes from the combustion (oxidation) of fossil fuels in which solar energy was captured and stored ages ago. However, the source of solar energy lies not in combustion but in nuclear processes. If the entire mass of the sun were composed of coal and oxygen in the appropriate proportions, it would have lasted for only a couple of thousand years at the present rate of energy radiation. Yet, the sun has been radiating energy at about the present high rate for billions of years, during which time solar mass has been continuously converted to radiant energy.

A number of different atomic reactions occur in the sun. By far the most dominant is the proton-proton reaction in which four hydrogen

*One calorie is the amount of heat necessary to raise 1 gram of water by a temperature of 1°C.

FIG. 1-3. The Great Spiral Nebula in the Constellation *Andromeda* (NGC 224). This, the closest of the great galaxies, is still nearly one million light years (6 billion billion miles) distant. Our own system of stars—the Milky Way system or Galaxy, would look something like this if seen from a great distance. (Photograph from the Hale Observatories.)

nuclei, each containing one proton, are synthesized into one helium nucleus (Fig. 1-4). The atomic weight of the hydrogen nucleus (one proton) is 1.0073. Although four protons should have an atomic weight of 4.0292, that of the helium nucleus is only 4.0015. Thus, in this nuclear fusion process matter is transformed into energy with a loss of 0.0277 parts of atomic mass.

An idea of the tremendous amounts of energy liberated from even this small conversion of mass to energy can be gained by substitution of appropriate values into the now-famous Einstein equation relating energy to mass:

$$E = mc^2$$

where E is the energy, m is the mass being converted, and c is the velocity of light (3×10^{10} cm/sec). (Notice that 10^{10} means 10 followed by ten zeros, etc.)

For each gram (about 1/15 of an ounce) of helium that is formed, 0.0277 grams of atomic mass will be converted to energy, or

$$E = 0.277 \text{ g} \times (3 \times 10^{10} \text{ cm/sec})^2$$
$$= 0.0277 \text{ g} \times 9 \times 10^{20} \text{ cm}^2/\text{sec}^2$$
$$= 0.25 \times 10^{20} \text{ g cm}^2/\text{sec}^2 \text{ or } 0.25 \times 10^{20} \text{ ergs*}$$

This is about equivalent to the energy released by the complete combustion of some 22,000 pounds of coal—all from 0.002 ounces (0.0277 g) of atomic mass! The sun is composed mostly of hydrogen, which is thus the fuel for the solar atomic furnace. In the transformation of hydrogen to helium the sun is losing mass at the fantastic rate of nearly five million tons every second. It has been doing so for billions of years in the past and will continue to do so for billions of years into the future. Despite this high rate of loss of mass, the sun is not obviously wasting away—a witness to the huge amount of matter it contains, an amount equal to 2×10^{27} tons.

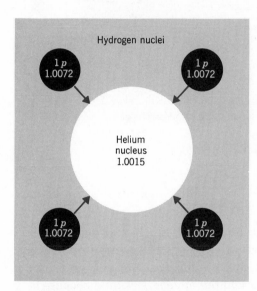

FIG. 1-4. The sun's energy is maintained by the fusion of four hydrogen nuclei, each containing one proton, into a single helium nucleus. Since four protons have an atomic weight of 4.0292 and a helium nucleus of only 4.0015, 0.0277 parts of atomic mass are lost in this atomic process. The conversion of this mass to energy provides the energy of the sun.

Size and Distance

One of the fundamental quantities in the solar system is the mean distance between earth and sun, known as the *astronomical unit*. When this is known, many other planetary distances can be found. One way of determining the size of the astronomical unit is to determine first the earth's average speed in its orbit about the sun. This is done by very carefully measuring the earth's motion relative to particular stars. The resulting

*An erg is a unit of work or energy in the metric system. It is the work done when a force of 1 dyne is expressed over 1 cm; 1 dyne is the force necessary to accelerate a mass of 1 g 1 cm/sec².

speed ($18\frac{1}{2}$ miles per second) when multiplied by the number of seconds in a year, gives the circumference of the earth's orbit. Because the earth's orbit is very nearly circular, we can use the formula:

$$\text{circumference} = 2\pi r$$

to solve for r and obtain for the radius of the orbit or mean distance between earth and sun, the value of 93 million miles.

From this information we can immediately calculate the size of the sun as in Fig. 1-5. The angular size of the sun in the sky is about a half degree (more exactly 31′59″). The simple computation (Fig. 1-5) then shows that an object at a distance of 93 million miles that has an apparent size of $\frac{1}{2}$° must be 864,000 miles in diameter—a diameter more than 100 times that of the earth.

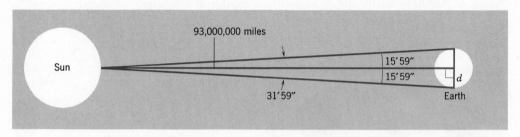

FIG. 1-5. Measuring the size of the sun. The angular size of the sun seen from the earth is 31′59″ (just over $\frac{1}{2}$°). If we split this angle to make two right triangles, then by simple trigonometry, the tangent of 15′59″ = d/93,000,000 miles, and d = 93,000,000 × 0.004646 = 432,078. But the diameter of the sun is 2d as drawn or is just over 864,000 miles.

Sunspots

When viewed telescopically (with appropriate light filters), the apparently unblemished surface of the sun is often seen to be marred by irregular dark gray to black regions called *sunspots* (Fig. 1-6). Strangely, these areas are actually quite hot and bright but appear dark only because they are less hot and bright than the surrounding solar surface. Even an electric light would seem dark if viewed against the sun. Although the origin of sunspots is uncertain, they have been established to be huge magnetic storms that must extend from the solar surface deep into the interior. The magnetic fields of force within each spot eject huge masses of electrons that are magnetically focused into beams. When the earth encounters these electron beams at times of sunspot activity, brilliant atmospheric auroras occur. These are luminous displays produced by the ionization of the upper atmosphere from the impact of high-speed electrons. The same ionization thoroughly disrupts long-range radio transmission.

The number of sunspots varies periodically with a cycle of 11.1 years. At times of maximum activity the sun actually shows a slightly increased level of radiation. Despite the somewhat cooler and less radiant region of the sunspots, the remainder of the surface of the sun has an increased activity. Attempts have been made to correlate many terrestrial phenomena with this sunspot cycle—for example, economic depression cycles, climate changes, sea-level variations, and tree growth. But no clear relationship between sunspots and anything terrestrial has ever been demonstrated, other than increased activity in the ionized layers of the upper atmosphere.

Because the sun rotates, sunspots appear to be carried across its surface. Measurements of their daily change of position enable astronomers to establish the sun's rotation rate—about 25 earth days.

THE PLANETARY SYSTEM

To a casual observer on earth, planets (described in Fig. 1-8) which are visible only through their reflection of sunlight, appear to be little different from the myriad of stars that dot the sky. To a more careful observer, the planets visible to the unaided eye (Jupiter, Saturn, Mars, Venus and Mercury) characteristically differ from the surrounding stars: planets are brighter than nearly all stars, they shine with a steady light usually unbroken by the twinkling typical of starlight, and perhaps most obvious of all, their positions among the stars change from night to night.

The planets all orbit the sun in ellipses of varying degrees of eccentricity or elongation. Because the sun always occupies the position of one of the foci of each of the elliptic planetary orbits, each planet has a closest point to the sun, or *perihelion,* and a most distant point of *aphelion.* Most of the planets, except for Pluto, have nearly circular orbits or orbits of low eccentricity.

The fundamental order of the motion of the planets about the sun was discovered by Kepler, who carefully analyzed the work of Tycho Brahe. Brahe had previously carried out the most prolonged and complete set of naked-eye observations of planets ever made. During the course of years of interpretation of Brahe's observations, Kepler developed his classical laws of planetary motion that finally demonstrated the order prevailing in the solar system. The laws are known numerically according to their sequence of deduction.

First law: Planets travel in elliptical orbits about the sun, which is located at one of the foci of the ellipse.

Second law: The radius vector of each planetary orbit sweeps out equal areas in equal times. This can be seen in Fig. 1-7 where the lines Ra to Rd (radius vectors), in order to sweep the equal areas (shaded), must traverse the two arcs 1–2 and 3–4 at different speeds—slower from points 1 to 2 and faster from points 3 to 4. As a consequence of this effect, the earth does not move the same distance about the sun each day. The resulting nonuniform motion of the earth has greatly complicated our method of keeping time as explained in Chapter 2.

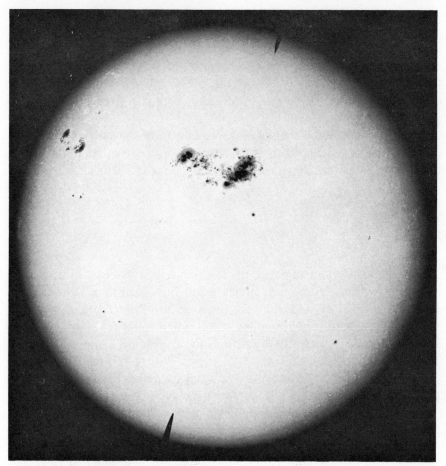

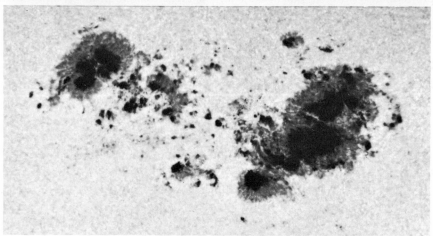

The Sunspot of April 7, 1947

FIG. 1-6. The great sunspot of April 7, 1947. This is one of the largest sunspot groups ever observed. The lower photo is an enlarged view of the main group. (Photo from the Hale Observatories.)

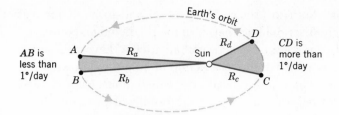

FIG. 1-7. Schematic view of Kepler's second law. The shaded regions on opposite sides of the sun represent the equal areas swept out in equal times by the radius vectors **Ra, Rb, Rc,** and **Rd.**

Third law: For any pair of planets the squares of their periods of revolution are proportional to the cubes of their mean distances from the sun, or

$$\frac{P_1^2}{P_2^2} = \frac{R_1^3}{R_2^3}$$

where P_1 and P_2 represent the periods of revolution of the two planets in years, respectively, and R_1 and R_2 their mean distances from the sun. The latter is expressed in terms of the earth's mean distance from the sun, which is designated as "1" astronomical unit. Because we know the earth's period and mean distance from the sun, observation of any other planet's period allows a quick calculation of its distance. For example, Jupiter is observed to have a period of 11.86 years. Because the earth's period in years is 1 and its mean distance is 1 (astronomical unit), substitution in the equation yields a mean distance of 5.2 astronomical units for Jupiter. Its distance in miles would be the earth's mean distance (93 million miles) times 5.2.

It is interesting to note that the heliocentric theory of Copernicus, stated in 1543 in his monumental work *De Revolutionibus Orbium Coelestium* (The Revolution of Celestial Bodies), was essentially a revival of an idea originally advocated by ancient Greek astronomers. The detailed observations and conclusions of Tycho Brahe, Galileo, and Kepler, confirming the theory, came afterward. The geocentric theory of Copernicus is an example of how science advances through pronounced and often logically hazardous leaps of the imagination on the basis of relatively scant data.

A summary of the size and distance relationships among the nine planets of the solar system appears in the scale drawings in Fig. 1-8. The asteroids are not included in this presentation.

Despite their relatively large size, Uranus and Neptune are too distant to be seen by the naked eye, in contrast to the nocturnal brilliance of the closer planets. Uranus, the brighter and closer of the pair, was discovered by direct telescopic observations. The discovery of Neptune stands as a milestone both to precise observation and to mathematical astronomy. Very slight discrepancies between the predicted and observed orbital motions of Uranus led a French mathematician, Leverrier, and a Briton, Adams, independently to the prediction of a then-unknown outer planet whose gravitational attraction caused the deviations. Neptune was then found in the sky within a degree of the predicted position. Pluto, the outermost of the planets, was found in much the same way.

With the exception of the inner planets Mercury and Venus and an outer planet, Pluto, the remaining six, including the earth, all have their own families of objects called satellites circling them. These number from 1 for the earth to 12 for Jupiter. It was in fact the observation of the motions of Jupiter's four largest satellites by Galileo, soon after his invention of the telescope, that confirmed Copernicus' earlier heliocentric model of the solar system, thereby refuting the long-held geocentric model. By means of his telescope Galileo could observe four of the largest of Jupiter's satellites revolving about the mother planet. For the first time it was shown beyond doubt that some of the celestial objects revolved about a body different from the earth.

Galileo then turned his telescope on the planet Venus, the third brightest object in the sky after the sun and moon. He observed that Venus

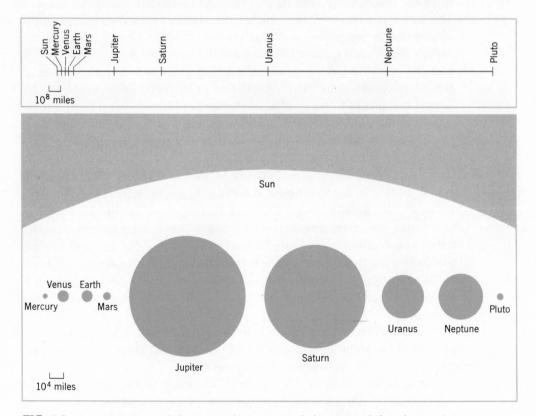

FIG. 1-8. Scale drawings of the mean distances and the sizes of the planets (to a different scale) including an arc of the sun.

    THE SOLAR SYSTEM

exhibited phases (Fig. 1-10) just like those of the moon (see p. 27). The only explanation for these phases was that Venus must revolve about the sun. With the conclusion that one of the seven major celestial bodies of the ancients revolved about the sun, not the earth, the supporters of the geocentric theory were further shaken.

The final crushing blow to the geocentric concept followed the demonstration by the physicist Bessel that the closest star (Proxima Centauri) described a small oscillation in the sky during the year. The principle

FIG. 1-9. Mars, Jupiter, and Saturn as photographed with the 100-inch telescope on Mt. Wilson, California. Pluto, photographed with the 200-inch telescope on Mt. Palomar, California, is so small and far away that it appears as a mere dot, no more prominent than the stars. The Martian polar cap, that waxes and wanes with the change of seasons on Mars, is very prominent. Jupiter is characterized by latitudinal bands and Saturn by its beautiful ring system. (Photographs from the Hale Observatories.)

of *parallax* that is involved is quite simple. If you hold up your finger at arm's length and shift your head from side to side, your finger appears to move right and left against the more distant background. The farther you hold your finger, the less is its apparent displacement. Relative to nearby (but more distant) stars, Proxima Centauri appeared to move to and fro. This effect, repeated each year, is only explainable by assuming that the earth moves in an orbit about the sun. When viewed from nearly opposite sides of the orbit, even very distant stars oscillate slightly, just as your finger shifts from motions of your head.

A number of pertinent facts regarding the planets are summarized in Table 1-1 and Fig. 1-8. Not included are the asteroids, a swarm of tens

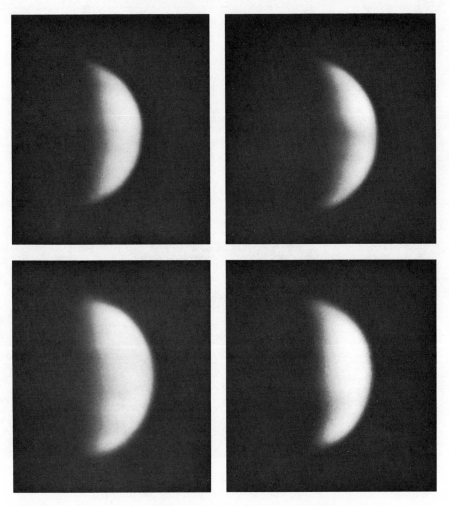

FIG. 1-10. Photographs of Venus between the "crescent" and "quarter" phase (from right to left). Initial observations of the changing phases of Venus by Galileo proved that Venus must revolve about the sun and helped crush the long-held geocentric theory of the universe. (Photographs from the Hale Observatories.)

TABLE 1-1 Planetary Data

Properties	Mercury	Venus	Earth	Mars	Jupiter	Saturn	Uranus	Neptune	Pluto
Diameter (miles)	3010	7610	7918	4140	86,900	71,500	29,500	26,800	3700?
Mean solar distance (millions of miles)	36.3	67.8	93	141	483	888	1790	2800	3680
Mean density (gms/cc)	4.8	4.9	5.5	3.8	1.3	0.7	1.3	2.2	?
Number of satellites	0	0	1	2	12	9	5	2	0
Rotation rate	88 days	?	23.93 hours	24.62 hours	9.83 hours	10.23 hours	10.82 hours	15.67 hours	6.4 days
Period of revolution (years)	0.24	0.62	1	1.88	11.86	29.46	84.02	164.8	248.4

of thousands of small planetoids, varying in size from small fragments to a few hundreds of miles in diameter. Although most asteroids lie in orbits between Mars and Jupiter, a few have paths so eccentric as to range from within the orbit of Mercury almost to that of Saturn. Other inhabitants of the solar system include comets (Fig. 1-11) and meteors. The latter are often called *shooting stars* from their luminous appearance when heated to frictional incandescence during passage through the earth's atmosphere. The very same effect is observed when a space vehicle makes its fiery reentrance into the earth's upper atmosphere. Frictional heating reaches about 7000°F on the outer surface of the returning vehicle. At times meteors are large enough to withstand their scorching trip through the atmosphere and actually impact on the earth's surface. They are then called meteorites and have provided us with very important clues to the age of the earth (Chap. 7) and the composition of the solar system.

THE MOON

Our moon once enjoyed the uniqueness of being the earth's only satellite. Since the advent of the "space age" it is now one among many and its uniqueness lies in being the only natural satellite. Our moon has many other distinctions, however. Compared with the satellites of other planets in the solar system, the moon is unique in being so large relative to the earth. All other satellites are minute in comparison to the planets they circle. One of Jupiter's satellites is the largest in the solar system but has only 1/10,000 the mass of Jupiter. Our moon has $\frac{1}{81}$ the mass of the earth. Also, whereas all but one of the many moons in the solar system revolve in the equatorial plane of their respective planets, our moon revolves about the earth in an orbit that cuts across the earth's equatorial plane at an angle of $28\frac{1}{2}°$.

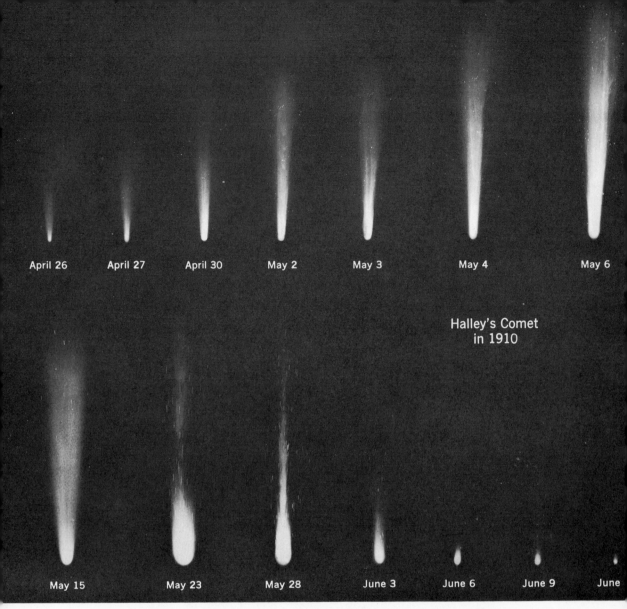

FIG. 1-11. Fourteen successive views of Halley's Comet between April 26 and June 11, 1910. The comet increased in size and brightness as it approached the sun and then decreased after circling the sun and receding far into space. The thin tail, through which stars are plainly visible, is driven out by the sun's radiation, hence is most pronounced when the comet is closest to the sun. (Photographs from the Hale Observatories.)

These facts pose serious problems to theories of lunar origin. One theory proposes that the moon originates from the same material as did the earth (see the end of this chapter) and in much the same manner; another proposes that the moon originated somewhere in the solar system far from the earth and because of a disturbance to its orbit, passed near the earth and was captured. A difference in history and composition is

FIG. 1-12. The moon as we knew it from powerful telescopes prior to the age of space exploration. This view brings the moon to a distance equivalent to about 50 miles. Compare this to the later close-up views in Figs. 1-13 to 1-16. Numbers refer to the locations of these close views. (Photograph from the Hale Observatories.)

implied by each of these theories. Modern lunar exploration may resolve much of the problem. Despite these speculative aspects of the moon, many important details are well known, as we shall see below.

Size and Distance

The diameter of the moon is very nearly 2160 miles, just a little more than one quarter that of the earth. Its small size and relatively low density give the moon a mass about $\frac{1}{81}$ that of the earth. These factors all combine to make surface gravity on the moon about $\frac{1}{6}$ the value of gravity at the earth's surface. Someone with a weight of 180 pounds would thus weigh

FIG. 1-13. View of the Sea of Fertility taken from Apollo 8 as it circled the moon. The largest crater in the foreground, which is 45 miles in diameter, is crossed by prominent rilles. The longest rill crossing the center of the crater from left to right must be younger than the crater through which it cuts. This view is at location 1 in the full moon view of Fig. 1-12. (NASA photo.)

only 30 pounds at the moon's surface. Without this advantage the heavy life-support packs of lunar explorers would be a severe and almost impossible burden.

The mean distance between the earth and moon is close to 238,857 miles. The initial distance determinations were made by standard surveying triangulation procedures in which the earth's diameter was the baseline. The procedure is similar to that in which the diameter of the sun was determined (Fig. 1-5), only now the distance is the unknown factor. Accuracy was improved with the application of radar reflections from the moon's surface. In this procedure the time lapse between the emission and return of the radar signal (which travels with the speed of light) multiplied by the speed of light gave the two-way distance to the moon. Further refinements are now being made by space probes, which already have narrowed the accuracy to a few hundred feet. Instrumentation left

FIG. 1-14. The bright crater, Tycho (location 2, Fig. 1-12), from Lunar Orbiter 5 140 miles above the moon's surface. This young, large crater was formed by forces so violent that material from it was ejected across much of the face of the moon. (NASA photo.)

at Tranquility Base by Apollo 11 astronauts Armstrong and Aldrin have made possible a distance determination with the remarkable accuracy of 6 inches!

The lunar orbit is also elliptical, like that of the planets; it has a closest distance to the earth (perigee) of about 226,000 miles and a greatest distance (apogee) of about 252,000 miles.

Lunar Period and Phases

The period of revolution of the moon about the earth is one of the most accurately known quantities in all of science. After many years of careful observation it has been established to be 27.32166 . . . days. This interval is known as the sidereal month because it is the time between successive passages of the moon across a line between the earth and a given star,

FIG. 1-15. A portion of the crater Copernicus (3 in Fig. 1-12), one of the most prominent features of the lunar surface, transmitted from Lunar Orbiter 2, 28.4 miles above the surface. From the upper horizon to the bottom of the photograph is about 150 miles, with the distance in the background, beyond the northern rim of the crater being greatly foreshortened. The mountains rising from the flat floor are 1000 feet high. That on the upper left (northwest) horizon is 3000 feet. (NASA photo.)

FIG. 1-16. (*a,b*) Apollo 11 astronauts at work on the lunar surface in the Sea of Tranquility (4 in Fig. 1-7). Note the small rocks and powdery surface into which the astronauts feet sink. Their corrugated footprints may well remain for billions of years since the moon has no water nor atmosphere to erode them. (NASA photos.)

as in Fig. 1-17. Here, the relationship shown by the line E,M,S, (earth, moon, star) is repeated $27\frac{1}{3}$ days later (E_1,M_1,S_1) as the moon orbits the earth. The synodic month, which is the basis for our monthly calendar, requires $29\frac{1}{2}$ days because it involves the alignment between the earth, moon, and sun. The reason for the difference can be seen in Fig. 1-17, where the moon is initially opposite the sun in the line, E,M,S. But $27\frac{1}{3}$ days later the moon has not yet reached the line connecting sun and earth, because of the earth's orbital motion. After $29\frac{1}{2}$ days (the synodic month), the moon is once again aligned with the earth and sun, as Sun, E_2, M_2.

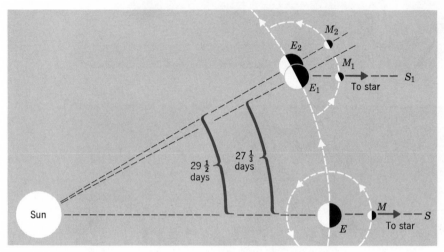

FIG. 1-17. Explanation of the sidereal and synodic months (see text).

The waxing and waning of the moon's phases and eclipses of the sun and moon are also more closely related to the synodic than the siderial period. The reason for the phases exhibited by the moon is apparent from Fig. 1-18. The portion of the illuminated half of the moon that can be seen from the earth determines the phase. Between the position of new moon (which is invisible from the earth) and first or last quarter, the moon shows a crescent shape—a very thin crescent when near the new-moon position and a broad crescent when near the first- or third-quarter positions. To an observer on the moon, the earth, too, would show phases. Note the earth as photographed from an artificial lunar satellite (facing Chapter 2).

Eclipses

Eclipses of the sun are among the most beautiful and awe-inspiring manifestations of nature. The beauty of the total-solar-eclipse phenomena is equaled by its scientific value. On these occasions the atmospheric

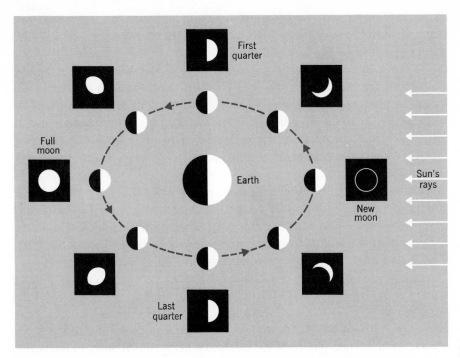

FIG. 1-18. Explanation of the phases of the moon. The inner set of circles shows the illuminated and dark halves of the moon. An observer on the earth would only see the halves perpendicular to the line of sight. The resulting appearance of the visible part of the moon is then as shown in the outer set of phases.

features of the sun, such as solar prominences, corona, and the like can be photographed and studied as at no other times. The Einstein effect, in which starlight passing the sun is deflected by the sun's gravitation, is reobserved and remeasured with each eclipse. This important observation was predicted by Einstein's *theory of relativity*. Rare atmospheric phenomena also occur at these times from the sudden chilling of the shadowed portion of the earth's atmosphere.

The relationship of the sun, earth, and moon at times of eclipses is shown in Fig. 1-19. When the moon's shadow falls on the earth at new moon, a solar eclipse is visible to those within the shadow; when the earth's shadow falls on the moon at full moon, a lunar eclipse is visible. Note that the moon barely obscures the sun during solar eclipses so that a very small region of the earth's surface "sees" totality. The earth's shadow is so large however, that the entire moon can be eclipsed and can remain so for a relatively long time compared to the maximum of a few minutes for a total eclipse of the sun. We can see from Fig. 1-19 that a lunar eclipse is therefore visible from the entire dark half of the earth. Even when totally eclipsed, the moon continues to shine with a very subdued copper-orange light from sunlight refracted or bent into the earth's shadow by the atmosphere.

If the orbits of the earth and moon were in the same plane in space, eclipses of the sun and of the moon would occur at each new and full moon, respectively. Because the moon's orbit is inclined to the plane of earth's orbit (at about 5°), eclipses are relatively rare. Although two partial eclipses of the sun must occur each year, there may be a maximum of seven, occurring as four solar and three lunar or five solar and two lunar eclipses.

Of the two, the total solar eclipse is certainly the more spectacular and memorable. At the instant of totality the sky darkens as at night and the stars become visible but may be overlooked as the remarkably delicate and beautiful corona (solar atmosphere) bursts forth. To those who have seen it, the effect is quite overwhelming and never to be forgotten.

The Moon's Surface

Although ideas about the structure and composition of the moon have played an important part in interpreting the origin and history of the earth, the subject has become even more vital with the era of manned lunar landings.

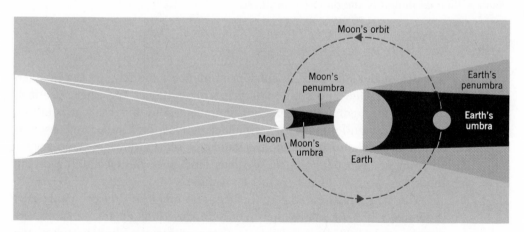

FIG. 1-19. Geometry of eclipses of the sun and moon. Sizes and distances are distorted for diagrammatic purposes. During a total eclipse of the sun, places on the earth's surface within the moon's umbra experience a total eclipse; places in the penumbra experience a partial eclipse, the degree of which is in proportion to the distance from the umbra. During a total lunar eclipse, the entire moon enters the umbra so that the eclipse is visible all over the earth's surface. When the moon is in the penumbra, so much sunlight is refracted into this zone by the earth's atmosphere that no apparent diminution of moonlight occurs.

Because the lunar period of rotation and revolution are equal, the moon always presents the same side to the earth but, artificial lunar satellites and astronauts orbiting the moon have returned pictures of the moon's far side (Fig. 1-20). The lunar surface we see is about half covered with large dark areas, called *maria* (sing. *mare*), seas (the dark areas in Fig. 1-12) and half with bright regions that are much more mountainous and pockmarked than the smoother gray regions. Superimposed on these broad regions are mountain ranges, individual cratered mountains (some of very great size), and circular craterlike depressions in the surface. Bright systems of *rays* radiate from a number of large craters like those from the crater Tycho (Figs. 1-12 and 1-14), reaching out for more than 600 miles. *Rills* or cracks (Fig. 1-13) also cover much of the moon surface. Some of the linear depressions seem to have the same meandering and sinuous patterns characteristic of terrestrial rivers. The true nature of these features await further lunar landings and exploration.

From pictures transmitted by Surveyor spacecraft and actual observations and photographs by astronauts (Fig. 1-16), we now know that the lunar surface is covered by fine dust and fragmental material of all sizes. Much of the latter was probably ejected from lunar craters at times of large meteor impacts. Some may be fine meteoric debris that is rarely seen on the earth because small-size meteoric dust completely vaporizes from the heat produced by friction in the atmosphere.

Perhaps the most striking aspect of the lunar surface is the stark, rugged appearance of its relief features. Because the moon has no atmosphere, neither wind, clouds, rain, nor storms occur. No erosion has occurred to wear down the high places and fill the low regions. Aside from the much smaller effects of expansion and contraction of its surface rocks from the intense heating during the two-week lunar day and extreme cooling during the two-week lunar night, little change takes place on the surface. Footprints made by the first men on the moon from Apollo 11 will remain for countless millennia. We can predict this because a small soil sample brought back from the moon by the crew of Apollo 12 was determined (as described in Chapter 7) to be 4.6 billion years old. No rock of similar antiquity has ever been found loose on the earth's surface, and only a small handful of rocks with an age of 3.5 billion years has ever been found anywhere in the earth's crust after 150 years of geological exploration. But some of the rocks scooped from the lunar surface by Neil Armstrong (of Apollo 11), the first man to walk on the moon, were dated at 3.5 billion years while soil samples were 4.6 billion years old. This small, random sample of lunar material thus tells us that the moon is at least 4.6 billion years old. These extremely old lunar samples vindicated the lunar voyages. It was argued by supporters of the expensive Apollo program that our knowledge of the earth would be improved. And so it has! We will see in Chapter 7 how information from terrestrial rocks had earlier been interpreted by rather subtle but sound procedures to indicate an earth age of at least 4.7 billion years. No rocks of such age have ever

FIG. 1-20. Photo of the far side of the moon. This is the side we
never see from the earth but which was photographed by Lunar
Orbiter 5. (NASA photo.)

been found. We will also see, later in this chapter, that we believe that
the earth and moon formed at about the same time. Hence the 4.6 billion-
year age of the lunar surface confirms earlier inferences about the age
of the earth.

It certainly seems paradoxical to realize that we know the surface
of the side of the moon facing the earth, and much of the far side, better
than we know the surface of the earth itself. Actually this is not so
surprising. About 71 percent of the earth is covered by visually opaque
seas and water bodies. Other large areas are masked by nearly impene-

trable jungles. Even astronauts, with their broad global views, are hampered by the widespread cloud cover; this is always absent on the moon, which has no atmosphere. We might in fact expect the gaps in knowledge to widen further as continued lunar exploration refines our knowledge of its surface.

THE EARTH AND ITS ORBIT

Motions of the Earth

A reader of this book, sitting quietly in his chair, may feel that he and the earth are at rest. This is an illusion brought about by our inability to sense uniform motion. We are hardly aware of the motion of a smoothly moving vehicle of any sort—only of the sudden stops or changes in speed. This is even more true for the numerous terrestrial motions. For example, while sitting quietly, we are whirling about the earth's axis at a rotational speed that varies from nearly a thousand miles an hour at the equator to zero at the poles and is about 700 miles per hour in the midlatitudes. At the same time, the earth is revolving about the sun at a rate of 66,600 miles per hour, the speed necessary to complete the annual cycle in 365.24 days (the length of a year).

In addition to these purely terrestrial motions, the sun with its family of planets and smaller bodies is hurtling through space at about 40,000 miles per hour toward the constellation *Hercules*. Further, the sun belongs to a system of some billions of stars tied into a gravitational system known as the *galaxy*, which is both rotating and moving through space at remarkably high speeds. We are hardly at rest.

The Ecliptic and the Seasons

The earth's path about the sun is known as the *ecliptic* because the moon must cross this orbit at the times of new or full moon in order for eclipses to occur. The plane of this orbit, called the *plane of the ecliptic*, is a fundamental reference surface in space science. As was noted earlier in this chapter, the earth's orbit is an ellipse of small eccentricity. Its perihelion point, which changes slowly with time, is passed about January 3, and is close to $91\frac{1}{2}$ million miles from the sun; its aphelion point, passed about July 4, is $94\frac{1}{2}$ million miles, resulting in a mean distance of 93 million miles. Interestingly, the earth is closest to the sun when the Northern Hemisphere has its winter season and farthest, in summer. Simple solar distance cannot, therefore, play a direct role in determining seasonal temperature changes.

The seasons are, however, related to the earth's motion about the sun as explained in Fig. 1-21. At the time of the summer solstice (June 21) the direct rays of the sun fall on the Tropic of Cancer at 23½°N in the Northern Hemisphere because the earth's axis is tilted 23½° from the normal to the plane of the ecliptic. Six months later, at the winter solstice (December 21), the sun's direct rays fall on the Southern Hemisphere at

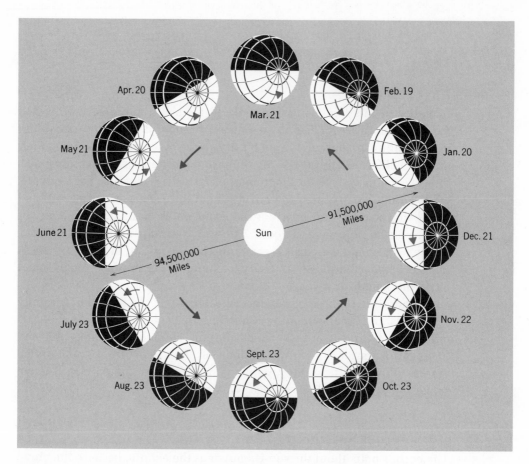

FIG. 1-21. The seasons are caused by (1) the inclination of the earth's axis, (2) the revolution of the earth about the sun, and (3) the parallelism of the earth's axis during the year. Thus, the direct solar rays fall on the Northern Hemisphere in summer and on the Southern Hemisphere in winter. From September to March, the direct rays fall on the Southern Hemisphere producing summer there and winter in the Northern Hemisphere. Notice that the line between the dark and illuminated halves of the globe (*the circle of illumination*) is just tangent to the Arctic Circle on June 21 and December 21. On June 21 the area within the Arctic Circle is fully illuminated; on December 21 it is completely in shadow. Between these times it is in partial illumination.

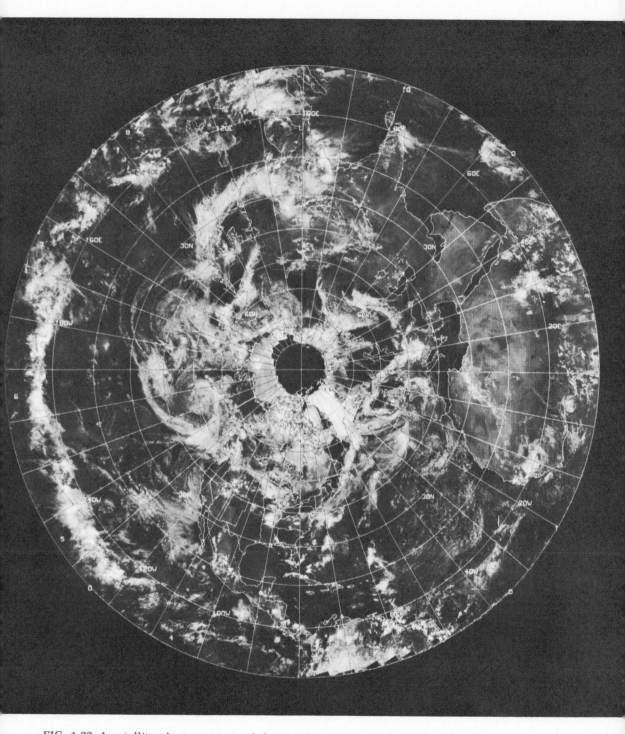

FIG. 1-22. A satellite photo—mosaic of the Northern Hemisphere on June 21 showing the "fully" illuminated polar region. Most of the white areas are cloud patterns related to storms. Although fully illuminated the central polar area appears dark because the satellite transmits its photo images when passing over the poles and does not record then.

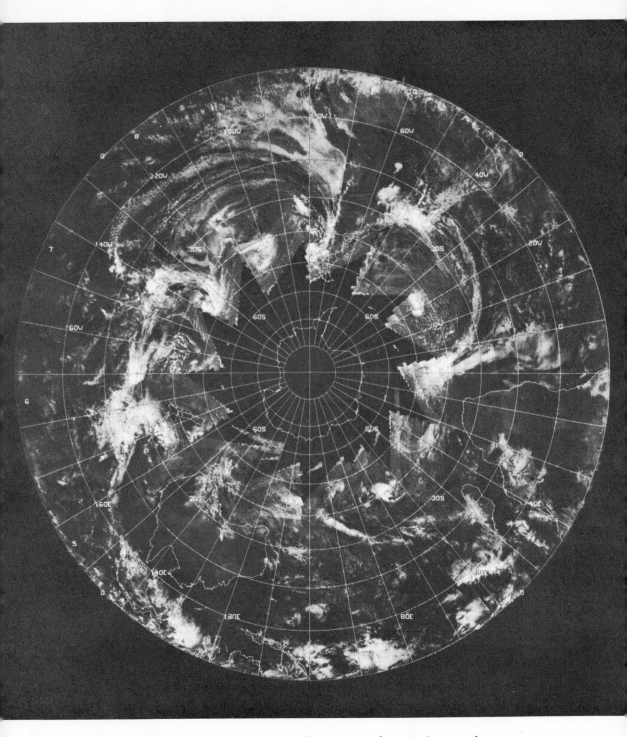

FIG. 1-23. The photo-mosaic of the Southern Hemisphere on June 21 shows the region south of the Antarctic Circle to be in darkness. (National Weather Satellite Center.)

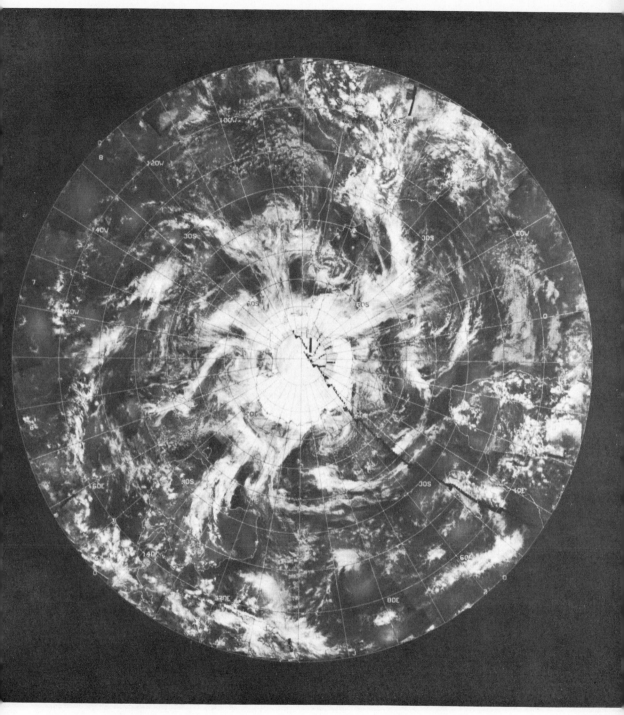

FIG. 1-24. On December 21 the entire south polar area is in sunlight as shown by the illuminated ice-covered Antarctic continent.

FIG. 1-25. On December 21 the Northern Hemisphere north of the Arctic circle is in darkness.

exactly $23\frac{1}{2}°$ S (the Tropic of Capricorn). At the times midway between the solstices, about March 21 and September 23 (the spring and autumn equinoxes), the sun's rays fall directly on the equator. The seasons are thus the result of (1) the earth's axis being tilted away from a perpendicular to the plane of the ecliptic, (2) the revolution of the earth about the sun, and (3) the self-parallelism of the axis at all times in the year. Variations in the illumination of the Northern and Southern Hemispheres caused by the earth's revolution are evident in the satellite photomosaics in Figs. 1-22 to 1-25.

In the course of the earth's revolution about the sun, an observer on the earth sees the sun shift its position with respect to the stars. Also, the noontime position of the sun undergoes a changing altitude angle with respect to the horizon and to the celestial equator—the projection of the earth's equator onto the sky. According to Fig. 1-21, we see the sun directly on the equator at the time of the equinoxes, and $23\frac{1}{2}°$ above and below the equator at the summer and winter solstices, respectively.

Precession of the Equinoxes

Among the motions described in the beginning of this section, a relatively slow but important one was omitted. The earth is wobbling, as it spins on its axis, somewhat like a spent top. The wobble, whose period takes nearly 26,000 years, is illustrated in Fig. 1-26. As a consequence, in 13,000 years Polaris, the North Star, which has long been the celestial beacon of navigators, will surrender this honor to the much brighter star, Vega. In twice this time, Polaris will again shine at the north pole of the heavens. All of the stars and constellations similarly undergo this slow cyclical displacement.

The effect of the axial wobble can be seen in an examination of the pyramids of ancient Egypt. Tubular opening through these structures were used for the purpose of telling time by noting when a particular star crossed the starsight. No longer can the intended stars be seen through the primitive sights.

An examination of Fig. 1-26 shows that another, climatically very important change must occur from this cyclical displacement of the earth's axis. When the axis is tilted in the opposite direction in 13,000 years, the orbital positions at which summer and winter solstices occur will be reversed, as will be true for the equinoxes of spring and fall.

At present the Northern Hemisphere experiences summer when the sun is near aphelion. But in 13,000 years, when the axis is tilted in the opposite direction, summer will occur near perihelion and will be warmer than at present. At this time, winters will be colder, occurring at aphelion, rather than perihelion as at present. This slow displacement of the equatorial points along the ecliptic caused by the axial wobbling is called the *precession of the equinoxes*.

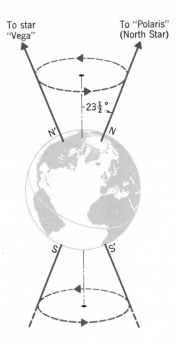

To star "Vega"

To "Polaris" (North Star)

$23\frac{1}{2}°$

N' N

S S'

FIG. 1-26. Precession of the equinoxes is caused by the wobbling of the earth with its inclined axis. At present, the axis is oriented so as to point approximately to the star, *Polaris,* the "north star," but in 13,000 years, after one-half a precessional orbit is completed, the bright star *Vega* will be the north star.

ORIGIN OF THE EARTH AND SOLAR SYSTEM

We have now seen that the earth, sun, and planets of the solar system constitute a unified system dominated by the gravity of the sun's enormous mass. Also, astronomers have shown that most of the elements composing the planets have already been observed in the sun. When all of these related facts are considered, it seems that the entire solar system had a common origin. This happened so long ago that most of the details are shrouded in the mists of the past—more than five billion years ago.

How do we know when it all began? One line of evidence comes from conclusions about the age of the earth's crust. The oldest rocks so far discovered have been dated at 3.6 billion years. Also, it has been estimated that known geological processes (erosion, deposition, etc.) involved in the formation of these rocks were going on as early as four billion years ago. And some of these processes involve the presence of rivers and oceans. From all of this we can reason that the earth's crust existed at that time much as we know it today. Further reasoning based on knowledge of the earth's interior suggests that the earth as a whole must have accumulated at least a billion years earlier, giving it a minimum age of five billion years.

Another indication of such a great antiquity for the solar system comes from observations of meteorites and lunar rock samples. Some meteorite ages have been dated by methods of radioactivity (Chapter 7) at about 4.7 billion years. Because most meteorites are regarded as fragments of a disrupted primitive planet, this date converges with our esti-mate of five billion years as the minimum age of the earth. This great age is also supported by the age of 4.6 billion years for the lunar rock and soil samples referred to above.

Then, from theoretical reasoning astronomers have been able to work out many details of the sun's evolution. The time scale for the develop-ment of the sun with its present size, composition, and luminosity also suggests a beginning compatible with that inferred from observations of rocks and meteorites.

Having established a reasonable, albeit approximate, minimum age for the solar system, we can now return to the problem of origin, and a problem it is. Once the general nature of the solar system was estab-lished, a number of very attractive theories of origin were proposed. But as our knowledge of the solar system increased, serious obstacles arose for each in turn. Although details of the most current views are quite technical, we can review the important features of modern thought on the subject.

It is not difficult to imagine that the origin of the solar system must be related to the evolution of the sun in view of its dominant mass and energy. At some time prior to five billion years ago our sun was vastly different. It was a huge, dark, cold, and diffuse mass of gas that more than filled the entire volume of the present solar system as far as the orbit of the outermost planet. Slowly at first, and then more rapidly, gravitational contraction occurred and with it the development of a slow rotation of the *protosun*. Gravitational contraction causes energy to be released in the form of heat. By the time the sun contracted to approximately its present size, after some hundreds of millions of years, it had warmed to about one million degrees centigrade. At this high temperature nuclear reactions (as described previously) began to occur, causing a much greater increase in temperature until the sun developed its present characteristics.

Left behind during the contraction was some five to ten percent of the original protosun. From this huge gas cloud or *solar nebula* the planets and satellites evolved. This concept of the evolution of the sun and the surrounding solar nebula seems now to be well established. However, the mechanism whereby the planets and satellites formed is less resolved. Alternative theories have been developed that describe how localized gravitational centers could have formed the primitive planets and satel-lites within the primordial nebula. One of the points of great philosophical significance about these modern concepts is that the development of a solar system appears to be a rather normal phase of the evolution of a star. Hence, countless other planets may circle other suns, with some having the proper composition and lying at just the proper distance to permit life as we know it to be supported.

Let us finally focus attention on the very primitive earth that at the time of complete accumulation from the solar nebula consisted of a mass of cold particles having a very mixed composition. As with the sun, gravitational contraction occurred, with heavy material settling to the central region and releasing heat as a consequence. The energy released by radioactive material also accumulated as heat. Most of the heat developed internally by both processes was retained in the interior because of the poor heat conductivity of terrestrial rock material.

Ultimately the central region reached the melting point, resulting in even more rapid segregation of heavy and lighter materials. The earth's heavy molten metallic core formed in this way. Above the core lay the mineral and rock material of the earth's mantle, which also became stratified into layers with the densest material at the bottom. The uppermost, light rock separated into the earth's crust which, as we have seen, was already well formed at least four billion years ago.

In later chapters we will reexamine the earth's interior and consider the methods by which scientists have discovered its nature.

STUDY QUESTIONS

1.1. Contrast the main point(s) of the geocentric and heliocentric concepts of the solar system.

1.2. What are several arguments (at least three) that supported the heliocentric theory?

1.3. Explain the process by which the sun develops its high temperature.

1.4. How much energy would be released by the complete conversion of 10 grams of mass into energy? How much coal must be burned to release an equivalent amount of energy?

1.5. The angular diameter of the moon in the sky (viewed from earth) just about equals that of the sun. As the mean distance of the moon from the earth is 240,000 miles, calculate the diameter of the moon by use of the information and procedure given with Fig. 1-5.

1.6. How do sunspots permit us to determine the sun's speed of rotation?

1.7. Define perihelion, aphelion, perigee, apogee.

1.8. The period of revolution of Neptune about the sun can be observed to be very nearly 165 earth years. By means of Kepler's third law of planetary motion calculate the mean distance of Neptune from the sun in astronomical units.

1.9. Which objects in the solar system shine by being self-luminous? Which shine only by reflected sunlight?

1.10. Explain why eclipses of the sun and moon do not occur at least once each month.

1.11. Draw a diagram to show the positions of earth, sun, and moon that would result in an annular eclipse of the sun—an eclipse in which a thin ring of the sun surrounds the eclipsed portion.

1.12. Explain the difference between the synodic and siderial lunar months.

1.13. What is the fundamental reason why the moon's surface appears very different from that of the earth's?

1.14. Why do we always see the same side of the moon?

1.15. What is the relationship of the "ecliptic" to eclipses.

1.16. Summarize some of the motions a person "at rest" on the earth actually undergoes.

1.17. Explain the "precession of the equinoxes."

chapter 2

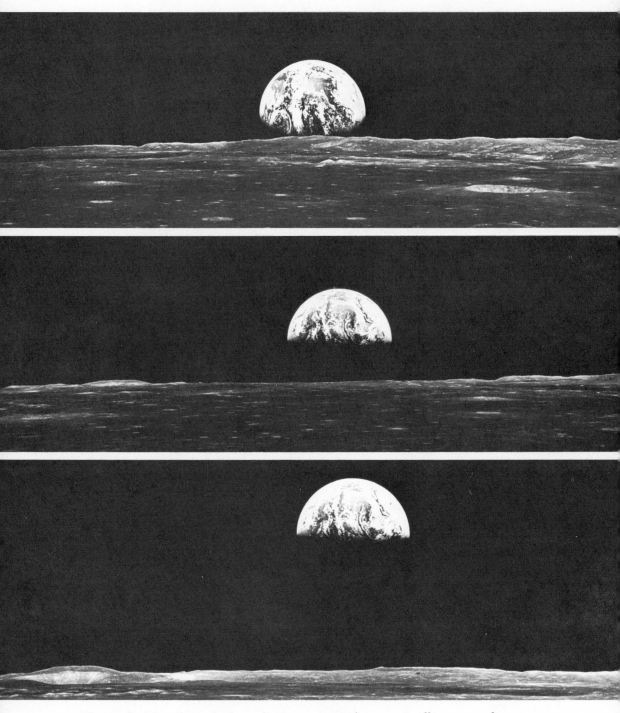

The earth rising above the lunar horizon as seen from an Apollo spacecraft. As the moon revolves about the earth, an observer on the moon would see the earth exhibit "phases" similar to those of the moon seen from the earth. (NASA Photo.)

The Planet Earth

To an observer with an unobstructed view of the horizon, the small part of the earth that is visible certainly appears to be essentially flat. But the concept of a flat earth was already obsolete with many of the scholars of ancient Greece. These thinkers who regarded the sphere as the most perfect of forms concluded that the earth must be spherical. Certainly the sun and moon appeared to be quite circular and a circle is but the cross-sectional view of a sphere.

The idea of the earth's sphericity is implicit in the records of the epic voyage of Pythias about 325 B.C. This remarkable Greek astronomer and mathematician traveled from the Mediterranean coast of southern France through the Straits of Gibraltar and then up along the western coast of Europe. He appears to have circled through the North and Baltic Seas and then reached as far as the coasts of Norway and Iceland (Fig. 2-1). His penetration into high northern latitudes is indicated by the increased length of the daylight period. This was reported to reach 22 hours of continuous sunshine with only a couple of hours of darkness. These observations, together with Pythias' observations of the changing positions of the sun and stars, would not be explainable by other than latitude changes on a spherical earth.

Despite the very early breakthrough by ancient scholars like Pythias, Plato, Aristotle, and Eratosthenes among others, the concept of a spherical earth and what it implied became submerged in the intellectual darkness

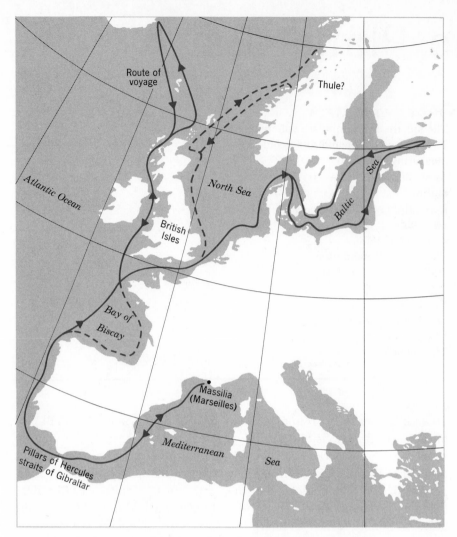

FIG. 2-1. The voyage of Pythias, about 325 B.C. Details of this voyage no longer exist but historians have deduced possible routes based on accounts of other early writers. A synthesis of these routes is shown here.

of the middle ages. Not until their reexposure by the ideas of Copernicus and Galileo and the magnificent voyage of circumnavigation of Magellan beginning in 1519, did the modern idea of the earth's true shape take form.

SHAPE AND SIZE OF THE EARTH

The Sphere and the Oblate Spheroid

A statement of the earth's shape can be made to different degrees of accuracy (all of them correct) for different purposes. We all know now that the earth is essentially spherical. For most everyday purposes, it is adequate to state that the earth is a sphere, as believed by the ancient Greeks, and that its diameter is about 8000 miles.

For more refined scientific purposes this statement is not sufficiently accurate because the earth's real form is not quite that of a perfect sphere but of an *oblate spheroid*—a sphere flattened at the poles and bulging at the equator. This first good approximation to the shape of the earth was deduced by Isaac Newton, who reasoned that forces resulting from the earth's rapid rotation should cause a slight decrease in the value of gravity at the equator compared to the value at the poles. The resulting oblate spheroid differs from a sphere in having a slightly compressed polar diameter (the axis of rotation of the earth) and an extended equatorial diameter, as shown in Fig. 2-2.

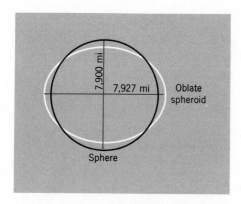

FIG. 2-2. Relationship between a sphere and the oblate spheroid of equal volume as seen in cross-section.

Recent accurate measurements of the size of the earth have established the equatorial diameter to be very nearly 7927 miles and the polar diameter, 7900 miles. The flattening of the sphere into an oblate spheroid thus amounts to only 27 miles out of 7927. The ratio of these numbers, 27/7927, or about 1/300, is used to describe the earth's oblateness or ellipticity.

It has been calculated that if the earth were composed entirely of water, its rotational shape would be very nearly that of the present spheroidal or ellipsoidal earth. This interesting conclusion suggests that the rock material composing the earth has actually responded to the mechanical stress of rotation as if it were a mobile liquid. In Chapter 7 we will return to this very interesting property of rocks whereby they respond as though liquid or plastic to the effect of a prolonged stress.

The Reference Ellipsoid and Geoid

Even the more refined shape of the oblate spheroid is not good enough for accurate map-making and locations of points on the earth's surface. The actual survey measurements neede ̄ ̄r maps are performed on the irregular earth's surface—for example, the surface shown in Fig. 2-3. Obviously the measurements must be referred to some more regular surface in order to make the computations that establish map locations. For this purpose, the *ellipsoid of revolution* was invented. An ellipsoid of revolution is simply the three-dimensional form that is developed when an ellipse is rotated about its short axis and is the same as the form called an *oblate spheroid*. If the earth had a smooth surface and uniform composition, its oblate shape would be the same as an ellipsoid of revolution—which is thus the shape of a fictitious earth. Geodesists (scientists concerned with the shape of the earth) have tried to develop a reference ellipsoid that would give the best fit to the earth's surface shape. The relationship of this ellipsoid to the surface is indicated in Fig. 2-3. To get this best fit, ellipsoids with ellipticities of 1/293 to 1/300 have been used for different parts of the earth's surface.

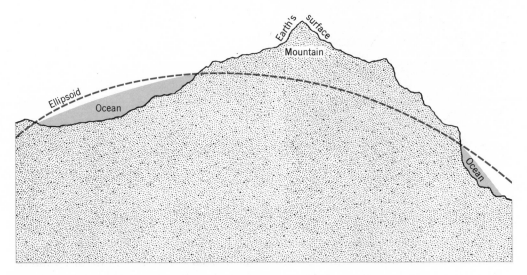

FIG. 2-3. Relationship of the earth's irregular surface to the uniform surface of a reference ellipsoid. Measurements performed on the irregular surface are transferred to the reference ellipsoid for the computations involved in map-making and the accurate locations of particular places.

Most maps with which we are now familiar are based on one or the other of these theoretical ellipsoids. But even the use of these refined shapes was not good enough, as shown by the sad fate of many aviators flying great distances over unmarked seas during World War II. Islands scattered through the vast Pacific Ocean were often located too inaccurately to be found at night or other times of poor visibility. Even places in Alpine regions of central Europe were inaccurately placed. However, there were enough surface reference guides and airline beacons to provide for successful locations of places within continents. Cartographers came

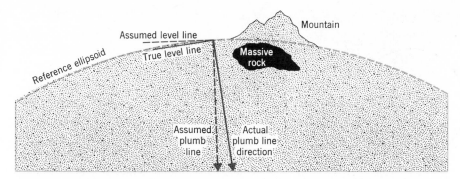

FIG. 2-4. Effect of more massive subsurface material or of an adjacent mountain in disturbing the plumb line from a direction parallel to the radius. The true level surface is normal to the plumb line direction rather than the radius.

to realize that the surface of the theoretical reference ellipsoids (or oblate spheroid) of the earth was simply not satisfactory for the construction of high-quality maps.

The trouble lay in the fact that the reference ellipsoid was fitted to the earth's topographic surface instead of to the true level surface of the earth. When measurements are performed on the earth's surface, the surveyor's level is parallel not to the topographic surface and not to the reference ellipsoid surface, but to the earth's true level surface or *geoid.* A level surface is one that is everywhere perpendicular to the force of gravity.

As water always spreads out to a level surface, the geoid corresponds to the position of mean sea level over oceans. On land, the geoid can be imagined as the form that would be shown by the surface connecting a network of narrow sea-level canals that crisscrossed the continents. This is really an undulating, irregular surface compared to the smooth reference ellipsoid because the level is locally deflected by inequalities of mass beneath the surface as well as by topographic irregularities, as shown in Fig. 2-4.

The determination of the geoid, which is not yet complete, has been carried out in two ways. One method is to measure the variations in gravity over the earth's surface. From these measurements the level or geoid surface is determined. Another procedure is based on the study of the orbits of artificial earth satellites. For example, when *Vanguard I,* the first United States artificial satellite, was lofted into orbit, some very interesting observations were made. The theoretical orbit differed from the observed orbit. To correct the former, a "pear-shaped component" had to be added to the geoid, as illustrated in Fig. 2-5.

In Fig. 2-6 we have a summary of the relationship of the earth's topographic surface, the reference ellipsoid and the geoid. Note how the latter corresponds to the surface of the ocean, which is below the reference ellipsoid.

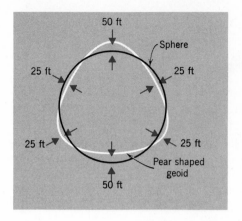

FIG. 2-5. The "pear-shaped" aspect of the geoid and a sphere of equal volume. In the latter, the value of the departure between the two surfaces is shown. (After R. Jastrow, *Missiles and Rockets, Transactions, American Geophysical Union.*) Compare this diagram to that in Fig. 2-2.

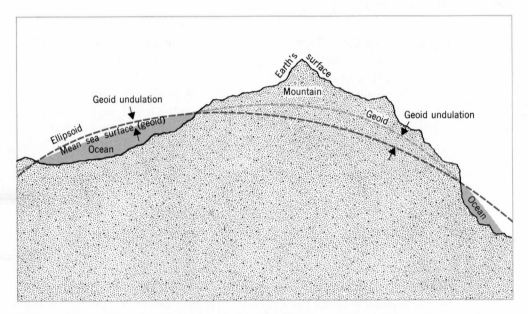

FIG. 2-6. The geoid surface is now added to the earth's surface and the reference ellipsoid shown in Fig. 2-3. The map computations performed on the reference ellipsoid must be corrected for the difference between this surface and the geoid which is the best figure obtainable for the true "level" shape of the earth.

As the shape of the geoid becomes established through gravity measurements and satellite observations, the results are plotted as in Fig. 2-7. This shows the actual difference in elevation between the reference ellipsoid and the geoid at any point. Positive values mean the geoid surface is above the ellipsoid surface and negative values, the reverse. From maps such as these, final corrections can be made to the positions of points located through the use of a more uniform reference ellipsoid.

In summary, the earth, which is a fairly good sphere, can be described

FIG. 2-7. Map showing the departure of the geoid from the reference ellipsoid as developed by the Applied Physics Laboratory of Johns Hopkins University. Continuous curved lines are contours of equal departure shown in meters. (A meter is equal to 3.28 feet.) Positive values lie above the reference surface and negative values below.

better as an oblate spheroid. The theoretical reference figure of the earth is the ellipsoid of revolution that best fits the earth's irregular surface. In cartography, maps constructed with respect to the reference ellipsoid must be corrected to the surface of the geoid—the undulating level surface of the earth.

Measuring the Earth

Strangely, although it was not until the days of the European Renaissance that the earth's spherical nature was firmly established, the assumption that it was a sphere was implicit in the earliest measurements of its size. The earliest accurate determination, by *Eratosthenes* in Alexandria, Egypt, about 200 B.C., can be duplicated by any schoolboy who knows the distance between two points in a north-south line.

Eratosthenes followed the two assumptions of Greek scholars of his time that the earth was spherical and that the sun's rays were parallel to one another (because of the sun's great distance). He knew that on June 21 (the summer solstice) the noon sun was vertically overhead at Syene, Egypt, where its image was seen directly below in a narrow well. At Alexandria, about 5000 stadia (an Egyptian measure) north, it was observed that a vertical rod cast a short shadow at noon, forming an angle of 7°12′ (7.2°) between the sun's rays and the rod.

The geometry related to these assumptions and observations is shown in Fig. 2-8, where the rods are seen to converge at the earth's center if continued down below the surface. Therefore, the angle of 7°12′ between the sun's rays and the rod at Alexandria must equal the central angle (latitude angle) between the radii for Syene and Alexandria. Also, because a central angle is measured by its arc, we can solve for the earth's entire circumference by the proportion:

$$\frac{5000 \text{ stadia}}{7.2°} = \frac{x \text{ stadia}}{360°}$$

$$x = \frac{5000 \times 360}{7.2} = 250,000 \text{ stadia}$$

This solution is thus the first known realistic determination of the circumference of the earth. Although stadia of different lengths were in use, one of the reasonable conversion factors results in a circumference of close to 26,000 miles. This is surprisingly accurate in view of the known limits of measurement at that time.

The present method of determining the earth's circumference has also been to extrapolate from the measurements over small distances. In this procedure the mean distance for a degree of latitude (69.055 miles) is multiplied by 360° to get the polar circumference of 24,860 miles. The equatorial circumference, which is obtained from the mean distance between degrees of longitude (69.172 miles) measured along the equator, is again multiplied by 360, giving 24,902 miles.

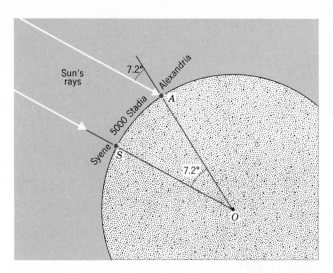

FIG. 2-8. How Eratosthenes made the first measurement of the size of the earth. The angle at the center of the earth, AOS, was determined to be 7.2°. The circumference is found from the proportion 7.2°/5000 stadia = 360°/x, where x is the entire circumference.

LATITUDE AND LONGITUDE

Terrestrial latitude and longitude make up the basic reference grid of the earth. They are as important in studying earth science as are street and road maps in finding our way about a city or countryside, to say nothing of their vital part in marine and aerial navigation.

Latitude

One way of expressing the latitude of a place is to refer to its distance in degrees north or south of the equator as measured along a north-south arc (meridian) on the earth's surface. The latitude of the equator is thus 0° and of the north and south poles, 90°N and 90°S, respectively. Lines connecting places of the same latitude must be circles that lie in planes parallel to the equator and to each other, with the equator being the largest of these circles, whose size becomes progressively less toward the poles. Of the latitude circles, only the equator is a *great circle*, one that connects opposite ends of a diameter of a sphere.

The concept of *geocentric latitude*, strictly applicable to a perfect sphere, is often used in science. It is the angle subtended at the center of the earth by lines from two points on a given meridian, one point being on the equator and the other at the position whose latitude is described (point "*p*" in Fig. 2-9). Notice that the geocentric latitude angle lies in a

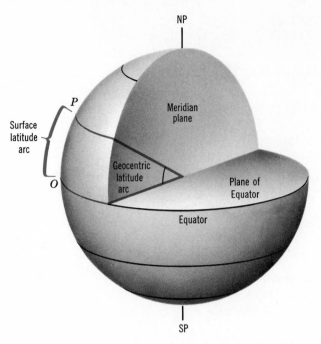

FIG. 2-9. The relationship between geocentric and surface latitude for a perfect sphere.

north-south plane (meridian plane) perpendicular to the equatorial plane.
If the earth were a perfect sphere, surface latitude would always equal
geocentric latitude. However, the oblateness of the earth causes the
amount of curvature to change continuously from equator to poles. The
curvature of the surface is greatest at the equatorial bulge and least at
the flattened polar zone.

Because of the decreased curvature—or increased flattening—with
higher latitudes, the distance between parallels of latitude a degree apart
increases progressively toward the poles. The reason for this is apparent
in Fig. 2-10. The heavy line represents a cross section of the earth from
the north to the south pole. If the lesser curvature of the polar regions
were developed into a complete sphere, it would then form the sphere
with the larger cross section, A. The greater curvature of the equatorial
region could be developed into a sphere of smaller cross section, B. A
degree of latitude as measured by equivalent central angles of each circle
intercept arcs of different lengths. The polar arc must therefore be larger
than the equatorial arc for the same latitude angle. These diagrams are
of course exaggerated; the actual variation in the length of a degree of
latitude from equator to poles is given in Table 2-1.

Determining the Latitude

In practice, latitude is measured by astronomical observation of the angle
above the horizon (altitude) of the sun or of a particular star at noontime,

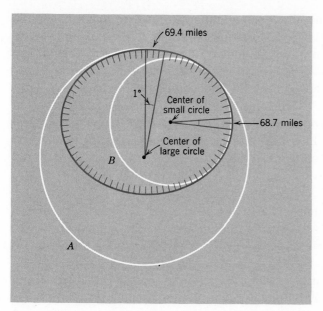

FIG. 2-10. The variation of the length of a degree of latitude from equator to poles results from the decreased curvature of the earth with increasing latitude. The flatter polar region matches the curvature of the larger circle with a longer radius, whereas the shape of the equatorial region matches that of the greater curvature of the smaller circle with a shorter radius. Equivalent central angles thus subtend arcs of different linear size, as shown. (The 1° angles are not to scale.)

TABLE 2-1 Variations in the Length of a Degree of Latitude for Each Ten Degrees from the Equator to the Poles

Latitude	Length of a Degree of Latitude (miles)
0°	68.704
10°	68.725
20°	68.786
30°	68.879
40°	68.993
50°	69.115
60°	69.230
70°	69.324
80°	69.386
90°	69.407

Source. From S. Gannett, U. S. Geological Survey Bulletin 650.

the time it crosses the observer's meridian. To use the sun, for example, some simple geometric relationships permit the determination of the observer's latitude once the altitude of the sun is measured at noon. This determination is least complicated on the equinoctial days (March 21 and September 23). At these times, the sun is directly overhead at noon at the equator. But at 1° north or south of the equator, the altitude will be 89°; at 2° north or south, 88°; and so on until at 90° north or south the altitude will be 0°. Thus, if the altitude can be measured, the latitude can be found, using the simple formula

$$L = 90° - A$$

where L is the latitude and A is the altitude of the sun. The geometry can be seen in Fig. 2-11. Because the sun is so distant, its rays are always parallel to one another as shown by the broken lines. The sun is directly overhead (altitude = 90°) at the equator and has an altitude, A, for the observer at 0. Because the point directly overhead is called the zenith, Z is defined as the *zenith angle*. Angles Z plus A equal 90°. Also, angles Z and L are equal (recall your elementary geometry). Hence, $L = 90° - A$.

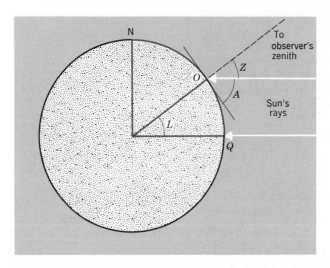

FIG. 2-11. The determination of latitude from the altitude of the sun on the equinoxes. L is the latitude, Q the equator, O the observer's position, A the noon altitude of the sun, Z the zenith distance, and N the North Pole. The arc NOQ is the observer's meridian.

At all times other than the equinoxes (as in Fig. 2-12), the sun is at some point in the sky either north or south of the equator; the angular distance from the equator is called the sun's declination, D. The declination of the sun for every day of the year (a critical value in latitude computation), can be found in the nautical almanac or other astronomical

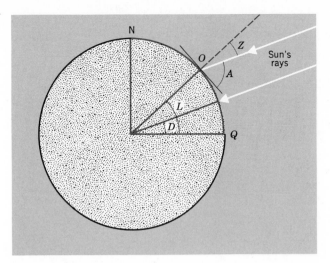

FIG. 2-12. The determination of latitude from the observation of the sun's altitude in the Northern Hemisphere when the sun has a positive declination, as indicated by the angle D. Other nomenclature is the same as in Fig. 2-11.

tables. When the sun is north of the equator, its declination is given as positive; when it is south, as negative. In the previous case with the sun overhead at the equator, the latitude was simply $90° - A$. But with the sun overhead at some latitude north of the equator, as in Fig. 2-12, $90° - A$ only yields part of the latitude angle, namely, angle Z. To get angle L we must add angle D, the sun's declination, to angle Z, which is equal to $90° - A$. Hence, when the sun is north of the equator, between March 21 and September 23, for a person in the northern hemisphere,

$$L = (90° - A) + D$$

In fall and winter, when the sun is south of the equator, we subtract D to get the latitude, or

$$L = (90° - A) - D$$

Figure 2-13 shows the setup if the sun's declination is greater than the observer's latitude, as would occur if the sun were overhead at 20°N and the observer were at 10°N. From the reasoning in the prior two cases, it is a simple exercise to show that

$$L = D - (90° - A)$$

Longitude

The other important element in our geographic grid system is longitude. Meridians, defined previously as north-south lines passing through the

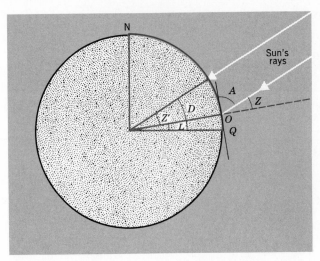

FIG. 2-13. The determination of latitude from the altitude of the sun in the Northern Hemisphere when the observer's latitude is less than the sun's declination. The nomenclature is similar to that in Fig. 2-11 and 2-12.

poles, are the reference lines used in longitude description. Longitude is the distance in degrees between the prime meridian (0°), which passes through Greenwich, England, and the meridian passing through a given point. The actual measurement can be performed along the equator or any parallel of latitude between the prime meridian and the given meridian.

It is clear from Fig. 2-14 that meridians must all pass through the north and south poles and thus converge to a point at the poles. Hence, the maximum linear distance between two meridians one degree apart is at the equator, where the value is 69.172 miles. If the earth were a perfect sphere, the decrease in this distance poleward would occur according to the cosine of the latitude. For example, at 60° (cosine = 0.5), the distance between meridians should be one half the equatorial separation, or 34.586. The actual value at 60° for the real earth differs very slightly, being 34.674 miles.

Note that all meridians are great circles because they connect opposite ends of the diameter of a sphere—the earth's axis. In the case of latitude circles, only the equator is a great circle.

A measurement of great value, particularly in navigation, is the length of a degree of longitude measured along the equator. Recall that this was determined to be 69.172 miles. If we multiply this number by 5280 feet (the number of feet in the conventional statute mile), we obtain 365,228.16 feet per degree along the equator. By dividing this by 60, the number of minutes in a degree, we get 6087.14 feet per minute of arc of the equator.

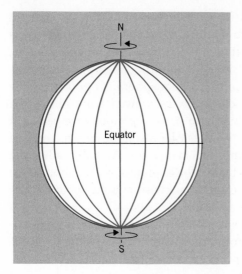

FIG. 2-14. The convergence of meridians of longitude from the equator to the poles. Curved arrows show direction of the earth's rotation.

This distance has been called the nautical mile and is equal to 1.15 statute miles. Quite good approximations of the linear distance between points can be made from any map or chart containing meridians by remembering that a minute along the meridian is about 1 nautical mile; a degree, 60 nautical miles; and so on. Then, multiplying nautical miles by 1.15, statute miles can be found. For example, mark the distance between two points on a map along the edge of a card. Then lay the card along the meridian and determine the number of degrees and minutes intervening between the points and compute quickly the number of nautical or statute miles involved. Your answer will be a good approximation of the correct distance.

In this book, the statute mile of 5280 feet is always intended unless the nautical mile is specified.

THE EARTH'S ROTATION

The spinning of the earth on its axis has tremendous importance in all of natural science. Any variation in the earth's rate of spin or rotation is so slow a change and so small an amount as to be negligible for practical purposes.

Although any point on the earth's surface (except at the poles) turns at a constant angular rate of 15 degrees per hour, the linear rate, as is evident from Fig. 2-14, is maximum at the equator and zero at the poles. The equatorial rotation rate is 1036 miles per hour (the earth's circumference, 24,860 miles, divided by 24 hours). This rate decreases poleward by the cosine of the latitude; at 60° north or south it is thus 500 miles per hour.

If we were to view the earth from some vantage point in space, it would rotate in the direction shown by the arrows about the polar regions in Fig. 2-14. In describing the direction of rotation of the globe as a whole, terms such as "westward" or "eastward" have little meaning. The movement shown by the arrow in the northern hemisphere is opposite to the movements of the hands of a clock. Thus, as viewed from above the northern hemisphere, the earth rotates in a *counterclockwise* sense; if viewed from above a point in the southern hemisphere, this movement is *clockwise,* as indicated by the appropriate arrows. Actually, all of the planets, when viewed from above their orbital planes, rotate about their axis and revolve about the sun counterclockwise similar to the earth, except for Uranus, which appears to rotate clockwise.*

Because the earth's rotation is so smooth and uniform, it is not possible for us to "feel" it directly. For this reason, the movement of the heavens about the earth, which is purely a consequence of rotation, was, as we discussed earlier, misinterpreted for so long as movement about the earth.

The Foucault Pendulum

There are many consequences of rotation: the diurnal revolution of the objects in the sky—sun, moon, stars, planets; the development of the oblate shape of the earth; the deflection of winds, currents, and other materials moving freely over the surface; the circulation of the atmosphere and oceans, and so on. But for persons living on the earth, there is only one good experimental proof or verification of rotation—the classical *Foucault pendulum* experiment. The principle of the Foucault pendulum is so fundamental, not only for the proof of rotation, but also for the understanding of winds and ocean currents, that we must grasp it in some detail.

This experiment is based on two factors: (1) as the earth rotates about its axis, any small surface area has a component of this rotation about a local vertical axis, and (2) once a pendulum is set in motion it tends to swing in the same direction in space. The first factor will be explained in detail because it lies at the heart of the principle of the Foucault pendulum.

Imagine that you can view the earth as in Fig. 2-15. The Line NS is the earth's axis of rotation running through the center of the earth at C. The plane at N represents a small horizontal portion of the earth's surface centered exactly at the north pole. NC, a radial axis from the earth's center to the pole point, coincides with the axis of rotation in this case. As the earth turns, it is easily evident that the plane at N will make one complete rotation about N in one day as summarized in the small inset to the right.

*The apparent clockwise rotation of Uranus can be explained by the overturning of its axis of rotation so that its north pole is below the orbital plane.

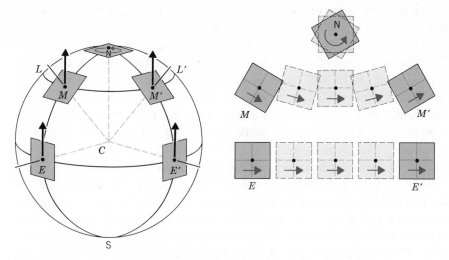

FIG. 2-15. Explanation of the Foucault Pendulum effect (see text).

Now consider the plane M at some latitude LL' between the pole and the equator. The plane at M lies along the meridian $NMES$, which in a given time rotates to $NM'E'S$. Plane M has been drawn with two sides parallel to this meridian for ease in reference. Line CM is the radial axis from the center of the earth to the point M in the center of the plane. Only at the pole, N, does the radial axis to the surface coincide with the earth's axis of rotation. This is most important. As the earth rotates, plane M is carried to M'. Note how the orientation of the plane has changed compared to the fixed arrow direction. This line, and in fact the entire plane (really a portion of the earth) has turned about the point M and the radial axis CM while the plane rotated from M to M' about the axis of rotation NCS. This turning of the plane is summarized in the inset to the right. In other words, if you stand anywhere between the poles and the equator, any small surface area of the earth is actually rotating about the spot on which you are standing. Although not obvious, this is evident from considering Fig. 2-15.

At the equator, as meridian NES rotates to $NE'S$, plane E is carried to E'. Note that the orientation of the plane has not changed nor has there been any rotation about the radial axis CE'. The orientation of the plane relative to this axis remains constant as summarized in the inset. At all times the plane E maintains the same orientation to the earth's axis of rotation, NCS.

In summary, while the polar surface turns 360° about the earth's axis of rotation in one day, a surface area anywhere else also rotates about a local axis that is different from the axis of rotation. The amount of surface rotation grows progressively less the greater the distance from the pole, until at the equator it is zero. The amount of daily rotation thus varies with the sine of the latitude angle (0 at the equator and 1 at the poles). For example, at 30° (sine of 30° = $\frac{1}{2}$) a local surface area will undergo one half of a full rotation about its local vertical axis in one day.

The movement of these reference planes could not occur without the earth's rotation. In 1851, L. Foucault, a French physicist, reasoned that if he could show such a motion, he could prove that the earth is rotating. Recall that once set in motion, a freely swinging pendulum tends to swing toward the same point in space. In Fig. 2-15, assume that the pendulum is set oscillating in the direction of the arrows. At M' the pendulum will still swing in the same direction but by this time the corner to the right is lined up with the arrow. An observer on the earth, not being aware of the actual rotation of the surface, sees only the relative motion between pendulum and ground. To him the direction of swing of the pendulum *appears* to change and in a direction opposite to that of the rotating surface. At the equator, no relative motion between arrow (pendulum) and the surface occurs.

On the basis of this reasoning, Foucault erected a large pendulum in the dome of the Pantheon in Paris. A heavy cannonball was suspended from the inside of the dome by a wire 200 feet long. Because the period of oscillation of a pendulum depends on the length of the support, the long wire provided a very long period with a very slow rate of swing, thereby reducing friction and permitting the pendulum to oscillate for more than a day. The cannonball hung just above a layer of sand. As the weight moved, a pin projecting from its bottom traced a fine line in the sand. The direction of the line changed slowly during the day in a clockwise direction at about $11\frac{1}{4}$° an hour. Actually, the pendulum direction remained fixed while the floor beneath rotated in a counterclockwise direction at a rate appropriate to its latitude. If we take 49° as the latitude of Paris, the daily rotation rate of the surface about the pendulum is

360° × sine of 49° = 271.2°

This simple experiment was, and is, the definitive experimental proof of the earth's rotation. The rotation of a given position of the earth's surface is most fundamental to the behavior of the atmosphere and oceans, resulting in very significant deflections of moving currents of air and water that will be discussed in later chapters.

Modern photographs and observations from space have provided the second and more striking observation of the earth's actual rotation. The five striking views (frontispiece) of the earth were taken by the NASA Applied Technology Satellite that maintained a fixed position 22,300 miles above South America near the mouth of the Amazon River. The satellite retained a fixed position because its speed was synchronous with that of the earth's rotation. The circle of illumination (or sunrise line) just crosses South America at 7:30 a.m. As the earth rotates, this line progresses westward until at noon, the entire half of the earth beneath the satellite is illuminated. Following noontime, the "sunset line" (the other side of the circle of illumination) moves into view and crosses South America at 7:30 p.m., completing one daylight cycle. Visible in the photos are all of South America and portions of North America. Africa, Europe, Greenland, and cloud-covered Antarctica. Several major storm-cloud patterns are also evident.

PHYSICAL PROPERTIES OF THE EARTH

A number of vital statistics should be noted when considering the earth as a planet, in addition to size and shape.

The surface area of the entire earth, which can be calculated from the other dimensions given previously, is very nearly 197 million square miles (197×10^6 miles^2). As the globe is about 71 percent water-covered, this area is close to 140 million square miles, while the remaining 29 percent of land equals about 57 million square miles. From our knowledge of the mean radius we can calculate the volume of the earth according to the familiar formula $V = \frac{4}{3} r^3$, where r is the mean radius of the earth.

If we use 3958 miles for r, the volume (v) in round numbers is 250 billion cubic miles (250×10^9 miles^3).

The mass, or amount of matter in the earth, is another important value. We cannot, of course, weigh the earth to determine this quality. The method of finding the earth's mass is of such significance in science as to be worth reviewing.

To do this, let us imagine that the entire mass of the earth is concentrated at the center. This is mathematically permissible for a perfect sphere, which the earth approximates closely.

Newton's classical law of gravitation states that objects attract each other with a force directly proportional to the product of their masses and inversely proportional to their distance between them or

$$F \, \alpha \, \frac{M_1 M_2}{d^2}$$

From this law we can express the earth's attraction on a small, unit mass at the surface as

$$g = \gamma \frac{M_e}{r^2}$$

where g is the acceleration of gravity, γ (gamma) is the experimentally determined constant of gravitation, M_e is the mass of the earth and r is the radius of the earth (the distance from the center to the unit mass). The mass, M_e, can be determined from

$$M_e = \frac{gr^2}{\gamma}$$

This equation is easy to solve if we use metric units of measurement wherein the acceleration of gravity, g, is 980 cm/sec^2; r, the mean radius, is 637,000,000 (637×10^6) cm; and γ is 6.67×10^{-8} dyne-cm^2/sec^2.

$$M_e = \frac{980 \text{ cm/sec}^2 \times (367 \times 10^6 \text{ cm})^2}{6.67 \times 10^{-8} \text{ dyne-cm/sec}^2}$$

$$= 6 \times 10^{27} \text{ grams}$$

or

6.6×10^{21} tons

From the volume and mass just determined, we can calculate the density of the earth. This basic property compares the weight of a body with the weight of an equal volume of water. Density is more fundamental a physical property than mass because it tells us how packed or compressed the material is. Mass alone is a function of both the size and the density and tells little about the basic physical makeup of an object. Density (D) is expressed as the ratio of the mass (M) to the volume (V) or

$$D = \frac{M}{V}$$

It is convenient to use the metric system in density measurement because the value for pure water is 1 gm/cc (4°C), which makes numerical comparison very simple.

If we use metric equivalents for mass and volume, the density of the earth is found as

$$D = \frac{5.98 \times 10^{27} \text{ g}}{1.083 \times 10^{27} \text{ cc}}$$

$$D = 5.52 \text{ g/cc}$$

The value just computed is the average density for the entire earth and means that the earth weighs 5.52 times as much as the weight of a sphere the size of the earth composed of pure water. This value, as is evident in Table 1-1, is the highest density of any planet in the solar system. The detailed variations in density of the different layers and portions of the earth, which depart significantly from the mean values, will be considered in later chapters.

TIME AND THE CALENDAR

One of the few things in nature that is essentially beyond the control of man is the inexorable passage of time. We can define it, use it, measure it, and relate it to all sorts of events, but we cannot alter, modify, or control it. The affairs of man would fall into complete chaos were it not for our common standards of time, to which all people refer.

We have two different standards of time, both related to motions of the earth. Daily "clock" time is based on the earth's period of revolution about the sun, with monthly subdivisions based on the moon's synodic period.

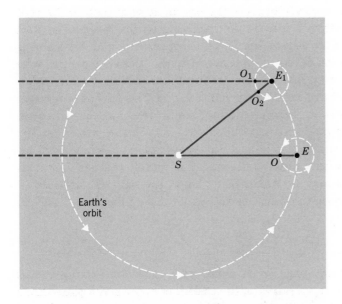

FIG. 2-16. The effects of rotation and revolution in determining the length of the solar day.

Clock Time

The length of the solar day is one of our universal units of time. The solar day is the time interval between successive passages of the sun across a given meridian. It is evident in Fig. 2-16 that the earth must rotate through an angle a little more than 360° (361°) in a solar day. When the earth is at E, the sun is on a meridian for an observer at 0. As the earth rotates, it also moves in its orbit, reaching E_1 when the sun is once again on the observer's meridian at 0_2. Note that the earth has turned beyond 360° in order for the sun to cross the same meridian. Because the period of revolution of the earth about the sun is very nearly 360 days, the arc 0_1–0_2 is very nearly one degree, so that the earth rotates very nearly 361° in a day. The time for this rotation (of 361°) is the solar day, which is divided into 24 hours; each hour is subdivided into 60 minutes and each minute into 60 seconds. A second is thus 1/86,440th of a solar day.

Although the earth's rate of rotation is very uniform, its rate of revolution is not. Recall that as a consequence of Kepler's second law,

(p. 13), the earth travels at different speeds in its orbit depending on its distance from the sun. Hence, the earth rotates for a little more than 361° between successive meridian crossings of the sun during some parts of the year, and a little less than 361° during others. In other words, the actual length of a solar day is a slightly variable quantity during the year. To remove the obvious difficulties that would arise from timekeeping by the real sun, we use a fictitious or mean sun that crosses the meridian at regular intervals. The resulting time, called *mean solar time*, is obtained by taking the average length of all of the *apparent* solar days in the year and dividing this interval into 24 hours, giving us the mean solar day.

Time based on the actual sun is called *apparent solar time* and is used by navigators and others in "shooting the sun." For such people, who must use the time when the real sun crosses the meridian, a relationship known as the equation of time has been developed. This tells us how much the real sun is ahead or behind the mean sun; for example, when an accurate clock registers 12 noon, the fictitious or mean sun is on the meridian but the actual sun may be a number of minutes before meridional passage, or may have crossed the meridian minutes earlier. Equation of time corrections may be found in tables in the nautical almanac or can be estimated from the diagram known as the *analemma*, Fig. 2-17.

Local and Standard Time

The *local time* for any place is its mean solar time. For example, local noon is the time the mean sun is on the meridian of the given place. Each meridian thus has its own local time. The time differences for different meridians can be realized from the fact that as the earth rotates, 360° of longitude turn under the sun in one day, or 15° in one hour, 1° in four minutes, and so on. Places 15° of longitude apart are also one hour in time apart. Because the earth turns from west to east, for each 15° of longitude to the west of a given place, time becomes progressively one hour earlier, and the reverse. Since the time as measured by the sun is different at meridians even seconds of an arc apart, we have adopted the convention of standard time zones. Accordingly, each meridian that is a multiple of 15° beginning with 0° is the center of a standard time belt that extends $7\frac{1}{2}$° to the east and west of the central meridian. This, the reference for Eastern Standard Time (EST) in the United States, is the meridian 75° W, and Eastern Standard Time is thus 5 hours earlier than the 0° or Greenwich Meridian Time (GMT). The boundaries of the time zones are often irregular in order to encompass a complete entity such as a city, an island, and the like. The distribution of time in relation to longitude over the earth is illustrated in Fig. 2-18.

For scientific and technical purposes it is necessary to have a single reference time that will be the same for all observers anywhere on the earth. For example, earthquakes (discussed in Chapter 10) are abrupt and

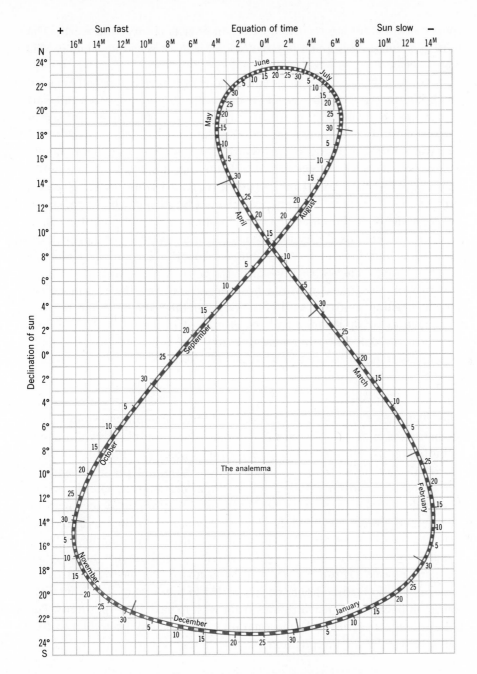

FIG. 2-17. The analemma. This diagram gives the difference between mean and apparent solar time as well as the sun's declination for each day of the year. For any day of the year, the time difference is found directly above along the upper horizontal margin, and the declination along the left hand margin. Thus on September 20, apparent solar time is 6.5 minutes ahead of mean solar time, and the sun's declination is 1.5° North.

special events that are recorded by seismologists all over the world. The time of occurrence is given in Greenwich Mean or Meridian Time (GMT) just described.

To avoid the serious time complications that have historically caused much confusion, the 180° meridian has come to play an important part in our time convention. Because the earth rotates from west to east at 15° of longitude per hour, time grows progressively earlier as we travel westward and later as we go eastward. If one were to travel completely around the world from west to east, losing one hour per 15°, an entire day would be lost. And if one traveled eastward from the same place, gaining one hour per 15°, an entire day would be gained, whereas a person not traveling from the same spot would notice no disruption of the calendar. Such a real possibility inspired the whimsical story of Edgar Allen Poe, *Three Sundays in a Week* and led to the happy ending of Jules Verne's *Around The World In Eighty Days*.

To avoid this time confusion, the *international date line* was invented. This line, which approximately coincides with the 180th meridian, departs from it to avoid splitting islands or groups of islands (Fig. 2-18). When you cross the date line from west to east, a day is immediately dropped; when crossing east to west you must add a day.

Navigation.

The relationship between Greenwich and local time provides one of the elements essential to celestial navigation. Navigation requires the determination of the latitude and longitude of the observer. Although accurate electronic navigation is now available to those ships and planes possessing the necessary instrumentation, conventional celestial navigation is still one of the fundamental procedures practiced. Latitude is determined by the method already outlined. Longitude is computed from the difference between the observer's local and Greenwich times. For example, suppose a clock that is always set to keep Greenwich time indicates 3 p.m. when the sun crosses a ship's meridian at local noon. The vessel must be on a meridian that is three hours earlier or 45° to the west of the Greenwich meridian because the earth rotates from west to east at 15° per hour. The ship is therefore at 45°W. Determination of latitude will provide the other value necessary for the position "fix."

The Calendar

Our year is based on the period of revolution of the earth about the sun. Many complications have developed in the use of this period because it is not a simple integral number but is 365.242 days. This time interval is known as the *tropical year,* as it is the time between annual passages of the sun across the point on the ecliptic defined earlier as the vernal equinox, or the point the sun appears to cross as it transits from south to north across the equator.

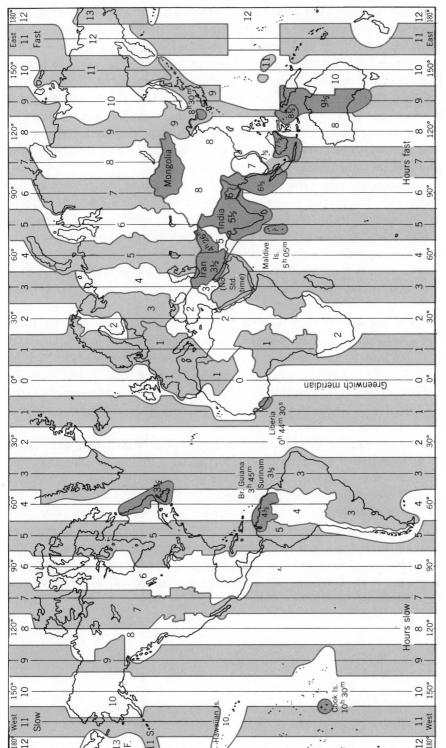

FIG. 2-18. The relationship between longitude and time. Since the earth turns 360° in 24 hours, it also turns 15° per hour. The standard time zones are centered on meridians divisible by 15, and extend 7½° on either side of the standard meridians. Over the oceans, the time zones conform exactly to the standard meridians but over land they deviate so as not to run through cultural centers.

One of the earliest calendars, known as the Roman calendar, was based on the synodic period of the moon. The year was designed to have only ten months from March to December. Confusion resulted, since seasons and festival dates fell at different times of each year. To correct this, two more months, January and February, were added. (February was arbitrarily made shorter by one day to give a total of 365 days in the year.) Because the synodic month is $29\frac{1}{2}$ days, the calendar was adjusted so that alternate months were 29 and 30 days long. It is believed that the number of days in a week derived from the number of planets then known, plus the sun and moon.

It soon became clear that the modified Roman calendar was also inadequate for proper date-keeping. In 46 B.C., Julius Caesar reformed the Roman calendar on the advice of Egyptian astronomers. The new *Julian* calendar was based on a year of 365 days, leaving an excess of almost a quarter of a day each year. To compensate, every fourth year was decreed to have 366 days; the extra day was added to February which was thereby increased from 28 to 29 days on years divisible by four (leap years).

The Julian calendar worked fairly well for a while, but over the centuries a significant error began to accumulate because the tropical year is a little less (by 11.2 minutes) than the average Julian year of 365.25 days. This difference adds up to a little more than three days in 400 years. As a result, events slowly shifted in the year so as to occur increasingly early. By 1582, for example, the vernal equinox, which should occur on March 21, actually fell on March 11.

To remedy the Julian calendar, Pope Gregory XIII took the advice of a consultant (the astronomer Clavius) and struck 10 days from the calendar, setting events back to their proper dates. And to prevent a similar occurrence, he further decreed that only the centuries divisible by 400 should be counted as leap years. Since every centurial year is divisible by 4 they would ordinarily all be leap years. This skipping of a leap year every hundred years except the fourth, removes most of the three-day error accumulated every 400 years in the Julian calendar. A slight error of less than one day in 3000 years is still accumulated—a problem to be dealt with in the future.

STUDY QUESTIONS

2.1. Define "oblate spheroid" and give the dimensions that describe the oblate shape of the earth.

2.2. Why must we determine the exact shape of the earth with a very high degree of accuracy?

2.3. Explain the "geoid."

2.4. Find the location of a ship giving its latitude and longitude from the following observations:
(a) On June 21, noon altitude of noon sun is 80°; Greenwich time is 2300 (11 PM); ship's local time is 1200 (noon).
(b) On a different day the sun's declination is 10° south and its noon altitude is measured as 50°; Greenwich time is 1700 (5 PM) at noon on the vessel.

2.5. Distinguish between "parallels" and "meridians." Which are always great circles? Why does the distance between successive parallels (say one degree apart) change between the equator and the poles?

2.6. List five consequences of the earth's rotation.

2.7. As briefly as possible, explain how the use of a Foucault pendulum proves that the earth rotates.

2.8. What are the only two places where a Foucault pendulum would show one complete rotation in a day.

2.9. How does the use of artificial earth satellites prove rotation?

2.10. Explain the Law of Gravitation—use both words and the formula with all terms defined.

2.11. What is the density of an object whose metric weight is 3.5 times the weight of an equivalent volume of water?

2.12. The average density of the earth is about 5.5 gm/cc. What can we infer about the density and even the composition of the earth's interior from this fact and the observation that crustal rocks have densities of about 3.0?

2.13. Distinguish between mean and apparent solar time. Which time is used ordinarily?

2.14. How are our basic time units (days, hours, minutes, seconds) determined?

2.15. Why do our standard time zones roughly cover 15 degrees of longitude? Why are the time zone borders actually quite irregular?

2.16. Explain what was wrong with the Julian calendar that led to the present Gregorian calendar.

chapter 3

A view of the earth's surface showing a variety of landscape features. (Photo by U. S. National Park Service, Department of Interior.)

The Face of
the Earth

In the previous chapters we examined the earth as an astronomical body in the solar system and considered the kind of information that relates to the earth as a planet. Much of this information could be obtained by an observer on the moon or on some other planetary body. In this chapter and the remainder of the book, we will examine the earth as a geologic body, including its atmosphere and oceans, and examine methods, principles, and facts only obtainable by an earthbound observer. The first close-up view of the earth is naturally of its surface. The earth's surface features are the results of geologic processes that have been going on for billions of years. Much of the rest of this book is concerned with the earth's composition, structure, and processes that have made the earth look the way it does today. Let us in this chapter see what the earth does look like at present, since we live on it and know it better than any other part of the universe, and because it is the present end product of the processes we plan to study.

RELIEF

Relief in the geologic sense is quite different from other definitions of the word—for example, the emotional feeling after completing the last chapter. By *relief,* here, we mean the difference in elevation between two points. It does not refer to absolute elevation. Thus, a plateau may have very high elevation, 10,000 feet above sea level, and yet be quite flat and featureless, having the characteristic of low relief. A rugged, mountainous area may have high relief from the jagged peaks rising thousands of feet above adjacent valleys, but the maximum elevations can be well below that of the high plateau.

When close up, mountains often look high and rugged and their valleys deep. When we fly over the same mountains at the great elevations of modern jet planes, their relief and ruggedness is much less evident as our view of their horizontal dimensions broadens. Similarly, in illustrations, mountains and valleys invariably appear very prominent in comparison with the lateral dimensions of a diagram. Within a relatively short distance, a relief of about 15,500 feet occurs in southeastern California from the floor of Death Valley to the top of Mt. Whitney. The relief visible is one of the greatest in the world, but on a scale-model region, it seems much less unimpressive. This is indicated in Fig. 3-1, which shows scale drawings of two rugged regions, one including Mt. Everest, the highest mountain in the world, and the other the region from Death Valley through the High Sierras. In a scale drawing, vertical and horizontal scales are equal, but when aspects of topography are emphasized, vertical exaggeration (in which the vertical scale is much greater than the horizontal) is used.

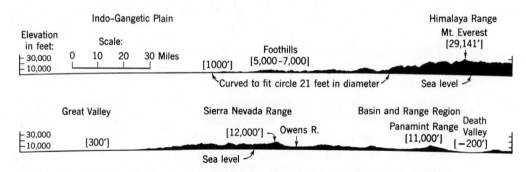

FIG. 3-1. Profiles drawn to true scale (no vertical exaggeration) of two regions having very high relief. (After A. N. Strahler from *Physical Geography*, John Wiley & Sons.)

CONTINENTS AND OCEAN BASINS

Although the surface of the oceans is the most prominent large-scale aspect of the earth's surface, the broad depressions containing the oceans, called the ocean basins, constitute the actual relief feature. The principal relief features of the earth are thus the continents and ocean basins. Interestingly, the effect of scale and distance again results in a distorted impression of the ocean basins. We normally imagine them as broad depressions flanked by the margins of the continents and presenting a form that would be concave upward. But if we draw the ocean basins to scale on a spherical section, it is immediately apparent that they are convex upward over their entire extent and follow the normal curvature of the spheroidal earth. It is our tendency to draw vertically exaggerated topographic profiles of the ocean basins relative to a flat rather than a curving surface that creates the impression of the ocean basins being concave hollows—they are really "convex hollows."

The ocean basins are deeper than the continents are high, and are also much greater in area. This applies to means as well as extremes; for example, the mean elevation of the continents is 2670 feet (about $\frac{1}{2}$ mile) while the mean depth of the oceans is close to 12,500 feet (about $2\frac{1}{4}$ miles). The greatest elevation on the continents, Mt. Everest, is a little more than 29,000 feet, whereas the greatest depth in the oceans, as now known, is in the Mariana Trench (southwestern Pacific Ocean) at about 36,000 feet below sea level.

The *hypsographic curve* provides a good presentation of the areal distribution of elevations on the continents and depths in the seas, as shown in Fig. 3-2. This diagram gives the information about depths and

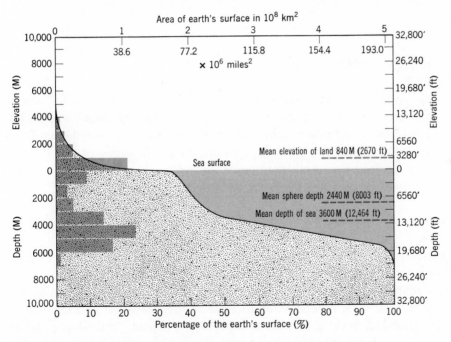

FIG. 3-2. Hypsographic curve of the earth's surface. This shows the area of the earth's surface above any level. The histogram along the left-hand side shows the frequency distribution of elevations (depths) at 1000 meter intervals. The two most frequent levels of +270 meters (886 ft) and 4420 meters (14,500 ft) are quite evident in this representation.

areas in two ways. Along the left side is a bar graph (histogram) showing the amount of area covered by each one thousand meters (3280 feet) of elevation on land and of depth in the seas. The percentage of the earth's surface as well as absolute areal dimensions are given along the horizontal axis of the graph (bottom and top, respectively). The continuous curve, the actual hypsographic curve, shows the amount of surface area *above* any elevation or depth value. For example, about 50 percent of the earth's surface is above a depth of 4000 meters (13,120 feet); about 29 percent is above sea level, and so on.

Both the histogram and the hypsographic curve reveal an interesting fact about the distribution of elevations and depths. Two preferred levels exist, one on land, with a mean elevation of about 270 meters (886 feet) and one below sea level at a mean depth of about 4400 meters (14,430 feet). These levels result from a fundamental difference in density between the rocks that form the continental and oceanic parts of the earth's crust.

It can be deduced from this diagram that the volume of the ocean basins below sea level is considerably greater than the volume of the continents above sea level—in fact, the ratio of the volumes is about 10 to 1. Thus, if the earth's surface were smoothed out so that no elevations or depressions existed, the oceans would then completely surround the earth as a uniform water layer almost 7900 feet deep.

FEATURES OF THE CONTINENTS

Superimposed on the broad continental platforms and broader ocean basins are a great diversity of relief features—some large, some small. The larger features fall into natural provinces that permit a logical classification. Although plains, plateaus, and mountains (features of the *second order*) are found on both continents and ocean basins, it will be more convenient to consider the latter two regions separately.

Plains

Plains are broad areas of low elevation and low relief, the latter often considered as less than 700 feet. This purely topographic definition is not adequate to exclude portions of the continents that were once mountainous but had been eroded away to areas of low elevation and low relief. A plain has a geologic as well as topographic aspect. To the topographic definition given above, we must add that a plain must be immediately underlain by layers of rock that are horizontal or nearly so. One of the common occurrences of plains is along continental margins where the surface of the plain dips gently seaward and extends beneath the ocean as the continental shelf. No break exists between the shelf and the coastal plain, which is often simply that part of the former continental shelf that has emerged from the sea. At times the coastal plain may become submerged by a rising sea level, in which case it becomes part of the continental shelf. An example of a typical coastal plain is found in Fig. 3-3. The eastern seaboard of the United States is an ever-widening coastal plain southward from New York City. The plain continues with still greater breadth around the southern Gulf Coast margin of the United States and then narrows into northeastern Mexico. The general distribution of this extensive feature can be seen in Fig. 3-4, a physiographic or landform map of the United States (North America).

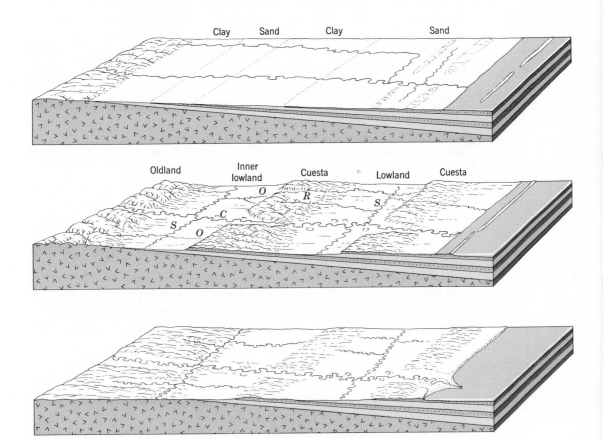

FIG. 3-3. Block diagrams of a typical broad coastal plain in different erosional stages. (From A. N. Strahler, *The Earth Sciences,* Harper and Row.)

Another type of plain indicated in this diagram is the interior plain or central lowland, which is again underlain by nearly horizontal rock units. The central plain or interior lowland of North America, through which the Mississippi River flows, slopes gently upward, north from the Gulf of Mexico. To the east the lowland merges with the Allegheny Plateau. Westward from the Mississippi Valley it rises slowly and then gives

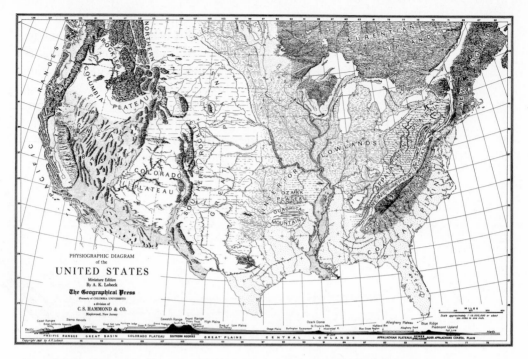

FIG. 3-4. Physiographic diagram of the United States. (From A. K. Lobeck, Courtesy C. S. Hammond & Co.)

way to the much higher Great Plains, which reach elevations of thousands of feet. Such upper-level plains are usually referred to as *high plains,* to distinguish them from the more typical lower forms. The maximum relief exhibited by plains occurs in the stream valleys of high plains. Because sea level limits the depth of downcutting by streams (see Chapter 6), the valleys of coastal and interior lowland plains are subdued, whereas the valleys of high plains can be cut rather deeply.

Plateaus

A plateau is a broad, flat, or gently undulating upland surface of relatively high elevation, often bounded along part of its border by a sharp topographic (cliff) drop to a lower level. To this topographic definition is again added the geologic requirement of horizontal rock units immediately underlying the surface. The rock forming the surface may be of sedimentary origin, that is, formed of rock deposited initially as sediment, or it may be formed by successive layers of lava or of layers of fragmental material ejected from one or more volcanoes. If the surface of the plateau is geologically young, it may be relatively unbroken by stream valleys. If older, the erosional work of streams will have carved valleys of great depth into the plateau rocks. If many such valleys exist, the plateau may be cut into numerous, smaller flat tablelands between streams valleys.

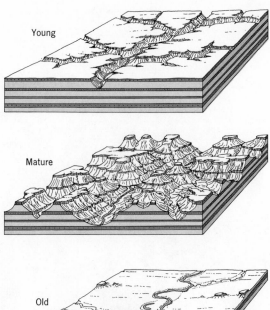

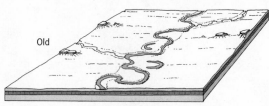

FIG. 3-5. Block diagrams showing the subsurface structure and topography of a plateau in different erosional stages. (From A. N. Strahler, *Physical Geography*, John Wiley.)

The entire region is called a plateau regardless of the amount of dissection by streams. Figure 3-5 shows a schematic view of a plateau in different stages of erosional history.

One of the largest plateaus in North America is the Colorado Plateau, centered approximately at the unique meeting point of the four states of Colorado, Utah, Arizona, and New Mexico (Fig. 3-4). The colossal Grand Canyon has been cut more than 7000 feet deep into the southwestern portion of this region. At least 5000 feet of strikingly colored and often weirdly shaped layers of sedimentary rocks are present on the walls of and on residual pinnacles within this awe-inspiring work of nature, illustrated very inadequately in Fig. 3-6.

The Columbia lava plateau, covering about 50,000 square miles in eastern Washington and northeastern Oregon, is an excellent example of a plateau built up by successive flows of molten lava. In places, the total accumulation is nearly 5000 feet. The Columbia and Snake Rivers have cut spectacularly deep valleys into the black rock formed by the cooling of the lava. This region is also shown in the physiographic map of Fig. 3-4. The Deccan Plateau, which covers much of western India, is a broad, high plateau region underlain by thick ancient lava flows far more extensive in area than the Columbia Plateau of the northwestern United States.

We will return to these great lava plateaus in a later chapter when the problem of the source of all of the molten material is investigated.

FIG. 3-6. Photograph of the Grand Canyon of the Colorado River the colossus of all canyons in Northwestern Arizona. (Courtesy, American Museum of Natural History.)

Mountain Chains

Although a great variety of topographic forms called mountains exist, this second-order feature on the continents really refers to chains of mountains, hundreds of miles broad and often thousands of miles in length. In the eastern United States, for example, the Appalachian Mountain chain extends from northeastern Alabama northeastward to the Gaspé Peninsula of Canada and then into Newfoundland. The Rocky Mountains are traceable all the way from Alaska to New Mexico. The Andes Mountains extend more than 5000 miles as a continuous chain along the entire western coast of South America. It is significant that the length is usually great compared to the width of the great mountain chains. It is also significant that the great mountain systems are rarely in the center of a continent but rather in the marginal regions. The world relief map in Fig.

3-7 emphasizes this fact of mountain distribution. One of the most extensive nearly continuous systems in the world is the Alpine, Carpathian, Caucasus, Himalayas Mountain system that stretches from southeastern Asia to southwestern Europe. The borderland characteristic of the great mountain chains will be discussed further in a later chapter on the origin and evolution of mountains.

Topographically, a mountain is an elevated region having a limited summit area bounded by relatively steep slopes. But the actual elevation of a mountain area may vary all the way from the high, craggy Himalayas to the low, undulating slopes of New England. The fundamental requirement of a mountain region is a geologic one in that the underlying rocks ment of a mountain region is geologic in that the underlying rocks must be deformed and distorted from their original structure or be of volcanic origin. The degree of this deformation is very broad. Some mountains are composed of rather gently undulating rock structures as in Fig. 3-8a and others of very markedly deformed rock units as in Fig. 3-8b (p. 82).

Superimposed on the large features of the second order are the myriad of details of the more minor features that constitute the geological scenery. These include, for example, the tremendous variety of erosional forms cut into the larger second-order features; also included are the smaller constructional forms produced by deposition of rock material in flowing or quiet bodies of water and the forms developed through volcanic action and the like. The smaller features of both lands and ocean bottom are sometimes referred to as relief features of the *third order*. Rather than attempt a description of this multitude of forms, we will refer to them as they are involved in the different processes to be described throughout the remainder of the book.

FEATURES OF OCEAN BASINS

Hidden beneath the widespread surface waters of the oceans is an almost fantastically rugged and varied topographic scenery, far higher in relief and grander in size and appearance than any equivalents on the continents. Our knowledge of the ocean floor has increased by giant strides from technological developments and applications following World War II. Prior to this time most of our understanding of the invisible world beneath the sea came from the use of lead-line soundings. They gave only widely spaced observations of inherent inaccuracy due to the drifting of the long lines from the effect of currents. A small number of observations were derived from echo sounders. These instruments emitted a sound in the water that traveled to the ocean bottom at a speed of 5000 feet per second and was then reflected up to a recorder on the ship. The time interval for the round trip was halved to get one-way time and then multiplied by 5000 feet per second to get the depth.

FIG. 3-7. Relief map of the world. (Courtesy B. C. Heezen.)

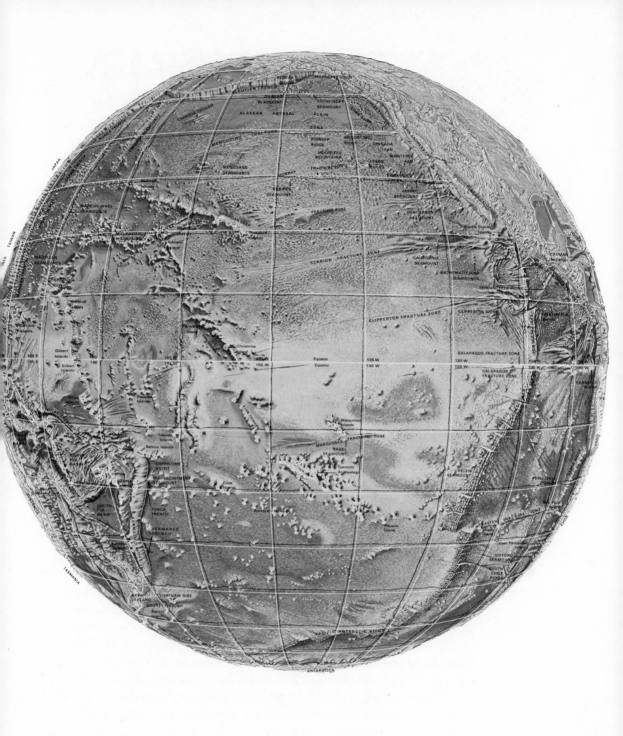

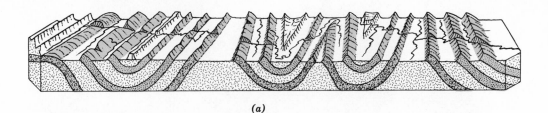

(a)

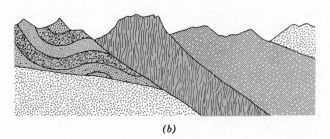

(b)

FIG. 3-8. Diagrams showing (a) mountain structure formed from simply folded rock layers and (b) intensely deformed rock units.

"Seeing" the Sea Floor

Although the first echo sounder was put into use in 1922, results between that time and World War II were relatively inaccurate and were generally limited to shallow water. Also, early echo sounders required the observer to listen for the signal with headphones and time the interval from transmission to return by watching a clock. By 1953, scientists of the Lamont-Doherty Geological Observatory of Columbia University developed the precision depth recorder (PDR), which makes a continuous record in any depth of water as the vessel cruises. The accuracy of the PDR is better than 1 fathom in 3000 (6 ft in 18,000 ft). As of 1970, millions of miles of precision depths have been recorded continuously by the research vessels of many institutions from many countries. Some examples of PDR records made by vessels of the Lamont Observatory are shown in Fig. 3-9.

Time marks are placed automatically on PDR records by an accurate chronometer. Then, by comparison with the navigational positions of the vessel, a control that has now become accurate to a few hundred feet through the use of navigational satellites, points on the PDR profile can be located on marine charts. The oceans are vast and despite the large numbers of crisscrossing tracks giving continuous depth data, our knowledge of the details of the ocean floor is certainly not comparable to that of the continents. In fact, it is in many ways not as complete as our knowledge of the surface of the moon. But the depth data already accumulated have revealed a surprisingly large amount of information about the features of the sea floor. Interpolation and imagination based on data have then filled in bathymetric features not directly observed or only partially observed.

Two methods of analysis of the depth data have been employed, as

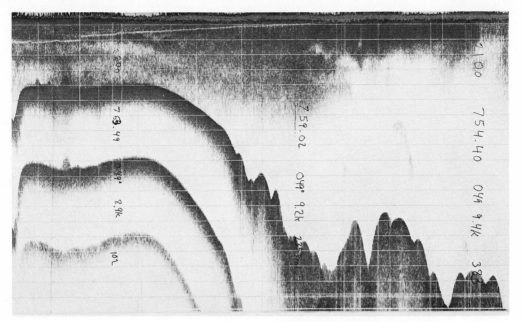

FIG. 3-9. Examples of a record from a precision depth recorder showing the topography where the continental shelf descends to the continental slope off Cork, Ireland. Each horizontal line from the surface down represents 20 fathoms (120 feet). The two lower (later) images represent later reflections between the bottom and the sea surface. (Courtesy, Lamont-Doherty Geological Observatory, Columbia University.)

shown in Figs. 3-10 and 3-11. The former shows a portion of a Japanese hydrographic chart of the western Pacific Ocean. Bathymetric contours are used to connect points having the same depth. To the skilled reader of contour maps, the three-dimensional aspects of the features contoured are immediately apparent. In addition, a three-dimensional effect is made visually apparent by means of careful shading. Precise elevations and depressions on the ocean bottom are given by the values assigned to the contours that are always drawn for fixed vertical intervals.

Another way of analyzing the depth data is somewhat more artistic and interpretational. This is the method developed at the Lamont-Doherty Observatory of showing ocean-bottom depths through the use of numerous points giving precise values as in Fig. 3-11, which shows a portion of the bathymetric chart of the South Atlantic Ocean. Mountains, cliffs, fractures, plains, and the like are clearly more distinct in this procedure, in which the creators of the chart, rather than the reader, have done most of the interpretation of the meaning of the depth values.

Marine geologists have delineated three very broad bathymetric provinces making up the ocean basin. These are the continental margins, the ocean-basin floor, and the midocean ridges. The distribution of the provinces for the North Atlantic Ocean is shown in Fig. 3-12. Below the

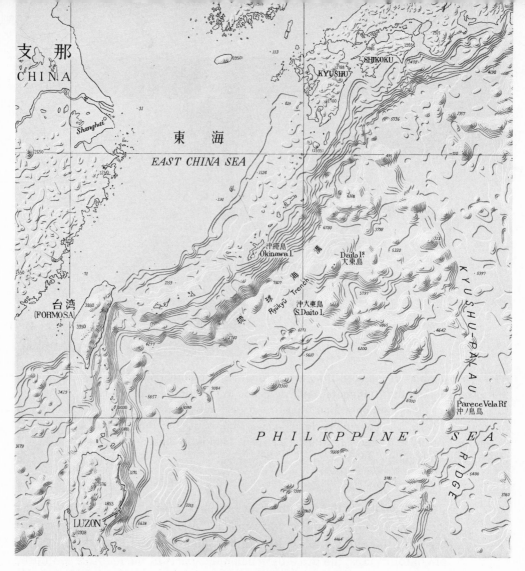

FIG. 3-10. A portion of a Japanese hydrographic chart of the western Pacific ocean. Relief is indicated quantitatively by depth and qualitatively by shading.

map is a typical profile across the ocean. Since the horizontal distance is several thousand miles and the vertical relief about three miles, a high degree of vertical exaggeration is required to display the bottom topography. Although the ocean ridges are topographically simple, the continental margins and ocean floors are composites of a large number of features, each with distinct characteristics.

If we draw a parallel with the second-order features on continents, the equivalent features for the ocean basins (Fig. 3-13) can be considered as (1) the continental shelves and slopes, (2) the continental rises, (3) midocean ridges with related rises, (4) abyssal plains, (5) marine plateaus, and (6) ocean deeps or trenches.

FIG. 3-11. A portion of the bathymetric chart of the South Atlantic Ocean. (From B. Heezen, and M. Tharp, Courtesy of the Geological Society of America.)

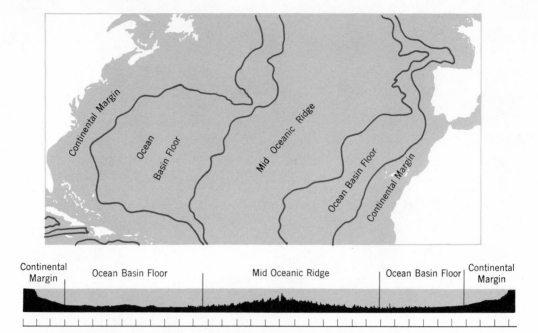

FIG. 3-12. Topographic provinces of the North Atlantic Ocean including a topographic profile from New England to North Africa. (Redrawn after B. Heezen, M. Tharp, and M. Ewing, Courtesy the Geological Society of America.)

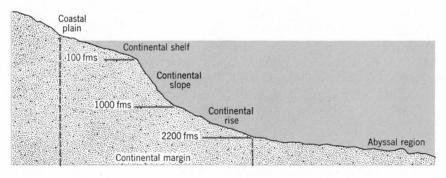

FIG. 3-13. Schematic (nonscale) drawing of the general features of continental margins and ocean basins.

Continental Shelf and Slope

The topographic features of continental shelf, continental slope, and continental rise are often combined into a single grouping called the *continental margin* as illustrated in the vertically exaggerated diagram in Fig. 3-13. In different regions the continental shelf varies from very narrow to extremely broad in width, but the depths of 400 to 600 feet reached at their outer margins, is typical of them the world over. For example, the continental shelf off Long Island is quite smooth and broad as shown in Fig. 3-14, where the 100-fathom (600-ft) contour marks the

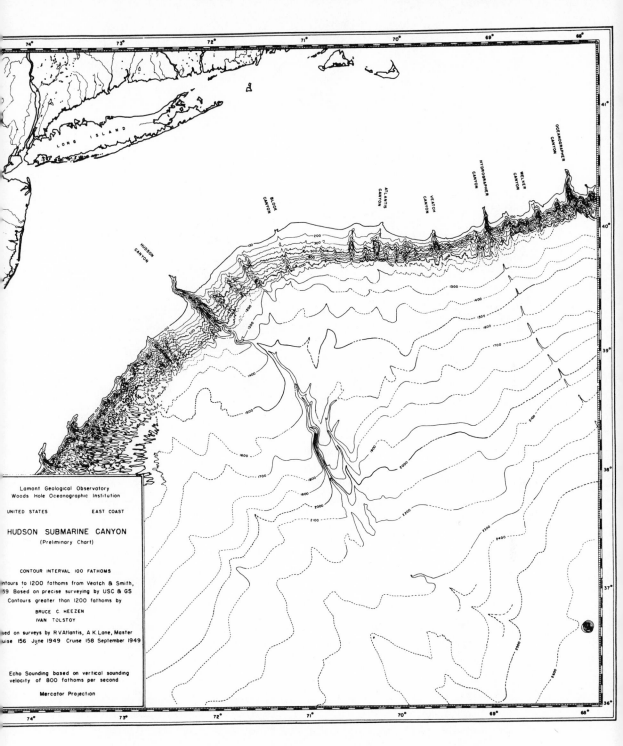

FIG. 3-14a. Bathymetric chart showing the continental shelf, continental slope (close contours), and continental rise south of New England. The submerged Hudson Canyon cuts the slope and rise south of Long Island. Bathymetric contours are drawn for depth intervals of 100 fathoms. (From B. Heezen, M. Tharp, and M. Ewing, Courtesy of the Geologic Society of America.)

Features of Ocean Basins   87

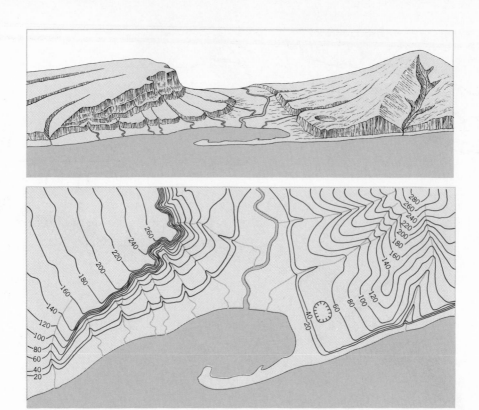

FIG. 3-14b. Illustration of the principle of *contour lines* used commonly to illustrate topography. Contour lines are lines connecting points having equal elevation. They are close together where slopes are steep and more separated where slopes are gentle. Note that in crossing stream valleys contour lines must bend to point upstream in order to remain at the same elevation. A typical V-shaped pattern characterizes steeply sloping valleys.

outer edge of the shelf. The width of this gently sloping surface (slope about $\frac{1}{2}°$) can be estimated from the scale in degrees and minutes along the north-south (meridian) boundaries following the method outlined in Chapter 2. The breadth of the shelf increases from about a degree from its southwestern end to at least $1\frac{1}{2}°$ at the northeastern part or from at least 60 to 100 nautical miles (69 to 115 statute miles).

The closely spaced contours from the 100- to 1100-fathom level depict the relatively steep continental slope, while the rest of the region to the southeast represents the continental rise. The V-shaped contours pointing northwest in the center of the chart shows the submerged Hudson Canyon, a valley of grand proportions. Presumably it was eroded in glacial times when sea level was several hundred feet lower than at present and when much of the shelf was exposed above water (Chapter 15).

Compare this continental margin with that off the coast of California, at Monterey Bay, as in Fig. 3-15. The continental shelf here is very narrow, barely 5 to 10 miles wide, before the steeper slope begins. Of interest again

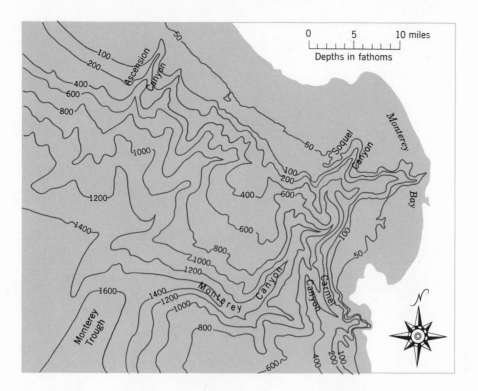

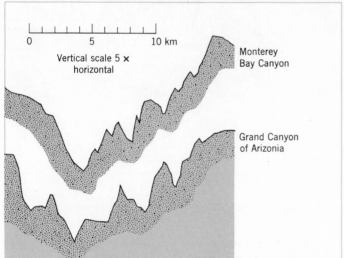

FIG. 3-15. Bathymetric chart of the sea floor off Monterey, California showing Monterey Canyon, which in the lower diagram is compared in profile view with the Grand Canyon of the Colorado River in Arizona. (Redrawn after Sverdrup, Johnson, and Flemming, *Oceanography*, Prentice-Hall.)

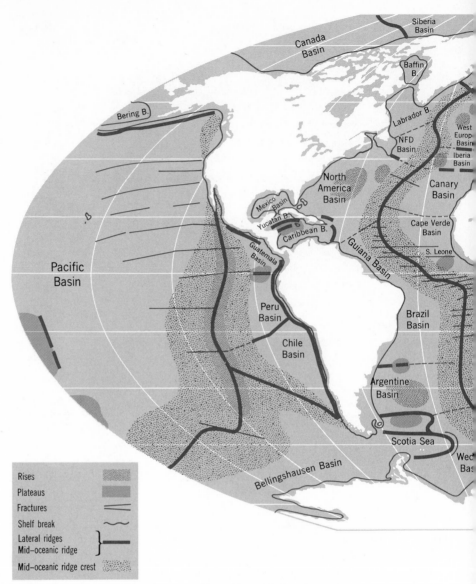

FIG. 3-16. Generalized topographic map of the world sea floor. (Courtesy B. Heezen and M. Tharp.)

here is the large submarine Monterey Canyon shown by the elongated V-shaped contours. In the lower part of the diagram, a vertical cross section across Monterey Canyon is compared with one across the Grand Canyon of the Colorado River in Arizona. The submerged canyon is actually somewhat more profound than the famous Arizona gorge and would represent a grand sight if the oceans receded. In depth and width, even the submerged Hudson Canyon rivals the Grand Canyon.

Midocean Ridges and Rifts

The great topographic swells, consisting of ridges and rises, are the most prominent and geologically significant features of the ocean basins. The

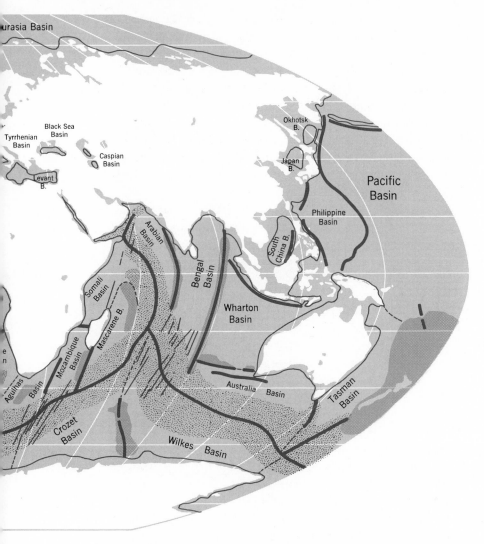

ridge axes experience frequent earthquakes and have a much higher heat flow upwards through the crust than the rest of the oceans. In Fig. 3-16, the global system of ridges is shown strikingly in the very generalized topographic map of the sea floor. The central region of the ridges is shown by the heavy line, which is flanked by mountains decreasing in elevation above the sea floor; the entire feature is often referred to as a *rise*. Perhaps the most striking is the continuous system that extends for 40,000 miles from the Arctic Ocean through the center of the North and South Atlantic Oceans (the Mid-Atlantic Ridge) and then bends around South Africa to the Indian Ocean. Here it splits into two sections, one going northward and the other continuing across the South Pacific Ocean. This portion continues as the East Pacific Rise, which extends northward to Baja California, off Mexico. Note how this extensive ridge system follows the shapes of the continents, particularly in the Atlantic Oceans, an effect related to the mechanism of sea-floor spreading to be discussed in Chapter 8.

The extensive submarine ridge system cannot be considered without important emphasis being given to the great *rift* that extends almost continuously along the ridge crests. Submarine rifts are relatively narrow deep clefts that would appear as great cracks in the earth's crust if the oceans were removed and the sea bottom viewed from a distance large enough to encompass much of the world in a single view. The rift atop the Mid-Atlantic ridge can be seen in the center of the profile in Fig. 3-12, where it abruptly interrupts the summit level of this ridge. Similar but less continuous rift valleys are found on continents such as those of east Africa, which now contain the deep, elongated lakes Tanganyika and Nyasa.

According to modern geologic thought, the ridge-rift systems are perhaps the single most important feature in the evolution of the earth's crust, as will be elaborated in Chapter 8. We will see there how the ridge-rift system in midocean regions and the ocean trenches (to be described soon) form the boundaries of a system of great crustal plates.

Submarine Plains and Plateaus

Abyssal plains are broad and remarkably flat regions that occupy the deep expanses of the ocean basins. PDR profiles made during crossings of these regions show them to be almost geometrically flat surfaces. Compare the rough mountainous areas in the east-central part of Fig. 3-17 with the unmarked, smooth regions (abyssal plains). Similarly, the smooth abyssal plain in the central part of Fig. 3-11 (South Atlantic) contrasts strikingly with the mountainous surrounding areas.

The true nature of abyssal plains was revealed by the use of the *seismic profiler,* a system whose sound signal could penetrate far below the bottom of the ocean floor. Profiler records quickly showed these deep plains to be formed of a thick carpet of sediments lying over the rough low portions of the sea floor. Further marine exploration led to a better understanding of the deep-sea process involved in the formation of abyssal plains. During millions of years, sediment eroded from the continents has been transported farther and farther seaward until the lowest portion of the sea floor was reached. Continued accumulation resulted in the blanketing of many topographic features (Fig. 3-18).

Superimposed on both the continental margins and the abyssal plains are broad elevated regions analogous to plateaus on land. In many cases they extend from the continental slope as broad, flat areas that then drop sharply to the floor of the deep ocean basin. A good example is the Blake Plateau, which lies just beyond the edge of the continental shelf to the east of Florida, as seen in Fig. 3-17. In other cases, plateaulike features may rise from a deep sea basin. Bermuda, an island group with only a small area above sea level, actually rises from an extensive volcanic platform known as the *Bermuda Rise* (Figs. 3-16 and 3-17). The *Melanesian Plateau* (17) to the east of Australia (Fig. 3-16) is one of the most extensive of the submarine plateaus.

FIG. 3-17. Portion of the bathymetric chart of the North Atlantic Ocean. (From B. Heezen, M. Tharp, and M. Ewing, Courtesy the Geological Society of America.)

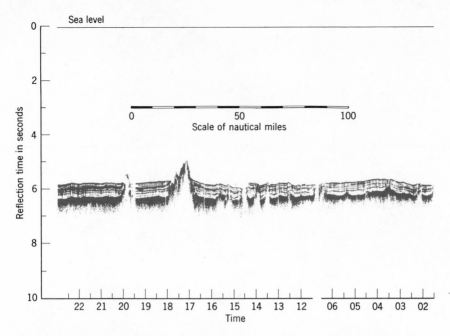

FIG. 3-18. Seismic profiler record of the floor of the Pacific ocean about 0°N, 140° W showing the sediment layers blanketing the rough topography beneath. Mountain peaks extend through the sediment in places and project above the floor of the ocean. Each time division on the vertical scale equals 2500 feet. Time on the horizontal scale is actual sailing time in hours. (Lamont-Doherty Geological Observatory record.)

Ocean Deeps and Island Arcs

The ocean deeps or trenches, like the ridges, are geologically very active regions. Deeps are frequently affected by strong earthquakes, which often reach catastrophic proportions (Chapter 10). Ocean deeps are long, narrow depressions extending to depths below 23,000 feet and for the most part lying to the seaward side of *island arcs*. The latter are arcuate chains of islands located offshore from certain parts of the continents; for example, the Caribbean Arc north of South America, the Aleutian Arc off Alaska, the East Indian Arc between Australia and Asia, the Philippine and Japanese Arcs of Eastern Asia, and so on. The islands of these groups (Figs. 3-7 and 3-16) occur as extended open arcs whose convex sides face away from the continents toward the open sea. Deeps and island arcs are both much more common along the margins of the Pacific than along those of the other oceans.

It is significant that ocean trenches always lie off mountainous parts of continents. In Chapter 8 we will see the important part that deep-sea trenches play in the evolution of the earth's crust. The earthquake activity of these regions becomes easier to understand in the light of the crustal evolution model that will be developed.

FIG. 3-19. Ripples formed in the sediment on the sea bottom by water motion at a depth of 13,000 feet off the southern tip of South America. Similar types of ripples, found at depths to 24,000 feet indicate relatively strong water movement at great depths. (From B. C. Heezen and C. Hollister, *The Face of the Deep,* Oxford Press. Photo by U.S.N.S. *Eltanin* of National Science Foundation.)

Fractures in the Sea Floor

The ocean basins contain other features of importance in the evolution of the crust. Figure 3-16 shows a number of east-west-trending lines off the western coast of North America (Pacific Basin). These represent gigantic crustal fractures wherein vertical movement has raised the floor of the ocean as much as thousands of feet above the floor on the opposite side of the fracture. In addition, lateral displacements also occur. Such displacement fractures are also evident in Fig. 3-11, where east-west-trending zones have produced displacements of the Mid-Atlantic Ridge.

One interesting aspect of all of these submarine features is their prominence when compared with similar features on land. Note especially the perfect conical forms of the volcanic mountains visible on the bathymetric charts of Figs. 3-11 and 3-17. On the continents such features are only so well-displayed when of very recent geologic origin. The reason for the profound difference between continental and marine relief features is the absence of strong erosional activity beneath the sea compared with the destructive effect of running water, glaciers, wind, and the like on the continents. On the moon the absence of an atmosphere has preserved features in all of their rugged grandeur from time immemorial. And under the oceans, the protection of the water has had somewhat the same effect, although many features in the deep basins have been covered by sediments.

We might, in passing, note the importance of deep sea photography to the understanding of processes affecting the bottom of the deep ocean. Photographs like that in Fig. 3.19 have revealed water motion and activity hiterto unsuspected.

We have made continued reference to the significance of the features described in this chapter. This will be elaborated on later in Chapters 8 and 9. It is safe to say, however, that our fundamental knowledge of the nature and evolution of the earth's outer zone, including the surface features, was quite in its infancy until the detailed investigation of the ocean bottom began with the end of World War II.

STUDY QUESTIONS

3.1. Imagine a region 10 miles across that has a maximum relief of 528 feet (0.1 miles). Suppose you must draw a topographic profile of this region from one side to the other. To make a scale drawing on a regular sheet of notepaper, what horizontal and vertical scale values would you use and what would be the size of the height and length of your diagram (in inches)? To emphasize the vertical elevation with a 10 times exaggeration, what values would be used?

3.2. According to the hypsographic curve, Figure 3-2, what is the area (in square miles and percent) of the continents above sea level.

3.3. How does the average depth of the ocean basins (below sea level) compare to the average elevation of the continents?

3.4. What are the two "preferred" levels and what is their significance?

3.5. Draw two labeled diagrams showing the similarity(ies) and difference(s) between plains and plateaus.

3.6. What is the significant difference between plains and plateaus on one hand, and mountains, on the other hand?

3.7. What are the features of ocean basins that are roughly comparable with plains, plateaus, and mountains on the continents?

3.8. If the sound used in a depth recorder takes 8 seconds to descend to the bottom and be reflected to the recorder, determine the water depth. (Sound speed in the ocean is 5000 feet per second).

3.9. Draw a simple contour map of a conical hill 210 feet high; use a contour interval of 20 feet. Draw a second contour map of a hill of the same size at the base but 370 feet high. Notice the difference in contour spacing. (Refer to Figure 3.14b.)

3.10. Describe the world mid-ocean ridge system.

3.11. Refer to the bathymetric charts in Figures 3-11 and 3-17. Why are some areas very rough while the low-lying plains are so remarkably smooth?

chapter 4

Crystals of calcite lining the inside of a geode structure (p. 118).

The Earth's Crust: Form and Mineral Composition

The earth is coarsely layered into four principal units. The outermost zone is the crustal shell, underlain by the mantle; this in turn surrounds the core in whose central region lies the inner core. Although the mass and volume of the crust are negligible compared with the dimensions of the interior units, it formed as part of the same general process responsible for the present structure of the earth. Hence a knowledge of the crust contributes tremendously to our understanding of the nature and history of the earth.

FORM AND DIMENSIONS

The general relationships among the crust, mantle, core, and inner core are shown in the scale diagram in Fig. 4-1. Even in this illustration where the crust is vertically exaggerated, it still appears quite miniscule in comparison with the entire earth. The crust is not a layer of uniform thickness but is quite variable as the result of prominent irregularities on its upper and lower surfaces. The lower surface seems to be a reflection of the major topographic relief of the upper surface. It descends into the mantle where topographic elevation is high and rises where topography is low. As shown in Fig. 4-2, it is thinnest beneath the seas, with a mean thickness of about 3 miles, and thickest beneath the continents, with a mean thickness of about 20 miles. The crust is still thicker beneath high mountainous regions, where the divergence of upper and lower surfaces produces a thickness

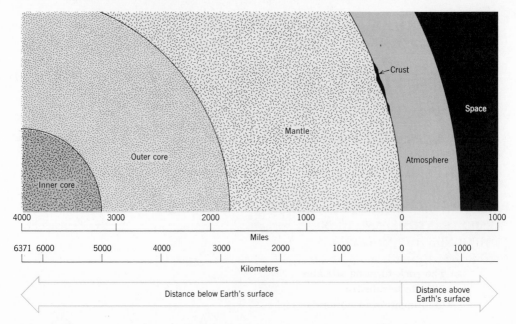

FIG. 4-1. Cross-sectional view illustrating the layered structure of the earth. The inner core, outer core, and mantle are drawn to scale; the crust and its features have a vertical exaggeration of three times.

of up to 40 miles. In Chapter 10 we will see how this and other knowledge of the earth's interior is acquired. We will also give more attention in later chapters to the fact that a striking difference exists between the composition and physical properties of the rock beneath continents and oceans. For the moment let us note the important fact that the rock immediately underlying the continents has a density of 2.7 g/cc; that beneath the oceans has the higher density of 3 g/cc.

In the previous chapter, it was emphasized that the continental surfaces are rather anomalous in covering a relatively small area of the earth's surface while projecting so prominently above sea level. We noted also that if the earth's crust were a uniformly thick zone, the continents would be deeply covered by the oceans, and we now see that there exists a most fundamental difference in density between continental and ocean-basin rock material. The continental rock shell is much lighter—about 10 percent less dense—than that of the rock beneath the oceans. This density contrast between the two portions of the earth's crust is one of the most—if not the most—significant facts about the continents and must be included in any final explanation of the origin of crust, continents, and ocean basins.

The earth's crust is a composite of a large variety of rocks. In places, a single rock type predominates and may occupy hundreds of thousands of cubic miles of crust; in other regions, rock types vary considerably in

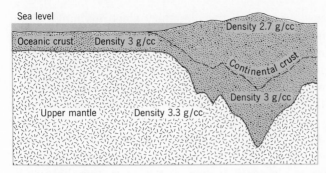

FIG. 4-2. Cross-sectional view of the earth's crust showing the thin oceanic and thicker, more variable continental portions.

relatively small areas and occupy rather small crustal volumes. Rocks are composed of minerals and cannot be understood until a knowledge of minerals is gained. Nearly all minerals are chemical compounds that are in turn composed of elements, the fundamental chemical entities. Thus, to understand the composition of the earth's crust, we must first review the nature of elements and compounds; then we can examine minerals and finally rocks.

ELEMENTS, ATOMS, AND IONS

The modern science of chemistry really began with the knowledge that chemical elements were unique substances in that they could not be separated—decomposed—chemically into other, simpler substances. Elements are characterized by specific chemical properties, including the way they unite chemically with other elements to form compounds. The smallest organized chemical unit into which an element can be reduced while retaining its chemical properties is the *atom;* this is the smallest part of an element that enters into chemical union with other elements.

The atom consists of a central *nucleus* composed of *protons* and *neutrons* surrounded by one or more *electrons* orbiting the nucleus. Protons carry a positive charge and electrons a negative charge, while as indicated by the name, neutrons have no charge. The chemical uniqueness of an element depends on the number of protons in the nucleus, this quantity being called the *atomic number* of the element. In the simplest of elements, hydrogen, the nucleus contains only a single proton whose positive charge is exactly balanced by a single planetarylike electron. As the atomic number increases, the number of nuclear protons and external, charge-balancing electrons increases. The electrons revolving about the nucleus conform to certain principles of atomic behavior, some of which are important to know for an understanding of the way elements combine to form compounds and minerals.

An atom is three dimensional; the planetary electrons move about the nucleus in orbits that can be imagined as lying in spherical shells rather than in a plane. According to atomic theory, specific numbers of electrons occupy specific shells, with a certain maximum number of electrons being permitted for a particular shell. The smallest number of electrons that can fill certain shells increases from 2 in the innermost shell to 32 in successively outer shells, and then decreases again. The higher the atomic number, the greater are the number of shells that have their full complement of electrons. Table 4-1 indicates the electron sequences for a number of different elements. It is important to realize that just as the solar system, with the sun and planets, is mostly space, so also is the atom. This is critical to the understanding of many processes that take place at high temperature and high pressure.

TABLE 4-1 Number of Electrons in the Shells of Some Elements

Element	Electrons In Each Shell
Hydrogen, H	1
Helium, He	2
Oxygen, O	2,6
Neon, Ne	2,8
Sodium, Na	2,8,1
Magnesium, Mg	2,8,2
Chlorine, Cl	2,8,7
Potassium, K	2,8,8,1
Bromine, Br	2,8,8,7
Krypton, Kr	2,8,18,8
Barium, Ba	2,8,18,8,2
Strontium, Sr	2,8,18,18,8,2
Radon, Ra	2,8,18,32,18,8,2

From the standpoint of the chemical reactions producing compounds peculiar to minerals, it is only the electrons in the outermost shell that are important. With the exception of hydrogen and helium, which have only a single shell filled by two electrons, most atoms tend to attain an outer shell with eight electrons. When atoms react with each other, they gain this stable number (called the *stable octet*) by either giving up one or more electrons so that the next underlying shell is the one with eight, or acquiring or sharing electrons to complete a shell. Whether they gain or lose depends on the electrical control of the positive nucleus. If it is strong, electrons are acquired from an atom with less nuclear binding of its outer electrons; if it is weak, the reverse occurs. As noted, reactions can occur by either sharing or transferring of electrons as shown schematically in Fig. 4-3a and b. Those atoms with few electrons in their outer

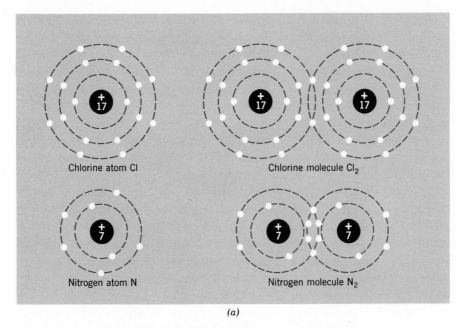

Chlorine atom Cl

Chlorine molecule Cl$_2$

Nitrogen atom N

Nitrogen molecule N$_2$

(a)

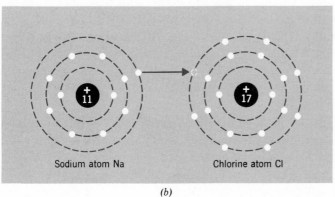

Sodium atom Na

Chlorine atom Cl

(b)

FIG. 4-3. Types of chemical bonding involved in the formation of molecules. (a) Illustrates electron-sharing in the case of chlorine and nitrogen. In the former, a pair (no particular pair) of electrons—one from each atom—is shared giving each outer shell a stable octet. The number of positive charges (protons) in the nucleus balanced by the negative electron charges is also indicated. In the case of nitrogen, three pairs of electrons are shared by two atoms in order to provide 8 electrons in each outer shell. Complex organic compounds of carbon usually involve similar electron sharing. (b) The *ionic* bonding is illustrated for the formation of the sodium chloride NaCl molecule in which one electron is transferred from the metal to the nonmental giving each outer shell the stable octet while retaining a net neutral electrical charge. Two or several electrons can be transferred in the case of other elements where more than one electron is involved to complete the stable octet.

shell tend to give them up in transfer reactions to achieve the stable octet; those with many electrons tend to acquire them to fill the incomplete shell. The former group constitute the metals and the latter, the nonmetals. If an element has its outer shell completed (contains two or eight electrons), it is chemically inert and with rare exceptions, never enters into a chemical combination. This explains the inert or noble gases, such as helium, neon, and argon.

When electrons are lost, the excess positive charges from the effect of the unbalanced protons in the nucleus give the atom a positive charge, the strength of which depends on the number of electrons given up. If electrons are gained in the transfer process, the atom gains a negative charge whose strength depends on the number of electrons acquired. In the sharing type of reaction, no gain or loss is involved, so that the combination of atoms remains neutral. Charged atoms are called *ions*. Metals, therefore, form positive ions and nonmetals, negative ions. Ionic behavior is of great importance in the forming and structure of the minerals to be discussed later in this chapter.

Although protons, electrons, and neutrons are the principal constituents of atoms, it should be noted that at least 30 subatomic particles have been identified by the methods of modern physics.

Of the 103 elements that have been observed or isolated, only 90 occur naturally in the earth's crust and many of these exist only in trace amounts, usually referred to as trace elements. And of the 90 elements occurring naturally, 8 make up the bulk of the earth's crust; the percentage composition by weight of these 8 elements is 98.59 percent. By volume, they form essentially 100 percent of the crust, since the bulk contributed by the small remaining quantity is negligible. In view of the importance of this small group of elements, their proportions in the crust by weight and volume are given specifically in Table 4-2.

According to this tabulation, two elements, oxygen and silicon, are by far the most abundant; oxygen, a gas, contributes nearly twice the mass of silicon, a solid. By atomic volume, oxygen is even more significant. Owing to the large size of this gaseous atom compared with the remaining

TABLE 4-2 Chemical Composition of the Crust

Element	Weight in Percent	Volume in Percent
Oxygen, O	46.60	93.77
Silicon, Si	27.72	0.86
Aluminum, Al	8.13	0.47
Iron, Fe	5.00	0.43
Calcium, Ca	3.63	1.03
Sodium, Na	2.83	1.32
Potassium, K	2.59	1.83
Magnesium, Mg	2.09	0.29

solid elements, oxygen by volume composes almost 94 percent of the crust. But we are not quite walking on air; oxygen in chemical combination with other elements does not retain its gaseous state. Aluminum and iron, two of the most valuable metals in our civilization, are the third and fourth most abundant elements. Despite this, their occurrence in the crust in commercially useful accumulations, or *ore deposits,* is relatively rare. Most of the iron and aluminum in the crust is in less-concentrated chemical combination with oxygen and silicon in the form of minerals from which they cannot be easily extracted.

COMPOUNDS AND MOLECULES

When chemical combinations of elements occur, the atoms of each of the elements involved become joined by the process of either electronic sharing or electron transfer. The combination of two or many atoms into a joined system is called a *molecule.* The physical and chemical properties of a molecule are quite different from those of the component atoms. The molecule thus becomes the smallest unit of a compound that still retains the properties of the compound.

The number of atoms of the different elements that combine with each other to form a molecule depends upon the number of electrons required to be given up, gained, or shared in the achievement of the stable octet. For example, hydrogen has one electron that it can surrender or share; oxygen, as seen from Table 4-1, requires two electrons to fill its shell to eight. Two hydrogen atoms are thus required to satisfy one oxygen atom when these gases combine to form water—hence the familiar formula, H_2O. Most of the compounds that compose the crust involve combinations of several elements, but the same principles of electron shuffling applies despite their more complicated-looking formulas.

Since the process of aqueous solution is so closely involved with ionic behavior and is so important in geologic processes, it is appropriate to examine the effects involved here. Let us consider the solution of sodium chloride, NaCl, the most common compound in ocean salts. Sodium has one electron in its outer shell; chlorine has seven. The salt compound thus contains sodium ions that have lost one electron each and are designated Na^+, and chlorine ions that have gained one electron each and are designated Cl^-. The opposite electrical charges cause a strong attraction between the ions that are combined to form the NaCl molecule. The strength of this attraction, known as *ionic bonding,* varies with different compounds and plays an important part in determining the structures and reactions of the mineral compounds in the crust.

In the electronic sharing that occurs among the one oxygen and two hydrogen atoms to form a water molecule, although the joined atoms develop a charge as though they have gained or lost electrons, the molecule itself has no net charge, as is true of molecules in general. By placing water in a strong magnetic field it has, however, been determined that the atoms are so arranged in the molecule as if the positive charges of the hydrogen were at one end of a rod and the negative charge of the oxygen were at the other, as illustrated in Fig. 4-4. In other words, the water molecules have an electric polarity. In pure water, the polar rods, often called dipoles because of the two unlike signs at opposite ends of the rod, tend to unite with positive and negative ends joining in a great variety of possible configurations.

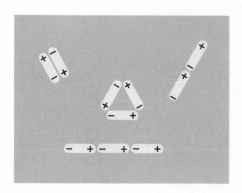

FIG. 4-4. Some possible combinations of *dipolar* water molecules that produce an electrically neutral fluid.

If grains of table salt, NaCl, are added to pure water, the ends of the dipoles begin to attach themselves to the oppositely charged Na^+ or Cl^- ions instead of to each other. Many water dipoles can attach to a single + or − ion. Gradually, the pull of the water dipoles separates the ion from the solid salt. As this process extends, the solid salt disappears because the small individual ions separate and go into "solution." The ease with which this dissociation occurs determines the difference in solubility of different substances in water. The solution process is illustrated schematically in Fig. 4-5.

Most of the nongaseous elements are sufficiently active chemically so that they combine readily with other elements and exist in nature only as compounds. A bare handful is found as free or native elements, the more common being copper, silver, gold, platinum, mercury, carbon, and sulfur.

A number of types of compounds are of particular importance in the earth's crust. The silicates, which are composed of silicon and oxygen in combination with one or more metallic elements, are the most important of the crustal compounds. They form minerals that make up more than 90 percent of the crust. Aluminum and/or iron are the chief metallic components of the silicates. Since oxygen and silicon make up 75 percent of the crust by weight (mass), it is not surprising that their compounds should be so important in crustal composition. Other compounds that are

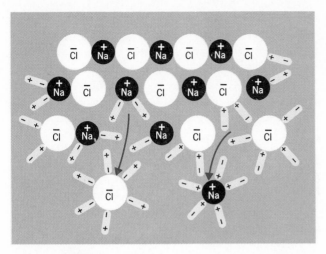

FIG. 4-5. Process of solution of sodium chloride by
water—a polar solvent. Negative ends of water dipoles
become attached to positive sodium ions in the NaCl
crystal and positive ends become attached to negative
chlorine ions. The Na and Cl ions become pulled loose
and disappear into solution. This occurs more readily
with substances that are easily soluble.

common in the crust but of lesser total bulk are oxides—compounds of
one, occasionally two, metals with oxygen; carbonates—compounds of
one, occasionally two, metals with carbon and oxygen; and sulfides—
compounds of one or more metals with sulfur and a variety of other
compounds of still-less-common occurrence. Most of our important ore
bodies are found as oxides, sulfides, and carbonates of economically
valuable metals.

MINERALS

Minerals are naturally occurring inorganic substances that possess specific
chemical compositions and particular physical properties by means of
which they can be identified. Chemically, nearly all minerals are compounds. Physically, they are mostly crystalline. Crystals are solid geometric forms whose boundaries or crystal faces always make the same
angular relationship with one another for the same mineral (Figs. 4-6 to
4-8). The visible external form of a particular crystal is controlled by a
very regular and characteristic arrangement of the ions or atoms of elements present into a geometric pattern called a *crystal lattice*. Chemical
composition of a mineral, determined by chemical analysis, tells us the

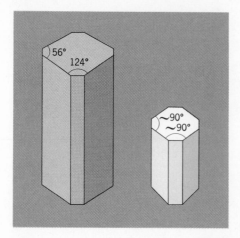

FIG. 4-6. Example of a typical hornblende crystal showing the characteristic angles of 56° and 124° for the four principal elongated faces.

FIG. 4-7. Example of a typical augite crystal development showing the nearly right-angle relationship among the four principal elongated crystal faces. The augite crystal is characteristically shorter or stumpier than that of hornblende.

kind and proportions of elements present but little about the internal structure. This is revealed by X-ray analysis, which shows both the size and the spacing of the internal particles. When photographed after passing through crystalline material, the X-ray beam images reveal the number, relative size, and spacing of these particles (see Fig. 4-9).

From X-ray analysis we have learned important facts about crystal chemistry. An ion of the same size can replace another in the crystal lattice. If their charges are not the same, however, additional substitutions are necessary for the crystal to remain neutral. Thus Mg^{++} can replace Ca^{++}, but if Ca^{++} replaces Na^+, further ion replacements are needed in which a metallic ion of lesser charge replaces another of higher charge. An understanding of this process has led in turn to a greater insight into the chemistry of minerals. Further application of the lattice structure of silicate minerals is given in the explanation of igneous rocks in the following chapter.

The chemical behavior of minerals is determined by their chemical composition; their physical properties are determined by the internal structure of the ions or atoms of the elements that compose them. More than 2000 different minerals, or mineral species, are now known, but the great bulk of them are quite rare. Only about two dozen are really common, and of these a bare handful make up the largest part of the earth's crust. And this handful, as we noted earlier, is silicates.

About seven elements occur in native form as minerals in the earth's crust. Of these only the nonmetallics, carbon and sulfur, occur in any large quantity. Carbon is found in two very different forms, soft graphite and the intensely hard diamond. Of the metals, gold, silver, and platinum sometimes occur in economically workable concentrations. Although they are rare, these three plus diamond have certainly played a disproportionately large part in the history of nearly all of mankind.

With the exception of the carbonates, calcite ($CaCo_3$) and dolomite $CaMg(Co_3)_2$, all of the other compounds except silicates are of relatively

FIG. 4-8. Examples of typical mineral crystal models and actual minerals that exhibit the forms shown on the wood models. (Courtesy American Museum of Natural History.)

little geologic importance. Those minerals belonging to the compounds called oxides, sulfides, sulfates, and the like that are not important as rock-forming minerals will be referred to as they are touched on in different parts of the book and will also be described in the identification tables given later in this chapter. For the moment we are concerned primarily with those rock-forming minerals of geological importance in the earth's crust.

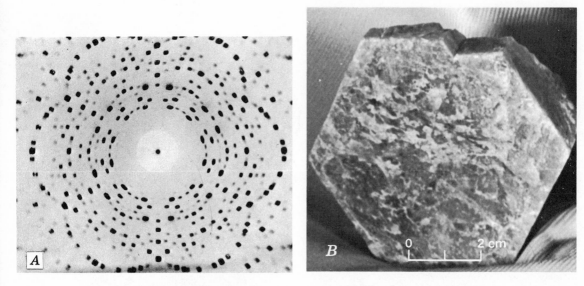

FIG. 4-9. (a) Pattern recorded on an X-ray photograph of the mineral *beryl*. A distinct six-fold internal symmetry is revealed. This internal particle organization is reflected in the six-sided symmetry of a beryl crystal in (b) (Photos by A. Hurlbut and B. Schaub, courtesy Longwell, Flint, and Sanders, *Physical Geology*, John Wiley and Sons.)

The Rock-Forming Silicates

This most important of all mineral groups is also the most complex chemically. Many of the silicates occur as very closely related members of a mineral group or family, one or more of which may be of particular rock-forming importance.

The feldspar group alone constitutes about 55 percent of all crustal minerals. The feldspars can be classified into two subgroups, orthoclase and plagioclase feldspars, on the basis of slightly different chemical and physical properties. Basically, they are all aluminosilicates in combination with potassium, sodium, or calcium. Details of the feldspar compositions and classification are shown in Table 4-3. A glance at their composition indicates where so much of the oxygen, silicon, and aluminum come from. Of the minerals listed in the table, orthoclase, albite, and labradorite are probably the most common. Notice that the composition of oligoclase, andesine, and labradorite is given in terms of the proportion of albite and anorthite present. The plagioclase group belongs to a *solid solution series* in which the two end members, albite and anorthite, are found mixed in a continuously varying proportion to form other members with intermediate compositions. Each of the three intermediate minerals actually covers a range of proportions of the end members of which the ratios given are the midpoints.

The *micas* consist of a number of individual minerals of which only two, *muscovite* and *biotite,* are common. Where disseminated as small grains in rocks, the micas are flakey in appearance, but where they occur as larger concentrations under special conditions, micas resemble "books" of closely spaced sheets along which they can be readily peeled by hand. It is the smooth, shiny faces of mica flakes that give rocks the sparkling, twinkling appearance when viewed under a direct light source. The common varieties, *muscovite* and *biotite,* are both complicated potassium-aluminum silicate compounds containing water in their molecular structure. In addition, the biotite contains iron and magnesium, which gives it a characteristic black color in contrast to the tan-to-white appearance of muscovite. The complete composition of these minerals is best seen from their imposing molecular formulas: muscovite— $KAl_3Si_3O_{10}(OH)_2$; biotite— $K(MgFe)_3(AlSi_3)_{10}(OH)_2$. When these minerals are heated intensely, water is given off by the decomposing mica flakes.

Olivine is of great importance in the denser, heavier rocks of the earth's crust, most of which underly either the oceans or the lighter rocks of the continents. Where visible in relatively large accumulations, olivine is composed of granular, olive-green grains, the color being always very characteristic. It often resembles somewhat a mass of compressed, olive-green, granulated sugar. Much of the reason for the high iron concentration in the earth's crust is found in the rather simple chemical composition of olivine, which is $(FeMg)_2SiO_4$.

Hornblende, the principal rock-former of the *amphibole* group, is a very complicated silicate of Ca, Na, Mg, Fe, and Al. It is a shiny, dark-green to black mineral often in the form of elongated grains with surfaces that form a cross section whose sides meet in oblique angles (of 124° and 56°) as in Fig. 4-6.

TABLE 4-3 Classification and Composition of Feldspars

Subdivision	Mineral Name	Composition	Color
Orthoclase	Orthoclase	$KAlSi_3O_8$	White
	Microcline	$KAlSi_3O_8$	Salmon to pink, rarely green
Plagioclase	Albite	$NaAlSi_3O_8$	White
	Oligoclase	Alb_3An_1	White
	Andesine	Alb_2An_2	Light grey
	Labradonite	Alb_1An_3	Grey to dark grey
	Anorthite	$CaAl_2Si_2O_8$	Black

Augite (the principal component of the pyroxene group) is also dark green to black, but is distinguished from hornblende by its stumpier crystals with a square cross section (Fig. 4-7).

Quartz, the last of the group of very abundant minerals, has by far the simplest composition, expressed by the formula SiO_2. The formula suggests that quartz is an oxide, but modern X-ray analysis has shown that it has the same internal structure as the silicate compounds in which silicon and oxygen are combined with one or more metals, as in those just described. This important internal effect will be discussed shortly. Quartz occurs in a very diverse array of colors, shapes, and sizes. When disseminated in rocks, quartz grains often appear as rather nondescript gray glassy patches. When in rich mineral pockets, quartz may occur as aggregates of beautifully colored and formed crystals, with single units commonly reaching a weight of many pounds. The compound SiO_2 in all of its forms is often referred to as silica, which is a chemical term merely indicating that the silicon is united with oxygen. Thus, all quartz is silica, but all silica is not crystalline (as is true of quartz), since it also occurs in nature in noncrystalline form.

Silicate Tetrahedra

All crystal forms are merely the reflection of a fixed internal arrangement of the atoms or ions that unite to form the complete crystal. The basic structural unit in silicates is the *tetrahedron* of silicon and oxygen. The tetrahedron is a geometric shape composed of four flat faces in the form of a pyramid (Fig. 4-10). In the silicon-oxygen tetrahedron, four oxygen ions are distributed as though at the four corners of the tetrahedron in Fig. 4-10. A silicon ion in the center of the tetrahedron actually makes contact with the oxygen ions, as in Fig. 4-11*a*. Figure 4-11*b* shows an opened view of the structure. The oxygen ions are coupled to the silicon ion by ionic bonds of great strength.

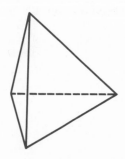

FIG. 4-10. Tetrahedron or tetragonal (4-faced pyramid). (The structure as shown is resting on one face; two faces are toward the front and one toward the rear.)

In the chemical reaction of the atoms to form this combination, silicon gives up four electrons and gains four positive charges (Si^{++++}). Each oxygen atom gains two electrons, becoming an oxygen ion (O^{--}). The tetrahedral unit thus has four positive and four times two or eight negative

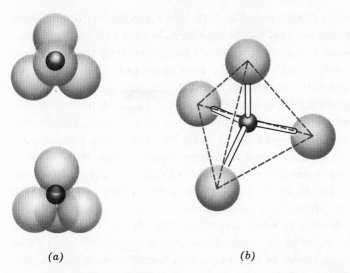

(a) (b)

FIG. 4-11. (a) Top and side views of the silicon-oxygen tetrahedron consisting of a central silicon ion and four oxygen ions. (b) An expanded view of the silicon-oxygen tetrahedron showing the geometric relationship of the complete complex ion.

charges, with a net charge of four negative units; it is designated $(SiO_4)^{----}$ or $(SiO_4)^{4-}$. This ion is the basic building block of the entire silicate system that forms most of the earth's crust. Most of the physical properties of silicates depend on the way the tetrahedra are joined together.

$(SiO_4)^{4-}$ is not a compound, since it has a net negative charge. To become a compound, it must unite with a source of positive ions in order to balance this charge. In some cases, the individual tetrahedra unite with appropriate metallic (positive) ions to form neutral mineral compounds. Olivine is an example of tetrahedra bound together by iron and magnesium ions that provide the positive units necessary to balance the negative SiO_4 ion and bind the tetrahedra together.

For most silicates, the tetrahedra form chains, sheets, or three-dimensional structures. In the case of chains, oxygen ions at opposite sides of the structure are shared with adjacent tetrahedra as in Fig. 4-12. Oxygen-sharing reduces the number of excess negative units in each of the silicon-oxygen ions. The chains of tetrahedra are then bound together by positive metallic ions that complete the electrical neutralization. Because the ionic bonding between the positive metallic ions and the tetrahedra are weaker in varying degrees than the strong ionic bonds within the tetrahedra, minerals composed of such chains form elongated, needlelike, or fibrous elements that can be separated or pulled apart, sometimes with great ease, as in the case of chrysotile (asbestos). In the case of augite, the chains are single with intervening positive ions; with hornblende, the single chains are combined, as in Fig. 4-13, into a double chain with three oxygen ions from one tetrahedron being shared with the silicon ions from three other tetrahedra. The net negative charge for each tetrahedron is reduced further, with positive metallic ions again uniting the chains.

The occurrence of the tetrahedra in a sheetlike arrangement is another common and important structure, as illustrated in Fig. 4-14. In this pattern, three oxygen ions of each unit tetrahedra are shared with three other silicon ions, and one oxygen is held uniquely by each silicon ion—or each tetrahedron is partially connected with three adjacent ones. If the positive and negative charges associated with the silicon and oxygen ions are added, it is seen that there is still a net negative charge for the sheet structure. Adjacent sheets can be connected by metallic ions that give a complete and neutral mineral structure. The sheets may be bound together as single or, as in the case of the micas, double sheets. In this structure, the "rough" sides of the sheets, in which the unshared oxygen stands above the plane of the shared oxygen ions (Fig. 4-14), are joined by positive ions to form a doublet. The smooth faces of the double sheets are bound by potassium (+) ions, but the bonding is so weak that the double units can easily be separated by the fingernail. The important

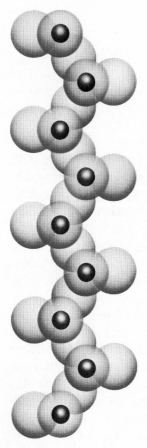

FIG. 4-12. Model of a single chain of silicon-oxygen tetrahedra. Successive tetrahedra are offset so that the oxygen ion between them can be shared.

property of *cleavage* in silicate minerals, in which a mineral splits along well-defined smooth surfaces, is dependent on the presence of weakly bound tetrahedra sheets.

The last important structure of the silicon-oxygen ion is that in which the entire neutralization is accomplished by the joining of the tetrahedra in a manner such that all of the oxygens are shared by other silicon ions rather than other positive metallic ions so that no net charge remains. In the structures considered above, the arrangements were either one-dimensional or two-dimensional (chain or sheets), so that some oxygen ions were not adequately neutralized unless other positive ions were added. In the complete oxygen-sharing, the structure is three dimensional,

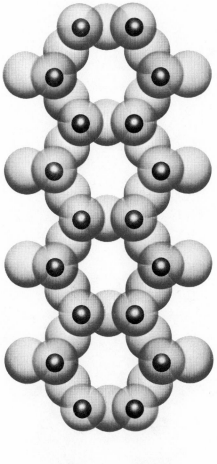

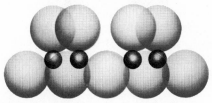

FIG. 4-13. View of a double chain model of silicon-oxygen tetrahedra. The oxygen ion between each single chain is shared by the inner tetrahedra.

FIG. 4-14. Model of a single tetrahedral sheet in which three oxygen ions are shared among three tetrahedra.

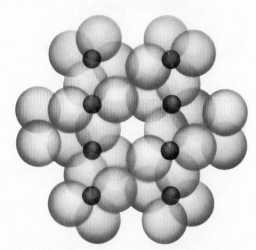

FIG. 4-15. Model of a tetrahedral network in which each oxygen ion is shared with adjacent tetrahedra as in the case of the internal structure of quartz.

FIG. 4-16. Quartz geode formed by precipitation of silica within a rock cavity. Compare with the calcite geode in the chapter opening. (Courtesy American Museum of Natural History.)

somewhat as indicated in Fig. 4-15 where each oxygen ion belongs to the silicon ions from two adjacent tetrahedra. Since each silicon ion is surrounded by four oxygens and each oxygen is coupled to two silicon ions, the total number of ions in such a completed and neutral structure is in the ratio Si_2O_4 or SiO_2 which is the chemical formula for quartz. Because the silicon-oxygen bond is so strong and so uniform throughout a quartz crystal, there is no preferred direction of splitting or breakage, as with the chain or sheet structures. Although chemically an oxide, from the atomic structure of the crystal, quartz is also properly classified as a silicate.

Accumulations of pure minerals, isolated from rock aggregates, are relatively rare. Most of the beautiful and ornamental specimens encountered have been taken from localized occurrences. For example, natural hollows or cavities are often present in rocks from a variety of causes. Water seeping through the rock will slowly dissolve mineral material that has been widely disseminated throughout the country rock of a region. On encountering an air-filled cavity, the water evaporates, precipitating a small mineral crystal on the rock wall just as a water solution of sugar or salt will precipitate crystals in a vessel after evaporation has occurred. Once a crystal has begun, it tends to concentrate further groundwater seeping into the void, causing enlargement of the original crystal or crystals that line the cavity. Such a structure, known as a *geode,* is one of the sources of fine mineral crystals, an example of which is shown on page 98 and in Fig. 4-16.

But in the earth's crust as a whole, minerals occur in less distinguished aggregates, forming the rocks to which we turn in the next chapter.

Identification of Minerals

The recognition of the common rock-forming minerals can be most satisfying. With a little experience we can usually learn to make sight identifications of the common rock-forming minerals very readily. The procedure for doing so, which is given more completely in the Appendix, simply makes use of a number of very apparent physical properties of minerals.

STUDY QUESTIONS

4.1. What is the fundamental physical difference between the rock that composes continents and that which forms the crust beneath the sea floor?

4.2. How may the facts in answer to question 1 help explain the elevation differences between continents and ocean basins noted in Chapter 3 and in answer to question 3.3?

4.3. Distinguish between "atomic number" and "atomic weight."

4.4. Explain the two methods by which chemical reactions occur according to atomic theory.

4.5. Why are the rare or noble gases so inert, chemically?

4.6. Distinguish between "atoms" and "ions."

4.7. Give several chemical compounds by name and formula and indicate, if possible, the electrical changes associated with each element involved.

4.8. What is the importance of the gas, oxygen, to the earth's crust?

4.9. What is a "silicate" and what is its importance to the composition of the crust?

4.10. When ions replace each other in a crystal lattice, what two conditions must be met?

4.11. List some common minerals with their chemical formulas.

4.12. Why are the silicates the most important of the mineral groups found in the crust?

4.13. Using plagioclase feldspars, as an example, explain the meaning of "solid-solution series." Refer to Table 4-3 for formulas.

4.14. What is the particular importance of the feldspars and olivine regarding rock composition?

4.15. Explain the silicon-oxygen tetrahedron.

4.16. Indicate the arrangement of the silicate tetrahedra in some of the important silicate minerals.

chapter 5

Sedimentary rocks in Bryce Canyon National Park, Utah. (Photo
by Edith B. Vincent.)

Rocks: Their Composition and Development

When we look at the earth's crust we rarely see pure minerals and never see the fundamental atoms and molecules that compose them. The earth's crust is actually composed of rocks that are mixtures or aggregates of minerals. Over much of the crust, the true rock, often called *bedrock,* is covered with a carpet of loose rock material or *regolith* (called *soil* when fertile) that forms from the natural breakdown of the solid rock. In this chapter we will be concerned with the bedrock of the crust and in Chapter 6, following, we will see how regolith and soil form as a consequence of rock weathering and erosion.

Rocks fall into three natural groups, depending on their origin: *igneous* rocks have been crystallized out of high-temperature, often molten material; *sedimentary* rocks are the consolidated products of deposition mostly by the seas but also by rivers, lakes, wind and glaciers; and *metamorphic rocks* are the rocks converted from initial igneous or sedimentary rocks after a history of subsequent heating or pressure deformation (or both). As rocks are the things of which the earth that we see or know is composed, a grasp of their nature and origin is essential to the study of earth science.

IGNEOUS ROCKS AND THE MAGMA

The bulk of the earth's crust is composed of igneous rocks, whose origin is suggested directly by the name, which comes from the Latin term for fire. With a significant reservation to be discussed later, we can define igneous rocks as those that have formed from the solidification of originally molten material called *magma* (from the Greek, meaning paste).

Very little of the earth's crust that we observe is primary—or is in the form and condition it was in at the time of the earth's formation close to five billion years ago. Considerable natural reworking of the crustal rocks has taken place, together with replenishment of new material from the underlying mantle. Igneous rocks that we see at present have mostly formed by cooling and solidification of different bodies of magma generated within or below the crust during the earth's long history. Magma consists of molten silicates containing a large amount of dissolved gases (mostly water vapor). When rock melts to form magma, its volume increases. Pressure from this volume increase as well as from the heated gases forces the magma up through the crust, where it follows initial natural breaks and fissures or creates its own pathways from the effect of the upward pressure. If the magma does not reach the surface, it solidifies into igneous rocks whose shape depends in part on the nature of the rock being invaded and in part on the nature and size of the magma body. If it reaches the surface, the magma may erupt in the form of a volcano or spread out over the surface as a lava flow.

Origin of Magma

Although the actual source of heat causing the liquification to form magma is not yet fully resolved, a number of heat sources are known that may together or separately generate molten rock.

Decrease of pressure at depth. The temperature within the earth increases from an average surface value of about 60°F to about 10,000°F within the core. The rate of increase of temperature is such that in the deeper parts of the crust (30 to 40 miles), the temperature can be above 2000°F. The high temperature at this depth might be adequate to melt the rock, were it at the earth's surface. But at a depth of some 38 miles, the pressure is about 19,000 times that at the surface, where rocks are subjected to atmospheric pressure only. Such high pressures preserve the rocks in a solid state. Hence, as the pressure increases, the temperature necessary to cause melting also increases. If movement or change in the overlying crust causes a large pressure reduction at depth, the temperature may then exceed the normal melting point of the rock at this level. But much more heat is required to melt a gram of rock than its specific heat (that necessary to raise its temperature 1°C). So that only a small part of the deep rock layer at best can melt from this cause.

Heat of radioactivity. It is now a matter of common knowledge that

FIG. 5-1a. Hand specimen of an igneous rock rich in the mineral *olivine* together with the slide containing the thin-section of the rock as prepared for examination in the petrographic microscope.

FIG. 5-1b. Photomicrograph of the thin-section in (a) as viewed in the petrographic microscope. The colors and other characteristics of the mineral grains are diagnostic for olivine.

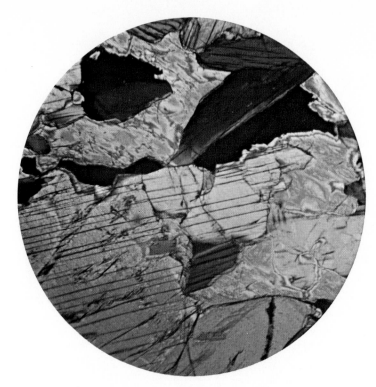

FIG. 5-1c. Photomicrograph of Manhattan Schist, one of the foundation rocks of the New York City region. The strongly colored mineral is hornblende and the parallel lines are the traces of the cleavage planes.

FIG. 5-1d. Photomicrograph of sedimentary hematite (iron ore) showing a variety of microfossils.

the atoms of certain elements, such as uranium, radium, and thorium, spontaneously decay into other elements by the process called *radioactivity,* discussed more completely in Chapter 7. The atomic disruption involved in radioactivity releases a relatively large amount of heat, which is absorbed in the rock containing the radioactive elements. Because rock material is a poor heat conductor, the heat released in radioactivity remains with the source rock, causing a continuous increase in temperature. Although the distribution of such elements is variable in the crust, good estimates can be made of their average occurrence and resulting heat output. If we take the crustal thickness to average 10 miles, the annual release of heat generated by known radioactivity is about 10 billion squared calories (or 10^{20} cal). Remember that a calorie is defined as the heat necessary to raise 1 gram of water $1°C$. Since rock requires much less heat than water for an equivalent temperature change, it seems clear that radioactivity can certainly provide the heat required for the generation of magma where local concentrations of radioactive elements occur.

It is important to realize that the radioactive elements have become concentrated in the crustal zone and particularly in the rocks beneath the continents. If the entire earth had the same distribution of radioactive materials as does the continental crust, the earth would be nearly all melted except for a skin on the top where the heat could be quickly dissipated. An idea of the heat available for geologic processes can be gained from the amount that escapes from the interior at the earth's surface. This annual heat loss amounts to about 40 calories for each square centimeter of the earth's surface, an enormous quantity when summed over the entire earth.

Composition and Behavior of Magma

Magma consists of molten or nearly molten masses of silicates occurring either within the crystalline rocks of the earth's crust or in the upper part of the mantle, beneath the crust. The magma chamber encloses the molten material often called a *silicate melt.* A large quantity of gases, particularly water vapor, are usually present in the melt in varying amounts. The high temperature and gaseous content of the magma make it possible to intrude into surrounding rocks, this being all the easier when pre-existing fissures or breaks are present. Magma that forces itself all the way to the earth's surface may spread out as lava, or erupt through a small vent to form a *volcano.*

Our understanding of the nature of magmas increased dramatically following the establishment of the Geophysical Laboratory in 1904 by the Carnegie Institution of Washington. Laboratory experiments on silicate melts were carried on in an effort to duplicate the high-temperature high-pressure conditions within a magma chamber. Laboratory results were especially fruitful when coupled to a careful field and laboratory analysis of rocks. The latter process involves the microscopic examination of a *thin section* of a rock specimen. A thin section is made by grinding and polishing a thin slab (about an inch square) until it becomes transparent at a thickness of one to two thousandths of an inch. When it is examined in a special microscope designed for rock studies (petrographic or polarizing microscope), a very specific determination can be made of the mineral composition and often of the rock history. When viewed in this way, mineral grains have a characteristic and often strikingly beautiful and colorful appearance. An example of a rock specimen and its thin section are shown in Fig. 5-1*a*, together with the view of the section as seen in the petrographic microscope (Fig. 5-1*b*). Examples of photomicrographs of other rock sections are included in this illustration.

Our studies have shown that natural magmas have much the same ionic composition as the solid silicates described in the previous chapter. Silicon-oxygen tetrahedra are present as negative ions that may be linked together by the one-, two- or three-dimensional networks in which oxygen serves as the bridge between the tetrahedra. The positive metallic ions that are ultimately required to join the tetrahedra in the formation of solid silicate minerals consist primarily of ions of sodium, potassium, calcium, magnesium, and iron. A major and important difference between the ions in magmas compared to those in solid silicates lies in their mobility, which may be quite high, especially for the positive ions.

Igneous rocks form from the crystallization of silicates as magma cools. The resulting rocks show a tremendous diversity in color and appearance because of a great variation in mineral composition and grain size (texture). The mineral composition and rock texture depend on both the composition of the magma and the conditions of its cooling and solidification. If the high temperature is quenched abruptly, such as by extrusion as lava on the surface, very rapid solidification occurs and the growth of large mineral crystals is inhibited by the inability of the ions to move through the melt. If cooling is slow or if the magma remains fairly fluid during the cooling process, large grains form because the positive ions retain a high degree of mobility, permitting large crystal growth.

The fluidity of different magmas varies at the same temperature. Some may be almost solid while others are still liquid. In general, the more complete the formation of networks or tetrahedra, the less fluid or more viscous is the magma. At the high temperatures of natural silicate melts, water tends to break the oxygen bridges that bind the silicon-oxygen tetrahedra into a network form. Hence, magmas containing much water tend to remain quite fluid as they cool and thus develop large mineral grains.

Magmatic Crystallization

The process by which liquid magma cools and crystallizes is called magmatic crystallization. From a study of artificial silicate melts and of natural rocks in the laboratory, N. L. Bowen of the Geophysical Laboratory deduced his classic *reaction series* as an explanation of the processes occurring in a cooling and solidifying magma. Bowen recognized two different genetic mineral series: one, involving silicate minerals rich in iron and magnesium, is called the *ferromagnesium* series; the second, involving only the plagioclase feldspars, is called the *plagioclase series*. These are called reaction series because later minerals form from the reaction of the preceding mineral with the remaining liquid.

In the ferromagnesium series, the first mineral to crystallize is olivine. The residual liquid then reacts with the olivine crystals to form augite and then with the augite to form hornblende and finally with the hornblende to form biotite. Recall from the preceding chapter that each of these ferromagnesium minerals have a different crystal structure, so that an internal rearrangement of ions must occur as crystallization proceeds. In the plagioclase series the first mineral to crystallize is the calcium-rich silicate, anorthite, and the last, the sodium-rich silicate, albite. Successive minerals involve only a substitution of sodium for calcium, without any change in the tetrahedral or crystal structure. The crystallization process of these reaction series in the same magma goes on simultaneously, with olivine and anorthite coming out of the melt at about the same temperature. The diagram in Fig. 5-2 indicates schematically the order of crystallization as the reaction series progress. Orthoclase and quartz are the last minerals to crystallize out of the melt. The final mineral composition of a rock must of course depend on the original magma composition.

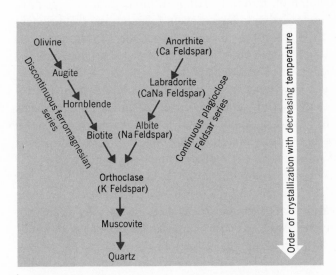

FIG. 5-2. The Bowen reaction series shows the sequence of crystallization of silicate minerals as a magma cools. Each mineral reacts with the remaining liquid portion of the magma. In the *continuous* or plagioclase feldspar series modification of the early minerals occurs; in the ferromagnesian or discontinuous series, the earlier minerals can disappear completely in the reaction to form the later minerals.

Mineral crystallization in the cooling magma is controlled by many of the principles developed in the preceding chapter. For example, in the ferromagnesium series, minerals with simple ionic bonding, like olivine, form first. As the temperature decreases, the energy of the ions decreases, permitting more complex coupling among the silicon-oxygen tetrahedra and the metallic ions that complete the crystal lattice. Because the lattice of each mineral in this series is different from the preceding one, forming entirely different minerals, the ferromagnesium series is called a *discontinuous-reaction series.*

In contrast, the feldspar series is a *continuous-reaction* series because substitution of metallic ions occurs within the same lattice as the temperature cools and "older" minerals react with the melt. The double-charged Ca^{++} ion is attracted into the lattice first. Reaction with the magma permits Na^+ ions to replace Ca^{++} ions because their ionic sizes are about the same. To compensate for the single-charged Na ion, modifications in the tetrahedral linkages occur.

The final Ca-Na composition of the feldspars found in a rock depends on how much of each ion was present in the magma. The plagioclase feldspars may range from anorthite, a pure calcium feldspar ($CaAl_2Si_2O_8$), to albite, a pure sodium feldspar ($NaAlSi_3O_8$).

The orthoclase variety of feldspar ($KAlSi_3O_8$), despite the similarity in composition, is not part of the continuous-reaction series because the K^+ ion is too large to substitute for either Ca^{++} or Na^+. In general, magmas are richest in silicate ions to which the metallic ions present become attached to form the common minerals. After all the crystal lattices have been completed, any excess silicate ions present form tetrahedral linkages with the proportion Si_4O_8 (SiO_2) as described for quartz in the previous chapter. The presence of quartz thus indicates that the original magma was very rich in silicate ions (or silica).

The order of crystallization can be deduced from a careful microscopic examination of many thin sections of rocks. The earliest minerals to form will have the best-developed crystal forms. As more and more of the melt crystallizes, less space is available for the full development of crystals. The last minerals to crystallize usually have very irregular shapes, filling whatever space remained at the end stages of solidification. Also, early-formed crystals may be all or partially enclosed by later crystals. The illustrations of different rock sections as viewed in the petrographic microscope, Fig. 5-3, show examples of crystal grains that are indicative of the sequence of crystallization.

Igneous Rock Types

Igneous rocks are usually classified in terms of their mineral composition and texture or grain size. Since these properties, once again, depend on the composition of the magma and the history of the cooling process, the recognition of an igneous rock type is important in deducing the geologic history of a region. Although a fairly complete rock classification and recognition scheme is given in the appendix, certain of the rocks are so important to discussions in later chapters that we must consider them here.

One major group of igneous rocks has an essential composition of quartz plus orthoclase and albite (feldspars). These light-colored minerals rich in silicon (Si) and aluminum (Al) form the generally light-colored *sialic* rocks. Granites are the most important example of sialic rocks because granites and the closely related granodiorites (see Table B-1 in the appendix) form about 95 percent of the mass of the continents (Fig. 5-4).

The other major group of igneous rocks is formed of minerals like augite, olivine, and plagioclase feldspars labradorite and anorthite. These dark-colored minerals, rich in magnesium (Mg) and iron (Fe), form dark-colored *mafic* rocks. Of the mafic rocks, *basalt* is the most important example.

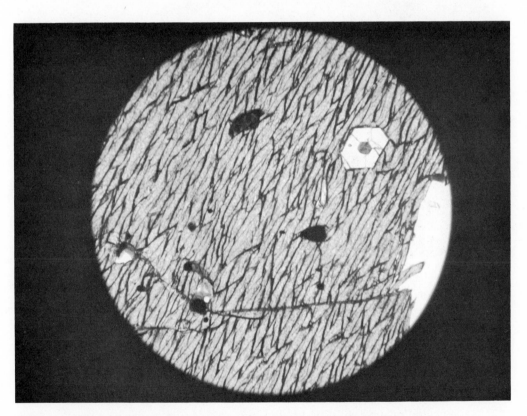

FIG. 5-3. A photomicrograph (×80) showing a hexagonal crystal of the mineral apatite completely enclosed by the later less-well formed crystal of hornblende. A crystal of still earlier formation lies within the apatite crystal. Crystals enclosed by others must have formed first. Also, those grains in a section which show the better geometric (crystal) forms, crystallize prior to those with irregular shapes.

Direct exploration of the sea floor by rock dredging has shown the ocean crust, where sampled, to be composed of typical dark to black fine-grained basalt. Indirect exploration by means of seismology (Chapter 10) has shown that rock with the properties of basalt constitutes most of the marine crust beneath the sediments. Also, most of the lava flows on the continents appear to be basalt that has come from a fairly great depth. The well known Columbia lava plateau of northwestern United States and the larger Deccan Plateau of India represent basalt flows thousands of feet thick and at least one hundred thousand square miles in area. We are thus led to conclude that basalt is a most important rock that not only composes the crust of the ocean basins, but extends as a globe-encircling layer beneath the continents. In Chapters 8 and 9 we will return to the problems related to the origin and history of the continents and ocean basins.

FIG. 5-4. Specimen of granite showing the typical granitoid or coarse crystalline texture. The dark grains include hornblende and biotite; the lighter include orthoclase and quartz. A microscopic section of the rocks is also shown.

Forms of Igneous Rock Bodies

Igneous rocks occur either as *intrusive forms,* which are bodies of congealed magma solidified below the earth's surface, or as *extrusive forms* in which the magma has erupted upon the surface.

Intrusive forms, often called *plutons,* depend on the nature and size of the parent magma and on the environment of intrusion. *Batholiths,* the greatest of these bodies, are invariably composed of huge masses of granite or granodiorite located within the cores of great mountain ranges. We see

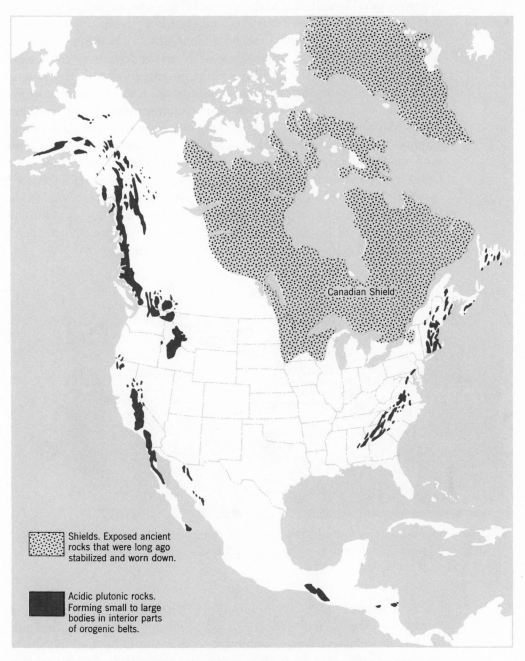

Shields. Exposed ancient
rocks that were long ago
stabilized and worn down.

Acidic plutonic rocks.
Forming small to large
bodies in interior parts
of orogenic belts.

FIG. 5-5. Map of North America showing the location of granite batholiths in
the interiors of mountain belts. In addition to these large masses of granite
rock that lie in the cores of great mountain ranges, note that most of eastern
Canada, and Greenland, the *Canadian Shield,* is occupied by a very extensive
region of granite formed billions of years ago. Each continent has a significant
granite shield area.

them today only because prolonged erosional scouring of the mountains has exposed their granite interiors. Although their surface exposure is generally parallel with that of the surrounding mountain range, batholiths spread out with depth so that their smallest horizontal dimension is at the surface. Several elongated, scattered batholiths mark the exposed core of the old Appalachian Mountain belt of eastern United States from northern Georgia and the Carolinas to the mountains of New Hampshire. In the United States the greatest batholiths lie in the mountains of the west. The Idaho batholith covers 16,000 square miles in the Northern Rocky Mountains of Central Idaho. The Sierra batholith extends for about 400 miles along the east-central border of California and averages 50 or more miles in width. But the greatest of all in North America is the granitic Coast Range batholith, which extends from southern Canada northward for 1100 miles into Alaska with a variable width of 80 to 120 miles. Batholiths occupy an important place in the modern theory of mountain building and will be given further discussion later in this and in Chapter 9. Figure 5-5 shows the batholiths of North America.

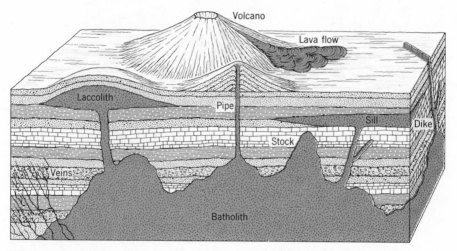

FIG. 5-6. Composite diagram illustrating the forms commonly taken by solidified magma. The problem of the batholith is discussed in detail later in this chapter.

A schematic relationship among the batholiths and other igneous intrusive as well as extrusive forms is shown in Fig. 5-6. *Dikes* and *sills* are tabular intrusive forms, occurring as thin sheets or thick layers. In the case of dikes, the body cuts across the preexisting rock structures. In the case of sills, the igneous material is intruded parallel or conformable with the enclosing rock units. Perhaps the most famous example is the Palisades sill, which extends for about 70 miles northward from New York Bay along the western shore of the Hudson River. This classical intrusion is illustrated diagrammatically in Fig. 5-7a.

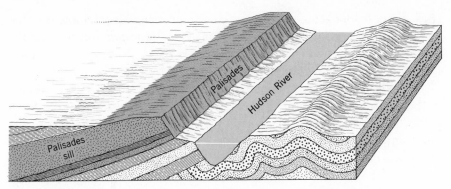

FIG. 5-7a. Diagrammatic view of the Palisades of the Hudson—a classical and world famous example of a sill which was intruded into the surrounding sedimentary rocks about 180 million years ago.

FIG. 5-7b. Basalt dike intruded into granite, Mt. Desert Island, Maine.

The name Palisades comes from the long four-and-five-sided columns that form the steep eastern cliff face of the Palisades, giving the entire structure the appearance of the high wooden stockade of colonial days. The columnar structure characteristic of the cliff actually extends throughout the body of the sill, as is also true for many other sills as well as lava flows. In the cooling process, shrinkage of the molten liquid occurs, forming tension cracks that radiate, under ideal conditions, at 120° angles from shrinkage nuclei. Drying mud is subject to the same forces, yielding the familiar polygonal mud cracks evident on the surface of a barren field after a rain. Under some conditions, columnar structures actually develop in fine mud materials as shown in Fig. 5-8. Columnar jointing, as the feature is called when exhibited by igneous rocks, forms from the intersection of the shrinkage planes.

If a sill-like intrusion domes-up the overlying rock layers as shown in the summary diagram (Fig. 5-6), the form is called a *laccolith*. Such forma, which may reach 25 miles or so across, can be reflected in the topography as small mountain units. The Henry Mountains in Utah represent a well-known cluster of laccolithic structures.

FIG. 5-8. Example of dessication or mud cracks in lime-mud in the Florida Keys.

Extrusive forms of igneous bodies include lava flows, volcanoes, and layers of fragmental material erupted explosively from certain types of volcanoes. We have already referred to lava flows, which are produced by the eruption of magma through fissures in the crust that reach to the surface. Although most lava flows are thin, some, as in the case of the Columbia and Deccan plateaus, accumulated thousands of feet of layered basalt, sometimes called *plateau basalt*. Volcanoes will be described further in Chapter 9.

SEDIMENTARY ROCKS

All of the rocks of the continents are subjected to continuous erosional forces (described in Chapter 6) that ultimately reduce them to loose sediments. After transportation by any of the agents of erosion (also to be described), the sediments are deposited, often far from the source, and subsequently become consolidated into sedimentary rock. We noticed earlier that the bulk of the earth's crust is igneous in nature—granite or granodiorite composing the continents and basalt composing the ocean basins. The sedimentary rocks form a veneer, which in the case of the contiguous United States, carpets some 80 percent of the land area. They are also the stuff out of which the great mountain ranges have been formed and in such cases are tens of thousands of feet thick. Perhaps most important to earth science is the fact that in their structure and composition sedimentary rocks preserve the history of the earth's crust to nearly 4 billion years ago.

TABLE 5-1 The Wentworth Scale of Clastic Sediment Size

Particle Name	Metric Size	English Equivalent (approximate)
Boulder	>256 mm	>10 inches
Cobble	64 to 256 mm	2.5 to 10 inches
Pebble	4 to 64 mm	0.16 to 2.5 inches
Fine Gravel	2 to 4 mm	0.08 to 0.16 inches
Sand	$\frac{1}{16}$ to 2 mm	0.0025 to 0.08 inches
Silt	$\frac{1}{256}$ to $\frac{1}{16}$ mm	0.00015 to 0.0025 inches
Clay	$<\frac{1}{256}$ mm	<0.00015 inches

Sediments: Clastic and Nonclastic

Sedimentary rocks form from the consolidation of sediments—the products of prior erosion. The nature of sediments depends to a large extent on their method of origin. When primarily mechanical in derivation, *clastic* (meaning broken) sediments and sedimentary rocks result. Although strong chemical alteration of the sedimentary grains may occur, the particles themselves were originally worn bits of preexisting bedrock. Clastic sediments range in size from boulders through cobbles, gravel, sand, and silt to mud or clay. Of these varieties, common observation shows the last or finest to be the most common form of sediment. Most of us know roughly what the above names refer to, but for uniform description and classification, geologists commonly use Wentworth sediment size scale described in Table 5-1.

As an example of sediment formation in the conversion of granite to clastic sediment, the erosional process may cause the breakup of a bedrock exposure of granite into initially large fragments that then become reduced in size by continued erosion. At some distance from the source, the boulders and cobbles become so worn and rebroken that the rock separates into the component minerals. Quartz grains are very durable, whereas the feldspar components (orthoclase and albite) become reduced to very fine particles that are chemically altered rather easily to clay. The ultimate erosion of a granite thus usually produces sand from the quartz content and clay from the feldspar content. Since the bulk of the granite is feldspar, the bulk of the resulting sediment is clay, which is therefore far more common in nature than sand.

Where close to the source, fragments of sand size and larger tend to be angular. In the course of transportation, sharp edges and corners are broken or abraded so that individual particles become smoother and more rounded.

The major group of sediments and sedimentary rock other than clastic is the *nonclastic*. This group includes material that is essentially organic and chemical in nature. As organic processes are really chemical, the nonclastics can be considered as resulting from chemical processes. Although some sediment may be precipitated directly from solution, a very abundant form of nonclastics usually consists of calcareous material composed partly of minute sea life or of sediment precipitated through biologic action. Calcareous muds, which ultimately become limestones, are the most important of the nonclastics. They often consist of worn and reworked skeletons of minute animals that lived in the upper levels of the oceans as *pelagic* (floating or suspended) sea life (Fig. 5-9).

A mixture of fine clastic muds with calcareous sediment often occurs, the amount of each present depending on their relative rates of production, which in turn depends on the environment in the region of deposition. In fact, some of the most significant history of the environment of the past is deduced from a study of the pelagic material recovered from the sea floor in the long sediment cores taken by oceanographic research vessels. The climatic interpretations from nonclastic sediments will be considered in more detail in Chapter 15.

Sediments to Sedimentary Rock

In the case of sedimentary rocks, *lithification* (rock forming) involves different processes, depending on the original sediment. Granular materials, such as silt, sand, and gravel, are converted to rock by the precipitation of natural cementing material in the pore spaces among the grains. Calcite, silica, and limonite constitute the most common of the natural cementing media and are precipitated from waters that percolate through the pores.

Mud or clay sediments become lithified by *compaction*. In this process the simple weight of accumulating overlying sediments squeezes the fine clay or mud particles together. As the water in which the sediment was originally deposited is driven out under pressure, the grains become

FIG. 5-9. Pelagic Foraminifera formed of calcium carbonate as recovered in a dredge haul from the central part of the north Atlantic ocean. (Courtesy A. Be, Lamont-Doherty Geological Observatory.)

bound by contact into a true rock. A reduction in 50 percent of original mud thickness can occur during compaction into shale, the resulting sedimentary rock.

In the case of clay or calcareous sediments, excessive compaction often leads to *recrystallization*, the third major form of lithification. In this process, larger grains of the same composition may form. Since a single larger crystal occupies less space than the original grains from which it formed, the total effect of recrystallization reduces rock volume further, as an accommodation of the pressure from increasing overlying sediment loads. Clay grains are converted to small flakes of mica, with the flakes being oriented parallel to the rock layers or perpendicular to the pressure. Grains of limy material recrystallize into single larger grains of calcite that although initially microscopic, may grow to be easily visible to the eye.

Types of Sedimentary Rocks

Although a classification and description of sedimentary rocks is given in Appendix C, a discussion of the principal types and their significance is necessary here in view of the importance of such rocks in the discussions in following chapters. Clastic sedimentary rocks, formed directly from the clastic sediments already described, include, in order of decreasing grain size, conglomerates and breccia, sandstone, siltstone, and shale. Limestone is the only important example of nonclastic rocks.

Conglomerates and breccias composed of pebble-sized fragments are relatively rare compared to the others and are distinguished from each other on the basis of the shape of the coarse fragments (Fig. 5-10). In conglomerates, the pebbles are quite rounded and in breccias they are relatively sharp and angular. Breccias thus imply a proximity to the source of the coarse material. The pebble components of conglomerates and breccias may be of quartz or fragments of an older rock.

Sandstone commonly consists of sand-sized quartz grains cemented by silica, calcite, or iron oxide. These rocks often have a characteristic gritty feeling from the effect of the quartz grains protruding above the cement. The rock *arkose*, an important variety of sandstone, is characterized by the presence of recognizable fragments of orthoclase and albite in addition to quartz. Since the feldspars tend to be destroyed quickly during transportation by erosional agents, their presence in a rock indicates deposition close to the source, and probably on land as a continental deposit. Arkoses also indicate the presence of a former close granite mass whose erosion provided the orthoclase and quartz. *Graywacke* is a "dirty" sandstone containing quartz, feldspar, and fragments of other fine-grained rocks, all embedded in a fine claylike matrix. *Siltstone* can be considered a fine sandstone.

Shale, the most common, by far, of sedimentary rocks, has a characteristic slabby appearance and earthy feeling, both a consequence of the compressed mud from which it formed. Shales tend to split or break in flat, roughly parallel thin slabs and have colors ranging from reds, greens, browns, and shades of gray to black.

Limestones exhibit a tremendous color variety from pure white to deep black and from extremely fine to quite coarse textures. If recrystallization has occurred in the lithification process, shiny flat faces of calcite may show prominently. Chalk is an example of a very fine white limestone.

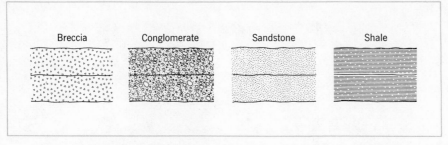

| Breccia | Conglomerate | Sandstone | Shale |

FIG. 5-10a. Diagrammatic examples of the four main types of clastic sedimentary rocks.

FIG. 5-10b. A slab of conglomerate showing rounded pebble inclusions.

FIG. 5-10c. Breccia showing angular fragmental inclusions.

The Environments of Sedimentary Rocks

Much of the history of the earth's crust, from the very recent to the very ancient geologic past, is deduced from a study of sedimentary rocks. The composition and structures of these rocks is a direct result of the environment of deposition, so that we can work backwards from a knowledge of the rocks to an interpretation of the past.

Foremost among the conditions of the past that we must "read" from sedimentary rocks is whether they formed on land or in the seas, and,

FIG. 5-11. Example of eroded stratified rocks in a horizontal (un-deformed) attitude in Bryce Canyon National Park, Utah. (Photo by Edith B. Vincent.)

if the latter, whether the seas were deep or shallow. Fossils and rock composition and texture are the best criteria for these interpretations. Climate at the time of deposition is another important aspect of the environment. After deposition and lithifaction, the shape and arrangement of the rocks are most revealing about subsequent deformations of the earth's crust. Although most of these interpretations are developed in subsequent chapters, the "tools" are best considered here with our discussion of the rocks.

Perhaps the most important single feature of sedimentary rocks is their *stratification* or layered appearance, a normal consequence of the deposition of sediments layer upon layer. Because the surfaces of deposition are almost always horizontal or nearly so, sedimentary strata originate in essentially a flat-lying or horizontal attitude. Stratification is emphasized by the presence of *bedding planes* between the rock layers. These planes provide the reference surfaces by which the structural arrangement of sedimentary strata can be recognized and interpreted. Figures 5-11 and 5-12 show how sedimentary rock structure is revealed by the stratification, without which the structural form would be difficult to discern.

Fossils, the remains of former plant and animal life, are one of the most important inclusions within sedimentary rocks. The kinds of fossils

FIG. 5-12. Example of stratified rocks exhibiting an anticline or upfold, Goshen, N.Y. (Courtesy American Museum of Natural History.)

Sedimentary Rocks   **139**

FIG. 5-13. Photograph of a sedimentary rock from Whitby, Yorkshire, England with numerous fossil cephalopods ($\times 1.5$). The specimen is an extinct invertebrate related to the squid, octopus, and nautilus. (Photo by G. Robert Adlington.)

present indicate whether the initial sediment was deposited in the sea or in water bodies on land. If deposited in the sea, the fossils present may tell whether the water was deep or shallow and whether it was warm or cold. Some of the fossil forms may be fairly large, like those in Fig. 5-13, or they may be so small that recognition and identification require the use of a microscope (Fig. 5-14).

Surface features formed on sediments almost at the time of deposition also contribute much to our knowledge of the local environment. For example, if deposited in water so shallow that frequent exposure to the atmosphere occurred, as on tidal flats and lake margins, mud-type sediments develop typical polygonal desiccation cracks. These features, which form from a volume shrinkage, resemble in appearance and origin the polygonal jointing of sills and lava flows described earlier. Examples of well-developed mud cracks were shown in Fig. 5-8.

The presence of sediment *ripples* often reveals something of the behavior of waves and currents that formed them as explained in Fig. 5-15a and b and Fig. 5-16a and b.

Cross-lamination is another important depositional feature bearing on the local environment. Movement of wind or water (ocean currents or rivers) often causes deposition of sediment grains in an inclined position. For example, wind-formed sand dunes commonly develop as in 5-17a, where sand is carried up the gentle windward face and falls down the steeper lee face, causing the dune to "march" downwind. Within the moving dune the sand grains are arranged in fine laminae as in Fig. 5-17b, and if dunes should form on each other in the course of time, the arrangement shown in Fig. 5-17c develops. Ultimately a layer of sand like that in Fig. 5-17d is formed with an internal cross-laminated structure. Since the inclination of the sand laminae is toward the downwind direction, sedimentary structures of this kind have been used to infer wind directions at times of deposition—often hundreds of millions of years ago. This enables comparison of past wind and climate belts with those of the present (Chapters 13 and 15). Similar structures that form in stream channels and in ocean beaches and sandbars are also helpful in interpreting flow directions of ancient rivers and ocean currents, respectively. An illustration of cross-laminated sandstone of the type described is shown in Fig. 5-18.

FIG. 5-14. Example of microfossils washed from the sediments of a deep sea core raised from the central equatorial Pacific Ocean ($\times$60). The fossils are about 50 million years old. (T. Saito, Lamont-Doherty Geological Observatory.)

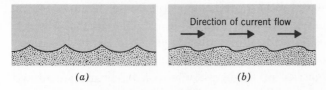

(a)   (b)

FIG. 5-15. Ripple marks: (a) oscillation ripples produced by waves and (b) current formed ripples. Since the former are always symmetrical and the latter asymmetrical, current ripples indicate the direction of flow of the current that formed them.

Turbidity Currents and Turbidites

Two very different types of observations established the existence of a most important depositional process in the deep sea not suspected prior to the middle of the twentieth century. The first event of importance occurred in November 1929. Following a small earthquake in the Grand Banks off Newfoundland, 13 telephone and telegraph cables connecting this country with Europe were broken. At the time the damage was ascribed to the shaking of the ocean bottom by the shock waves. But a serious question soon arose. None of the cables on the shallow and gently sloping continental shelf were broken. The 13 cables that did break lay along the steep southern margin of the continental slope, south of Newfoundland and the flatter sea floor beneath deeper water (Fig. 15-19).

FIG. 5-16a. Oscillation ripples visible in a slab of fine-grained sandstone. Ripples of this type, formed by the effect of wave action on sediment, produces ripples that are characteristically symmetrical, with sharp crests and broad troughs.

142   ROCKS: THEIR COMPOSITION AND DEVELOPMENT

Automatic recorders showed when each cable broke. Those high up on the continental slope broke almost at the time of the earthquake. The others broke in succession, and the 13th, out on the sea floor 450 miles from the tremor, gave way more than 13 hours later. Since seismic waves from earthquakes travel at speeds of many miles per second, the slow succession of breaks could not be explained simply by their passage. Also, the cable breaks involved at least 100 miles of broken sections.

A second crucial observation involved some of the early deep-sea cores taken by M. Ewing in the deep ocean beyond the continental shelf off the New York coastal area. These cores showed striking layers of relatively coarse sand, a deposit not thought capable of being transported from the land to the quiet waters of the deep ocean. Also, in some cores, the deep-sea sands contained shell material of small animals known to live only in the shallower waters of the continental shelf. Further, the ages of the shells in some of the sand layers, determined by radiocarbon dating (Chap. 7), turned out to be greater than the ages of shells in the underlying clay layers.

FIG. 5-16b. Asymmetric ripples formed in fine sand in the receding waters of an ebb tide. The asymmetry indicates the flow was from left to right. The trails are formed by crawling periwinkles. (Courtesy American Museum of Natural History.)

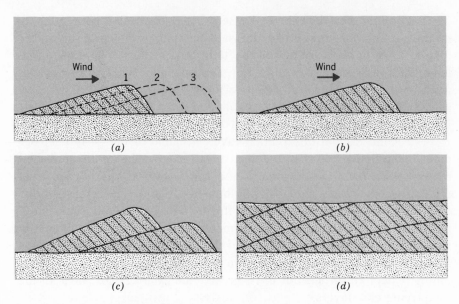

FIG. 5-17. Cross-section and profiles showing different views of sand dune structure. In (*a*) the downwind march of a dune is shown, with the crest being displaced from position 1 to 3, as sand grains are carried up the gentle windward slope and then slump down the steeper lee slope. In (*b*) the internal structure of the dune is shown as it would appear at some time after development. Sand laminae slope downward, "leaning" into the wind. In (*c*) one dune is shown overriding another. In (*d*), the result of several overlapping dunes is shown giving a cross-laminated structure within an otherwise horizontal sand layer.

These and subsequent observations were explained on the basis of *turbidity currents*, an effect known previously from laboratory experiments and observations in lakes but not previously applied to the sea. Turbidity currents (*turbid* meaning muddy or cloudy) are downslope movements of dense muddy water beneath normal clearer water. An illustration of such currents is given in Fig. 5-20. Once water is set in motion by increased density, the resulting current will have transportational and erosional capabilities analogous to streams on land. Turbidity currents can occur over a broad reach of the continental slope, unlike stream action, which is confined to well-developed channels.

In the case of the Grand Banks earthquake of November 1929, the immediate failure of the eight cables on the relatively steep continental slope were explained by B. Heezen and M. Ewing on the basis of the shock and landslides on the slope. The huge amount of sediment stirred into the water by the slides then created a strong turbidity current that surged down the continental rise and gently sloping ocean floor at speeds decreasing from 60 to 13 knots (nautical miles per hour). These speeds, indicated in the velocity graph of Fig. 15-19, were determined by dividing

FIG. 5-18. Photograph showing cross-laminated sand structure of wind blown origin in Zion National Park, Utah (Photo by Edith B. Vincent.)

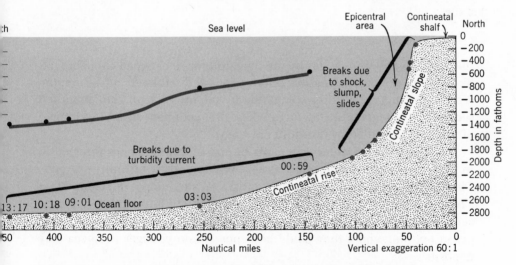

FIG. 5-19. Location of the submarine cables that broke following the Grand Banks earthquake of November 18, 1929. The upper eight cables failed from the earth shock and landslides that followed immediately. The lower five cables broke from the impact of turbidity currents that surged down the continental slope, rise, and sloping seafloor. The speed of the turbidity current at each of the five lower cables is shown by the curve above, whose speed scale is in the upper left margin. (After B. Heezen and M. Ewing, *American Journal of Science*.)

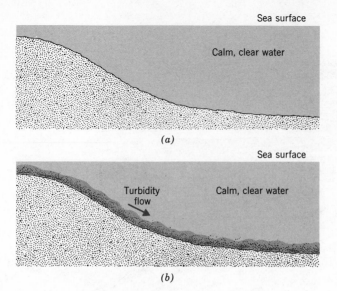

FIG. 5-20. Example of turbidity flow. In (*a*), still, clear water overlies a submarine slope. In (*b*), a disturbance has added mud to the water at the top of the slope causing it to become heavier than the clear adjacent water at the same level. The dense muddy, or turbid water surges down the slope beneath the clear water. Vast erosional and depositional changes take place on the sea bottom through this effect.

the distance between cables by the time difference between cable breaks. Subsequent to this explanation, photographs and sediment cores from the sea bottom supported this explanation by the presence of ripple-marked sediments evidencing strong current flow plus sediment structures typical of those expected from turbidity currents.

The problem of older sediments on top of younger ones referred to in the second set of observations is also explainable by turbidity currents. The flow of turbid water from the continental shelf down the continental slope to the deep ocean could erode relatively old and relatively coarse sand sediments from shallow water and transport them downslope to the ocean bottom, where they could be deposited on younger deep-sea clay deposits.

Since these historic observations and explanations, it has been well established that turbidity currents are a most important form of sediment transport in the ocean. By their action, sediments are continually drifting to the lowest levels of the oceans, where they finally form the broad flat depositional surfaces that characterize abyssal plains.

As turbidity currents slow after reaching the deeper sea, coarsest sediments drop first, followed by successively finer deposits. The decrease in sediment grain size from bottom to top is characteristic of turbidity deposits. Deposits formed by turbidity currents are called *turbidites*.

METAMORPHIC ROCKS

Strangely, rocks, like living things, must adjust themselves to their environment. This is accomplished by the process of metamorphism (meaning change of form). Metamorphic rocks were formed under conditions much more difficult to duplicate in laboratory procedures than igneous rocks, partly because of the slowness of the reactions involved and partly because of the complicated transfer of material to or away from the rock. Conditions and details of metamorphic rock formation have been arrived at primarily by deduction from careful field study and analyses of rock thin sections with the aid of the petrographic microscope, and secondarily by laboratory studies attempting to duplicate conditions in nature.

The Metamorphic Process

Igneous and sedimentary rocks that initially formed at or near the earth's surface developed under conditions of fairly low temperature and pressure. If this environment changes significantly because of greatly increased temperature or pressure, or both, metamorphism occurs. Solutions bringing in new mineral material may add further to the alteration. Throughout metamorphism the rock remains essentially solid. The metamorphic process is usually classified into regional and contact metamorphism.

Regional metamorphism involves far larger rock volumes and is thus of much greater geological importance. The environmental changes that cause regional metamorphism occur as a consequence of the mountain-building process (described further in Chapter 9), which often involves intense folding of originally underformed rock units. Associated with such deformation are increases in temperature and pressure that greatly change the environment, causing recrystallization of original mineral material into new minerals and often, new rock structures.

An idea of the factors involved in regional metamorphism can be gained from the temperature-pressure relationships for rocks at great depth, where the environment is far different from conditions nearer the surface. Let us consider first what we know about the increase of temperature with depth.

Observations in deep mines and oil wells carry our knowledge down for at most a few miles. But for most of the earth's radius the temperatures—and the temperature gradient—is estimated on the basis of both the amount of heat released through radioactivity and the heat conductivity of the crust and mantle. In Fig. 5-21 we see the vertical temperature gradient or changes with increase in depth or pressure. The pressure scale is in "atmospheres," a scheme often used with high pressure values. One atmosphere has a "weight" of 14.72 pounds per square inch so the pressure of 10,000 atmospheres (at about 20 miles—33 km) is equal to 147,200 pounds per square inch.

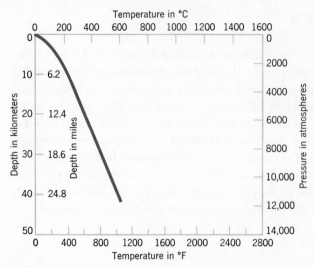

FIG. 5-21. The vertical temperature gradient to about 40 km (25 miles) in depth. The temperature-depth curve, which is a rough average of eight estimates based on different models of the earth's crust, is tentative, but nevertheless important in making further estimates. The primary units are in metric measure, but the nearest English values are also given. Depth is shown on the left and equivalent pressure on the right.

Now in Fig. 5-22, we have the temperature curve (broken line) plus another, the solid line that shows the temperature at which rocks with granite composition begin to melt, providing that some water is present (as it almost always is). It appears that the temperatures resulting from the normal vertical temperature increase will not reach the melting point of such rocks. Since metamorphism takes place in rocks that remain solid, the temperature of alteration must be less than that shown by the solid line and more than shown by the broken line, or in the region between the two curves. The grades of metamorphism will increase from the upper to the lower part of the "region" of metamorphism.

Solutions involved in metamorphism usually emanate from a nearby magmatic source. They may simply add new ions to change the original mineral composition, or they may replace ions with new ones, carrying away some original material. In all varieties of metamorphism, the presence of water greatly expedites the change.

Contact metamorphism is restricted to a fairly thin zone bordering an igneous intrusion. The contact aureole varies in extent from a few inches to a few hundred feet. Near the contact complete alteration of the country rock might occur, the alteration being in the form of baking and the introduction of new material from the magma. Further away the alteration may be barely recognizable. The presence of a baked zone above and below a sill helps distinguish such a form from a buried lava flow that only has a baked zone in the underlying sediments.

Metamorphic Rock Types

The way a rock changes in response to its environment depends primarily on its original nature and secondarily on the forces involved. Two broad groups called *foliated* and *massive* metamorphic rocks are recognized.

In the case of foliated rocks, when the composition permits, the original rock response to the high temperatures, pressures, and solutions results in both recrystallization and readjustment in the mineral grains so as to relieve the pressure. This is accomplished by flaky minerals such as mica and chlorite becoming oriented with their flat faces normal to the pressure. Linear crystals such as augite or hornblende become aligned with their long dimensions in the plane normal to the pressure. Embedded in these *planar* mineral structures may be new minerals that may be equidimensional but of greater density than the original material, thereby relieving pressures by decreasing the total volume. Garnets are a good example of dense equidimensional minerals that form in this manner. The parallel arrangement of platy or flaky minerals gives the rock a booklike or foliated appearance.

Initially, the newly formed metamorphic minerals are very small, in fact microscopic. Such rocks are of *low-grade* metamorphism and represent conditions just above the change threshold. With increasing intensity of the conditions causing metamorphism, an increase in the metamorphic "grade" occurs. Both increased grain size and particular mineral species distinguish the higher grades of metamorphism.

The classification of foliated metamorphic rocks depends on their grade of metamorphism as follows:

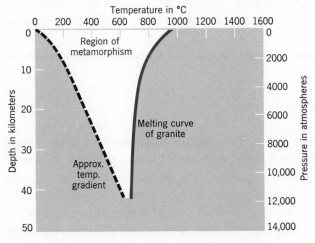

FIG. 5-22. Comparison of the approximate vertical temperature gradient curve (broken line) with the curve showing the temperature-pressure (depth) conditions at which sialic (granitic) rocks will melt providing water is present in the rock.

Slate is a very fine-grained, low-grade metamorphic rock composed mainly of microscopic mica flakes although some chlorite may be present. The foliation is so perfect as to give rise to the characteristic *slaty cleavage*—the property of splitting readily along closely spaced smooth, parallel surfaces.

Phyllite represents a somewhat higher-grade metamorphic rock having the general composition of slate. The mineral flakes, although larger than in slate, are still so small as to often be barely visible. But they are large enough to give the rock a characteristic satiny or silky sheen. The cleavage of a phyllite is less perfect than that of a slate.

Schists are the most common rocks formed in regional metamorphism. The platy minerals producing the foliation are usually clearly visible to the unaided eye and may consist of biotite, muscovite, talc, and chlorite. Although schist tends to break parallel to the foliated structures, the foliation is so distorted compared to the smooth quality of slaty cleavage as to be given the special name *schistosity*.

Gneisses are coarsely banded metamorphic rocks consisting of alternating layers of light and dark minerals. We include the gneisses under foliated structures, although the individual layers may not be truly foliated—that is, consist of parallel sheets of platy minerals. Quartz and feldspar characterize the light minerals, and amphibole, pyroxene, and biotite are most characteristic of the dark ones. Some gneisses that are typically granite in appearance except for the banding are termed *granite gneisses*.

The mineral content of the massive metamorphic rocks prevents their development of flaky minerals. Often a denser rock of the same mineralogy results. Thus, *marble* is the massive metamorphic equivalent of limestone. The many small grains of calcite in limestone become recrystallized into much larger grains in marble. In the case of a sandstone, quartz (simple SiO_2) grains also recrystallize to form the much denser quartzite.

A classification and recognition scheme for metamorphic rocks is given in Appendix D.

The Rock Cycle

Rocks exposed at the earth's surface undergo continuous change. They are eroded to sediments and then relithified, following which they may again be eroded to sediment or they may become deeply buried and metamorphosed. If then exposed at the surface, erosion to sediments will again take place and the entire cycle will recommence. Igneous rocks not exposed at the surface may become metamorphosed at depth, but after deep erosion and exposure, these will again begin a new cycle upon erosion and sedimentation. Table 5-2 indicates the direct and reversible aspects of the cycle for some of the standard rock types.

The Granite Controversy:
Is Granite an Igneous or Metamorphic Rock?

One of the fascinating and fundamental problems in earth science is the origin of granite. The problem has stirred a great controversy among geologists in recent years and has led to any number of conferences and publications. Although by no means solved, the question of granite and its relation to the genesis of the continents is so important as to merit examination here. The problem, simply, is this: is there a primary granite magma that has been the parent of the great plutonic (intrusive) granite batholiths? Having stated the problem, let us now see why the problem arose in the first place.

Basaltic-type rocks compose more than 98 percent of all extrusive igneous rocks. They also form the rock basement of the ocean basins. There is little doubt that great basalt magmas have been generated deep in the crust and upper mantle. Certainly they are required for the basalt of the great lava plateaus already described and for the great volcanic eruptions to be described in Chapter 9.

Also, granitic rocks make up at least 95 percent of all of the continental intrusive rocks. And they occur as the enormous batholithic intrusions that are revealed in the eroded cores of the great mountain chains. A major question in connection with batholiths is simply this: If the granite is truly the result of a magmatic intrusion, where are the rocks that have been invaded? Consider the volume of the Idaho batholith described earlier in this chapter; the exposed area is about 16,000 square miles. If the batholith is only 10 miles in depth, then without allowing for the normal batholithic spreading with depth, or that part already removed by erosion, the present minimum volume is 160,000 cubic miles.

TABLE 5-2 The Rock Cycle

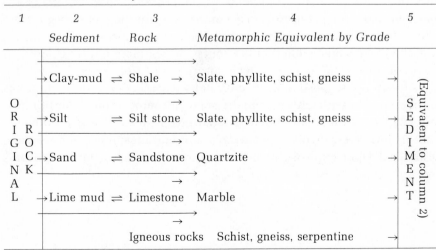

1	2	3	4	5
	Sediment	Rock	Metamorphic Equivalent by Grade	
ORIGINAL	ROCK			SEDIMENT (Equivalent to column 2)
	Clay-mud ⇌ Shale →	Slate, phyllite, schist, gneiss	→	
	Silt ⇌ Silt stone	Slate, phyllite, schist, gneiss	→	
	Sand ⇌ Sandstone	Quartzite	→	
	Lime mud ⇌ Limestone	Marble	→	
	Igneous rocks	Schist, gneiss, serpentine	→	

Where are the 160,000 cubic miles of rocks, presumably sediments, that were replaced by the granite? Volumes for the Coast Range Batholith in Canada would be even greater.

A second question involves the scarcity of extrusive sialic (granite equivalent) rocks in the mountain chains peripheral to the batholiths. If the tremendous magmas required for batholith intrusions existed, why are sialic lava flows so rare in mountain belts while basaltic flows are relatively common in these same belts?

A third point is the common occurrence of small granite and granitelike masses without any visible connection to a parent source. A fourth point is the actual rarity of deep-seated basalt or its coarse-grained equivalent, gabbro, in the cores of mountain chains.

Finally, the field observations reveal a large number of cases in which granite grades continuously into the sediments enclosing the batholith rather than showing the sharper contacts required by intrusions. To complicate the problem, there are also cases in which granitic batholiths do cut the surrounding rock quite sharply as though an actual intrusion occurred.

From these reasonings and observations the conclusion has been reached that much of the granites have formed through *granitization,* the process whereby preexisting sediments or metamorphosed sediments are slowly converted to granite. The exact nature of this process is still in doubt and may have involved the melting and fusion of sedimentary rocks deep in the heart of mountain chains from the high temperatures generated by both radioactivity and rock movements. Other methods involve a change in original rocks by a metamorphiclike alteration. According to the latter process, granite formed in the granitization process appears to be an extreme metamorphic rather than igneous rock. Both melting and solid-state metamorphism may have occurred.

The shales and sandstones that comprise the great bulk of sedimentary rocks are quite close to granite in composition—after all, they have been formed through the erosion of continental granitic rock to begin with. Thus, if sedimentary rocks composed of quartz, orthoclase, albite, micas, and the like become melted in the course of mountain building and then cool and recrystallize, a new cycle of granite can be generated. We are thus led to the suggestion of a continuous recycling of material that has been eroded to sediments essentially sialic in composition. After lithification, with or without additional metamorphism, deeply buried sediments that are intimately involved in the mountain-building process may become granitized. Upon exposure by erosion of overlying rocks, the cycle begins anew.

STUDY QUESTIONS

5.1. Why is none of the original crust of the earth now visible?

5.2. Explain how igneous rocks can be formed at or close to the earth surface.

5.3. Discuss the source of heat that produces magma.

5.4. Why do igneous rocks crystallize with a variety of grain textures? Define the main types of igneous rock textures.

5.5. Describe the Bowen reaction series distinguishing the plagioclase from the ferromagnesion branches. How does the composition of a mineral change as it progresses along each of these series in the process of cooling and crystallization?

5.6. How does the continuous reaction series explain the solid solution series described for the plagioclase feldspars in Chapter 4.

5.7. Describe the extrusive and intrusive forms of igneous rocks; use diagrams in your explanation.

5.8. Distinguish between clastic and nonclastic sedimentary rocks.

5.9. Construct a table showing sediment type and resulting sedimentary rocks.

5.10. By what natural processes are sediments converted to sedimentary rocks?

5.11. Why are shales the most common of sedimentary rocks?

5.12. What is the significance of an arkose compared to a pure quartz sandstone?

5.13. Explain the importance of "stratification" in interpreting the geology of region.

5.14. Define ripples, cross-lamination, and fossil.

5.15. How are relatively coarse sediments transported far from land and well beyond the zone where wave action stirs and disturbs bottom sediments?

5.16. Distinguish between "regional" and "contact" metamorphism.

5.17. What are the chief factors involved in metamorphism?

5.18. Construct a table showing sediments, sedimentary rocks, and the resulting metamorphic rocks.

chapter 6

Camel Rock, Wyoming. One of the remaining remnants of an original wide-spread layer of pale red sandstone that covered the entire region. In recent times atmospheric weathering has shaped the monolith into this interesting form.

Shaping the Continents: Erosional Forms and Processes

6A. THE STATIC AGENTS OF EROSION

WEATHERING
        Chemical weathering
        Mechanical weathering
        Factors that affect weathering
        Soils
MASS-WASTING
        Talus and the angle of repose
        Creep
        Earth flowage
        Landslides and avalanches
GEOLOGIC WORK OF GROUND WATER
        The hydrologic cycle
        Ground water and the water table
        Movement of groundwater
        Solution and precipitation

Since the time of its formation, the earth's continental crust has been subjected continuously to two oppositely directed processes, one leading to the growth of land forms such as plains, plateaus, and mountains and the other leading to their destruction through erosion. In Chapter 3 we considered the basic descriptions of land forms and in Chapter 9 we will give further attention to the evolution of continents and mountains. In this chapter we will examine the erosional forces that stem directly or indirectly from the presence of the atmosphere. Most of the features of the earth's physical scenery are the results of the erosional sculping of the lands by these forces.

As will be seen later, the source of energy for the upheaval of mountains is internal; that which drives the erosional machine is mostly external. All of these processes involve huge energy transformations. The internal sources of energy are both the earth's gravity and the chemical energy involved in rock decomposition; external energy is the radiation from the sun.

As we consider the chief erosional processes in this chapter, each will be related to the energy required. And notice once again that all of the mechanisms of natural rock destruction depend on the presence of the earth's atmosphere, and most particularly, its water content in all of its three states—vapor, liquid, and solid.

WEATHERING

Rock weathering consists of a number of separate processes, all of which reduce tough, solid bedrock to a soft, altered, loosened material that can easily be carried away by natural erosional agents, such as running water, glaciers, wind, and so on. In fact, without the presoftening action of atmospheric weathering, natural erosion would be almost negligible. Two broad types of weathering occur in nature, *chemical weathering,* or *decomposition* and *mechanical weathering,* or *disintegration.*

Chemical Weathering

In decomposition, the rock is chemically altered and weakened as new minerals form. In disintegration, the rock is simply broken into smaller bits. Although both processes may progress simultaneously, one or the other may predominate, depending on rock type and climate.

The energy involved in chemical weathering, like that in any chemical change, is related to the forces that bind atoms together to form molecules. If the electrons that take part in the chemical action are displaced to a higher orbit about the nucleus, they must do work to move out against the attractive force of the nucleus. Energy must be obtained from the environment for such a reaction. For example, nitrogen and oxygen, the chief components of the atmosphere, can combine to form the gas nitric oxide (NO) if an adequate amount of energy is available. The energy, in the form of heat, required for the formation of two ounces of NO is 44,000 calories. If this heat energy were not necessary, atmospheric nitrogen would long ago have combined with all of the oxygen present!

Chemical reactions may reduce the atomic binding energy of the electrons involved, thereby releasing heat energy. Ordinary combustion, in which oxygen combines with a fuel, liberates varying and often tremendous amounts of heat energy.

Chemical reactions in rock weathering mostly release heat energy or are *exothermic* because the original minerals formed in a higher temperature environment than is present at the earth's surface. And since the rock-forming processes described in the previous chapter all involve higher pressures than occur at the surface, chemical weathering under atmospheric pressure produces lower-density or higher-volume minerals than are present in the original rock.

The chemical weathering of minerals bears a striking and significant relationship to the Bowen reaction series (p. 125). We might expect that minerals that crystallize earliest—at the highest temperatures in the magma—would be the most unstable in the surface environment and that the last to crystallize would be the most stable. Field and laboratory observations have confirmed this and indicate the nature of the change in crystal lattices that causes decay. Recall that in silicate minerals, silica-oxygen tetrahedra are linked by metallic ions to form the particular lattice structure. Fe^{++} and Al^{+++} ions, among others, can be dissolved out of the lattice. Without the necessary positive ions to keep the tetrahedra linked, the lattice structure gives way, causing mineral decay.

S. Goldich essentially retitled the Bowen reaction series (Fig. 5-2) in accordance with mineral stability. If we consider the succession of minerals from higher to lower temperatures along the "Y" to have increasing stability, we then obtain the *Goldich weathering* or *mineral-stability series*. Quartz at the base of the Y, having no metallic ions to be removed, is thus the most stable of all common minerals. After the chemically weaker minerals have been weathered out of a rock, the remaining quartz is slowly reduced in grain size by mechanical weathering processes.

The solidity and coherence of the original bedrock is finally destroyed, and the loose decomposition products can be easily removed by the agents of erosion to be described in this chapter.

Water, the universal solvent, is of extreme importance in any type of weathering. In chemical weathering, water either takes part directly in the reaction or carries other corrosive agents to the scene, and often it does both. The solution action of water on rocks can be quickly appreciated from the salinity of sea water. By weight the oceans contain 3.5 percent of dissolved solids. Nearly all of this was carried in by stream water whose origin is in atmospheric rain and snow. This nearly pure water must have dissolved the present sea salts from the lands in the course of its ceaseless cycle of evaporation from the ocean, precipitation on land, and return flow to the ocean. The *hydrologic cycle* will be considered further in this chapter and in detail in Chapter 11 on the atmosphere. Detailed chemical analysis has shown that fresh precipitation contains a minute quantity of dissolved atmospheric carbon dioxide and sodium chloride, the latter picked up from ocean spray. After the water enters the ground, a relatively tremendous increase in the amount of dissolved matter occurs from the solution of mineral and organic matter.

Despite its importance, chemically pure water has far less of a solvent action on rocks than water that has acquired a weak acid composition. In chemical weathering the important acids in water result from the prior solution of carbon dioxide from the air and soil and from the solution of organic material. The former yields carbonic acid in the process.

$$H_2O + CO_2 \rightleftharpoons H_2CO_3$$

also written

water + carbon dioxide $\rightleftharpoons$ carbonic acid

in the ionic form as:

$H_2O + CO_2 \rightleftharpoons H^+ + (HCO_3)^-$
water + carbon dioxide $\rightleftharpoons$ hydrogen ion + bicarbonate ion

Both of the ions that form in this process contribute strongly to the reaction process, causing decomposition.

It is interesting at this point to speculate that if organic acids and carbonic acid are so important to chemical weathering and the erosion that follows, then billions of years ago when the vegetation was scarce or absent and when CO_2 may not yet have formed in our atmosphere, weathering and erosion must have proceeded at far slower rates than at present. If this is true, very important geologic implications regarding the evolution of the crust result.

Hydrolysis is perhaps the most important reaction in chemical weathering because of its effect on the feldspars—the most abundant of the rock-forming minerals. In the process, water in ionic form enters into the atomic structure of the original mineral, thus forming the new mineral. Recall from Chapter 4 that water can ionize weakly as shown by:

$H_2O \rightleftharpoons H^+ + (OH)^-$
water   hydrogen ion + hydroxyl ion

Many more H^+ ions are formed in the production of H_2CO_3, described above. Being quite small, hydrogen ions can penetrate into the crystal lattice structure of silicates and displace the metallic ions that help bind the silicon-oxygen tetrahedron, causing destruction of the silicate crystal.

To consider the typical feldspar, orthoclase, a mineral essential to the composition of granite, hydrolysis proceeds as shown by the following molecular equation:

$2KAlSi_3O_8 + 2H_2CO_3 + 9H_2O \longrightarrow$
$$Al_2Si_3O_5(OH)_4 + 4H_4SiO_4 + 2K(HCO_3)$$
orthoclase + carbonic acid + water $\longrightarrow$
Kaolin (clay material) + silicic acid + potassium bicarbonate

Plagioclase feldspars, containing calcium and sodium, hydrolyze even more readily than orthoclase, and all produce the important weathering products consisting of a clay mineral, a weak silicic acid that breaks down into H_2O and SiO_2 (silica), and a soluble bicarbonate of potassium, sodium, or calcium.

The resulting clay minerals are all-important and make up the bulk of the loose surficial weathered material, because feldspars are by far the most abundant of minerals. The soluble metallic ions in bicarbonate form provide important nutrient elements for vegetation. From this we can see the importance of weathering in the development of fertile ground from nonfertile bedrock. Much of the calcium bicarbonate, less used as plant

FIG. 6-1. Angular blocks of quartzite shattered by frost action. (Geological Survey of Canada.)

food, is carried into streams and thence to the oceans, ultimately forming limestones.

In the somewhat simpler *hydration* process, entire water molecules become attached to a mineral compound. In addition to the direct chemical effect, an increase in volume also occurs, causing further rock weakening. After hydrolysis to form clay minerals, hydration often follows, weakening the rock even further as the clays swell. Unlike hydrolysis, water of hydration does not enter into the crystal structure but is only loosely bound and can be driven off in a dehydration process by heating to only a few hundred degrees. For example, gypsum ($CaSO_4 + 2H_2O$) can be easily dehydrated by heat into simple $CaSO_4$—called plaster of paris. When water is added, the plaster is once again hydrated and the expansion occurs that makes it so useful in the manufacture of fine molds.

Oxidation, in which atmospheric oxygen combines with minerals, is a less important yet very widespread form of chemical weathering. Oxidation in combination with hydration is particularly important in the weathering of iron-bearing rocks. Most of the beautiful and varying shades of rock colors, particularly in arid and semiarid climates, result from surface weathering of iron minerals into one or more of the forms of hematite (Fe_2O_3) and limonite ($Fe_2O_3 + XH_2O$). Common rust is simply a limonite coating on metallic iron.

FIG. 6-2. Photo showing weathering of basalt from tree growth plus ice effects; Hook Mountain Park, N.Y.

Mechanical Weathering

In temperate latitudes and high elevations, where frequent alternation of freezing and thawing occurs, the *wedgework* of ice becomes a dominant disintegration process. Water, which normally freezes at 32°F (0°C), works its way into natural rock openings of all sizes, such as bedding planes, cleavage planes, or other natural fractures. In freezing, water crystallizes into a true solid and, contrary to most freezing processes in nature, increases by nearly 10 percent in volume when so doing. But if insufficient space exists, the water will not be able to freeze at 32° except for the topmost layer. If the temperature falls below 32°, the remaining water, in freezing at a lower temperature, exerts greater and greater pressure on the surrounding rock. At −7.6°F a maximum pressure of nearly 2100 tons per square foot can be exerted.

Although this maximum pressure may not be realized in nature, it seems clear that freezing water can generate pressures adequate to split rocks apart. The weaker the rock and the more numerous the weak zones, the more rapidly will wedgework disintegrate a rock (Fig. 6-1).

A second and related type of disintegration results from *root weathering*. It is certainly a rather common sight to observe trees and plants growing in most unlikely rock crevices. As roots and trunks grow, they, too, develop enormous splitting pressures on the walls of the enclosing rocks (Fig. 6-2). In addition to surface occurrences, there are the innumerable small roots and rootlets that grow down out of sight into bedrock below the soil. Plant weathering is not restricted by latitudes but is certainly more effective in moist than in arid areas.

Exfoliation, a process in which rocks scale off in sheets roughly parallel to the rock shape, is a form of mechanical weathering induced by a prior decomposition effect. Rocks are often sufficiently porous to allow any rain or snow water to penetrate a short distance below the surface. Although the surface and a thin layer below usually dry out rapidly, below this thin surface enough water may remain to promote hydrolysis and hydration, as in Figs. 6-3 and 6-4. The decomposition plus the associated volume expansion destroys the coherence of the rock as it separates the upper thin plate—usually a fraction of an inch to an inch or so in thickness—from the rest of the rock. The process is then repeated.

In *spheroidal weathering,* the combined effects of chemical and mechanical weathering tend to cause rounding of initially angular boulders or exposures of bedrock. In the extreme case of the rock cube in Fig. 6-5a, imagine weathering to proceed equally on all surfaces. Then because three surfaces being weathered are in contact, the rock at the corners will suffer the most. Hence corners and edges will round off, causing the rock to approach a spheroidal shape (Fig. 6-5b).

FIG. 6-3. Photo showing exfoliation in which a rock weathers into layers roughly parallel to the surface. (Courtesy American Museum of Natural History.)

FIG. 6-4. The "Old Man of The Mountain" in the White Mountains of New Hampshire is a product of weathering, mostly exfoliation. The side-view profile (a) is a fortuitous result of the exfoliation effect seen more clearly in the front view (b).

Factors That Affect Weathering

The ease and speed with which a rock weathers depends on at least two important factors. Foremost is the nature of the rock itself. The more soluble the rock (particularly limestones) or the chemically weaker its mineral composition and the more numerous the bedding or cleavage or other planes of weakness, the more rapid will be all forms of weathering. Shales with their closely spaced bedding planes nearly always succomb more rapidly to weathering than the more massive sandstones and limestones. Homogeneous crystalline rocks like granite are normally among the most resistant to weathering.

Climate is the most important external weathering factor. Clearly, both decomposition and disintegration proceed more rapidly in moist than in arid climates. Water is directly essential to hydrolysis and hydration, and plays a major part in all decomposition processes. It is also essential in ice wedgework, root weathering, and hydration. Interestingly, lime-

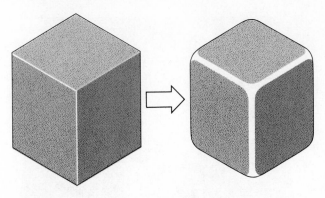

FIG. 6-5a. A cube of rock weathers most quickly at the corners, where three surfaces intersect and at a somewhat lesser rate along the edges where only two faces are exposed. Away from corners and edges weathering is slowest. The effect is to form a rounded boulder out of an initial angular one exposed at the earth's surface as seen in (b).

FIG. 6-5b. Rounded boulder of granite near the coast of Mount Desert Island, Maine. The more angular rocks visible along the top of the view have fallen from the cliff (not visible) more recently and have weathered less.

stones that are usually chemically weak, and valley-forming rocks in humid regions such as the eastern United States, are quite resistant in arid to semiarid regions, where they often form important mountain ridges.

Cleopatra's needle (Fig. 6-6a and b), in New York City, is a striking example of the effectiveness of weathering in a humid climate.

FIG. 6-6a. Cleopatra's needle in Central Park, New York City, a monument to the effects of different climate on weathering of rock. This granite obalisk was erected in Heliopolis, Egypt in 1600 B.C. and was stable in the arid climate of the desert. Soon after it was given to New York City near the end of the nineteenth century, rapid decomposition in the relatively humid climate has all but destroyed the hieroglyphics carved into the surface.

FIG. 6-6b. Close view of Fig. 6-6a.

FIG. 6-6c. The shape of this erosional remnant of red sandstone in the Garden of the Gods, Colorado contrasts strongly with that of the granite boulder in Fig. 6-6b. Although much older, the sandstone boulder is undergoing weathering in a relatively dry climate where angular weathering features tend to develop.

FIG. 6-6d. Balance Rock in the Garden of the Gods, Colorado is an interesting weathered remnant of red sandstone remaining in a rather precarious position after erosional processes have removed most of the surrounding rock.

Soils

The most important natural consequence of weathering is the "softening" of bedrock, without which most of the erosional processes, long as they may seem to us, would take almost infinitely longer. In the course of this softening, most of the lands become carpeted with a mantle of rock waste usually called *regolith*. Weathering is thus an integral part of the entire rock cycle referred to in the previous chapter, being a midprocess between the bedrock stage and the deposition of new sediment to form sedimentary rocks.

If chemical weathering has played a large part in the development of a regolith, soluble mineral materials, as discussed previously, provide the fertility necessary for vegetation. Once started, vegetation growth promotes further fertility through decay products and increased weathering. There is little doubt that soil is a nation's richest commodity. From it a nation develops its agricultural industry, providing food for both man and animals.

A soil profile, or section from the surface to the bedrock, is usually characterized by three layers or *horizons,* as indicated in Fig. 6-7. The true soil consists of the two upper, or *A* and *B*, horizons. The former is dark in color from the decayed and partially decayed organic remains.

FIG. 6-7. Soil profile visible on cut near Chatsworth, New Jersey. The A, B, and C layers described in the text are clearly shown by differences in color or shading. Scale is indicated by the vertical strap. (Photo by Leonard Zobler.)

By the process of leaching, water removes iron oxides and fine clay particles from the topsoil (*A* horizon) and carries these products down to the subsoil or *B* horizon, which is richer in mineral nutrients but lower in organic content than the topsoil. The *C* horizon contains a mixture of disintegrated rock, some decomposed rock, and some soil material from above. This layer, which will become soil, is transitional between the parent rock and the true soil.

We should not leave weathering without considering its impact on scenery. Much of the variety of the surface scene is due to the different types and conditions of the weathering processes. The shapes and colors of landforms are often a direct result of the type of weathering that has occurred. In humid regions, bedrock, in becoming decayed, converted to soil, and covered by vegetation, tends to develop rounded and "mellow" contours such as characterize the subdued hills of New England. In contrast, the absence of soil and vegetation in arid regions like the Colorado Plateau produces a covering of rock waste with little binding power. If steep slopes are present, the infertile regolith falls away from the bedrock, which thus maintains its fresh, angular bedrock topography. It is the skin-deep weathering of this bedrock surface that provides the variegated hues of dryland topography.

MASS-WASTING

Talus and the Angle of Repose

Mass-wasting refers to the downslope movements of soil and rock under the direct influence of gravity. The firm bedrock that may form a slope is quite "resistant" to the pull of gravity. But once the rock becomes weathered and thus converted to separate fragments, the slope can become unstable, and the components tend to fall to achieve the *angle of repose*. For a bedrock cliff this angle can be 90° or vertical. For loose material the stable slope has a much lower angle, depending on the size and shape of the particles and the degree of lubrication.

If we pour sand out of a container, the sand grains quickly roll over each other until the pile comes to rest, forming a low conical mound whose sides slope about 25° to 30° from the horizontal. On the other hand, fine, angular particles, such as clays or the more gritty *loess* of central United States, can accumulate with an interlocking structure that enables the mass to maintain a vertical slope.

Although the downslope movements of loose material result from the pull of gravity, the entire process is greatly enhanced by the presence of all-important water. We noted in Chapter 1 that weathering and erosion without an atmosphere would not occur—as on the moon. But weathering and erosion even with a dry atmosphere would be extremely retarded. We cannot overemphasize the importance of water in geologic processes.

In mass wasting, water acts as a lubricant that destroys intergrain friction and cohesion. Nearly all mass movements of earth materials at the surface are preceded by a fairly thorough saturation of the rock material. Depending on the slopes, rock or soil moves downhill at speeds ranging from imperceptible but significant rates to catastrophic slides and avalanches.

Wherever steep cliffs exist, their bases usually grade into an apron of fragmental rock material (talus). *Talus* accumulates slowly by the fall of individual rocks that have become loosened by weathering. After falling, the rocks achieve an angle of repose between 30° and 35° from the horizontal to form the *talus slope* (Fig. 6-8).

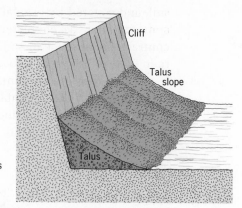

FIG. 6-8a. Diagram showing talus slope of broken rock fragments found at the foot of a cliff.

FIG. 6-8b. Talus slope at the foot of the Palisades Cliff along the Hudson River in New York State.

Creep

In the slowest of the movements, known as *creep,* a completely invisible downslope movement of either soil (soil creep) or rock (rock creep) occurs. Creep commonly results from the expansion and contraction of soil as a consequence of alternate freezing and thawing or wetting and drying. In either case, particles are heaved up as in Fig. 6-9 and after contraction they fall to a slightly lower elevation. Although the net daily migration is normally a fraction of an inch, when it is summed over all of the particles in the surface layer for a period of years, very observable effects occur.

The displacement of grass and soil particles is not very evident. But objects like trees, fences, and telegraph poles show characteristic tilting on hillsides because creep is strongest in the thin surface layer of soil and decreases rapidly with depth, as indicated by the arrows whose

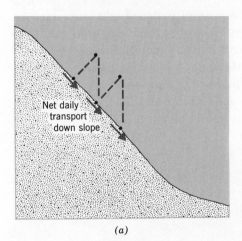

Net daily transport down slope

(a)

FIG. 6-9*a.* A major cause of creep is the freezing and thawing process in which a soil particle is lifted perpendicular to the ground surface by a small amount when soil water expands upon freezing. After the soil water melts, the particle slumps slightly down hill.

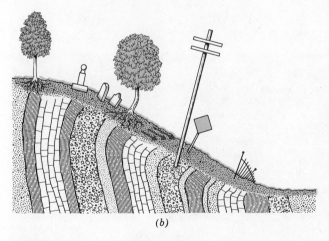

(b)

FIG. 6-9*b.* Illustration of the process of tilting of trees, fences, poles, etc. by creep whose speed, slow as it is, decreases downward causing objects standing in the ground to be rotated downhill.

lengths show the decreasing subsurface creep (Fig. 6-9*b*). Hence tall objects that are anchored deeply tend to be rotated as they migrate downslope as shown by the trees in Fig. 6-10.

Even cattle grazing on hillsides are known to cause creep as their hooves push soil downslope. Their effect is most noticeable after heavy rains when the ground is quite soft.

Earth Flowage

Although less continuous and widespread than creep, the different forms of flowage are locally more effective in the erosional process. In hilly humid regions where the soil is quite rich in clay minerals, saturation may cause excessive lubrication of the particles. Hydration and swelling of the clays further adds to the plasticity developed in the soils. Depending on the slope present, the angle of repose of material when dry may be exceeded when a particular degree of plasticity develops. At this time a long tongue of surface material will move downhill, forming an apron of slumped soil at the base of the slope and an elongated scar along the hillside. The speed of such an *earthflow* can vary from several feet per hour to many feet per minute depending on slopes, consistency, and amount of weathered rock that may be involved.

FIG. 6-10. Photo of creep effect showing rotation of trees toward the road compared to the vertical road-side sign-post.

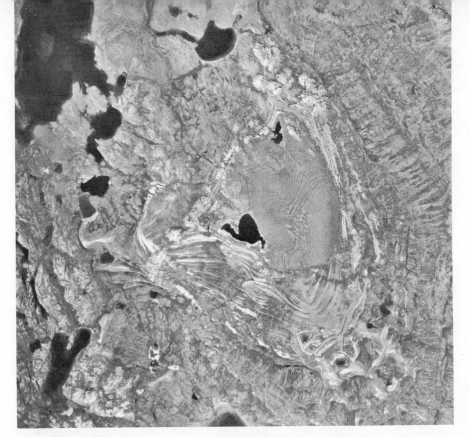

FIG. 6-11. Solifluction on a hill near Carr Lake, Northwest Territories, Canada. The hill was once an island in a large inland sea that covered this region when it was depressed by glacial ice-loading. As the hill was raised intermittently from the sea following the melting of the glaciers, concentric beaches were carved by the waves on what was once an isolated island. A solifluction tongue extending down over the series of beaches just south of the western (left) point of the island. (Royal Canadian Air Force Photo.)

In the cold lands of the far north, a form of earth flowage known as solifluction is one of the major causes of the lowering of the lands. Following the long cold Arctic winters, ice in the upper frozen soil and loose rock metls; the soil that is still frozen below the melted zone is impervious to meltwater. As the thaw continues, the watery rock waste begins to flow down even the gentlest of slopes over the frozen layer beneath. The newly exposed underlying rock and rock waste then experience further disintegration from weathering and thus become prepared for continued solifluction the following year. Figure 6-11 is an air photo showing a hill surrounded by a series of marine beaches raised through successive stages of crustal elevation. Tongues of solifluction-slumped material are clearly visible breaking through the geometric beach pattern.

Mudflows are the most rapid form of earth flowage. In arid to semiarid regions weathered rock is converted to soil too slowly to allow a protective cover of vegetation to form. Heavy rains frequently cause rapid saturation

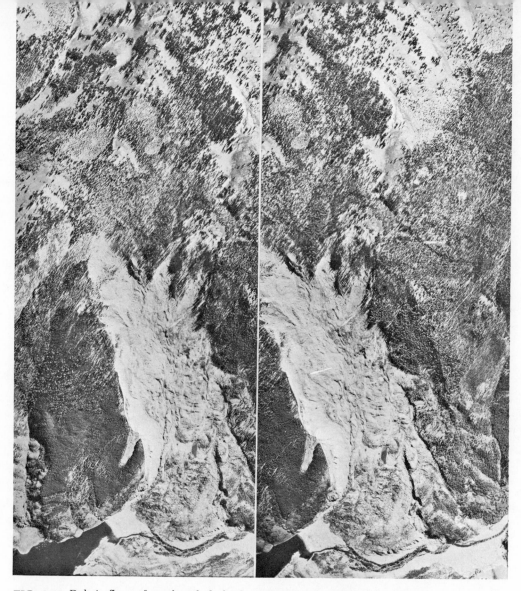

FIG. 6-12. Debris flow of mud and shale that slumped down the north slope of the Gros Ventre Mountains in Northwest Wyoming. Although the movement was slow, occurring intermittently in a two year period, topography was disrupted. The Gros Ventre River was dammed to form a lake (in lower left). The river can be seen beginning to recut its way across the slide. This photo is a stereo pair, and illustrates the way geologists often study topography. The use of a simple stereo viewer will bring the relief into sharp focus. (Courtesy University of Illinois.)

of the loose weathered layer, resulting in a continuous graduation from watery mud to muddy water. The entire sodden mass may move downslope, usually along a preexisting valley. The scale of this effect can vary widely from a slump involving a small amount of material to one of catastrophic proportions. A mixture of mud and saturated, weathered bedrock may flow downslope as a *debris slide* such as occurred in the Gros Ventre Mountains (Fig. 6-12).

Landslides and Avalanches

These are the most violent of downslope movements. Landslides usually involve actual bedrock (Fig. 6-13), which may contain internal fracture surfaces approximately parallel with a steep slope. After prolonged weathering, some of the widened fractures may become saturated enough to cause the rock to break loose. Huge masses of rock then hurtle downslope—often with great destruction to life and property. When viewed from a distance, wooded mountain slopes often show typical landslide scars where forest and soil have cascaded downwards. The striped landslide paths resemble a series of expert ski slopes on the wooded mountainside (for example, Fig. 6-14).

At least as violent as slides are the great *avalanches* in which immense masses of snow and ice, often mixed with rock, fall down snow—and ice—covered mountainsides. Although historical records are replete with accounts of slides and avalanches, one of the worst ever documented was observed by many in Peru on January 10, 1962. News records of the day indicate that some 3500 individuals lost their lives in a single catastrophic avalanche that began at the top of the peak known as Huascaran, over 22,000 feet in elevation. A block of ice falling from a high glacier snowballed into a tremendous mass of ice and rock debris. Millions of tons of ice and rock roared down the steep slope at nearly two miles a minute, stripping vegetation and soil from the mountain. The avalanche traveled about 12 miles and finally came to rest at the base of the mountain, where it covered an entire, well-populated village. In seven minutes the face of the mountain and valley were transformed into a scene of great destruction.

History repeated itself with still greater devastation on the slopes of Huascaran on May 31, 1970. On this day, a very strong earthquake of magnitude 7.8 (Chap. 10) occurred just off the coast of Peru. In addition to causing great damage and loss of life along coastal cities and villages, a huge avalanche of ice, water, rock, mud and other debris was shaken loose from the upper slopes of Huascaran peak. This material fell catastropically down on the villages of Yungay and Ranrahirca 7½ miles to the west and 13,000 feet down the mountain slope. Almost instantaneously the two villages were covered by 30 to 50 feet of debris. In Yungay alone, the population of 20,000 was lost beneath the avalanche. The total casualty list from the earthquake exceeded 50,000 persons. The tragedy at the foot of Huascaran Peak is recorded graphically in the illustrations of Fig. 6-15. (*P. 176–177*) (See also Fig. 6-38 for other examples of avalanche activity.)

GEOLOGIC WORK OF GROUND WATER

The Hydrologic Cycle

Water in the soil and rock below the surface is one element of the hydrologic cycle, described in detail later in Chapter 11. A preview of this

FIG. 6-13. View of landslide in Franconia Notch, New Hampshire. The photograph, taken just after the slide, shows the combined rock, soil, and vegetation debris.

important water regime is necessary at this point. An endless cycling of water occurs in nature, which we can imagine to begin with evaporation of water from the oceans. After transportation in the atmosphere, much of this water is precipitated on the lands mostly as rain or snow. Some of the precipitated water "runs off" in streams; much is evaporated directly back into the atmosphere; and some is absorbed from the soil by plants and either used in photosynthesis of starch or "transpired" back into the atmosphere. A large amount seeps deeply into the ground and becomes groundwater.

Ultimately groundwater finds its way into streams and back to the oceans. In high latitudes and at high elevations an intervening stage exists where water is trapped on land in the form of moving masses of snow and ice, whose meltwaters may enter the ground or flow into rivers or directly into the sea.

It may safely be said that most of the erosion of the lands is simply a consequence of the movement of water back to the sea after being

FIG. 6-14. Common form of landslide on steep, wooded slopes as photographed in the Green Mountains of Vermont.

transported to the land. In this section we will examine the action of water while it resides below the earth's surface. In Chapter 16 we will consider the important aspects of the use of such water as a water supply.

Groundwater and the Water Table

Water from rain or melted snow enters the ground at the earth's surface. In response to gravity, the water tends to move (percolate) directly downwards through the pore spaces between soil or sand grains. The actual amount of space available for water movement and storage in the ground varies from about 20 to 45 percent of the total rock volume (see Fig. 6-16), depending on the size, shape, and degree of compaction of the unconsolidated material. Even solid bedrock has a definite *porosity* (percent of open pore space), which is normally much lower than that of unconsolidated materials. In any kind of rock, the larger the pore space or openings, the greater is the ease of water motion, or *permeability*.

Most of the loose rock *overburden* is composed of clay minerals derived originally from the weathering of feldspars. Because clay soils compact readily under pressure, soil porosity decreases progressively with increase in depth. When bedrock is encountered the decrease is very abrupt. At a depth depending on the amount of water entering the ground as well as on the rock porosity, the available pore spaces become completely filled or saturated with water. The zone of saturation is the true

SHAPING THE CONTINENTS: EROSIONAL FORMS AND PROCESSES

FIG. 6-15. (*a*) The neighboring villages of Yungay and Ranrahirca prior to the avalanche of May 31, 1970. (*b*) The same view as 6-13(*a*) showing the slide debris covering the two villages to depths of 30 to 50 feet. (*c*) Large-area photo showing the entire mountain slope and source of the avalanche that devastated the villages below. (Photos courtesy National Aerophoto Service of Peru.)

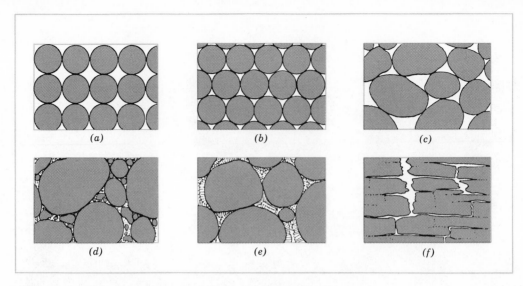

FIG. 6-16. Nature of porosity in rocks. (a) shows the maximum porosity from "open packing" of spherical sand grains. In (b) porosity is decreased from "close packing" of sand grains. (c) shows a more natural sand with high porosity because the grains are well-sorted or have nearly the same size. In (d) the porosity is lower because of poor sorting in which small grains fill some of the pore spaces between large grains. In (e), porosity is still lower because of natural cement that lithified sand into sandstone. (f) shows a limestone made porous by water solution along bedding and natural fractures.

groundwater zone. Above this is a zone where the pores contain air as well as any water still percolating downward from the last rain (Fig. 6-17). Between the two zones is the all-important *water-table* surface.

The depth of the water table is very variable from place to place depending on climate and thickness and type of overburden. In humid regions the water table usually lies close to the surface—within a few feet to 10 or 20 feet. In arid regions the depth is normally much greater and may be 100 or more feet below the earth's surface. Even in the same place the depth of the water table is quite variable. It is highest in late winter or early spring and lowest in late summer.

Between periods of rainfall, when the ground may dry out from surface evaporation, a small upward flow of groundwater may take place. Water from below the water table can be drawn up into fine open pores by the process of "capillary action," in just the same way that water or ink is drawn into a dry blotter. This process can provide surface vegetation with water during dry spells when little or no direct rainwater remains in the soil.

Movement of Groundwater

Groundwater tries to live up to the expression, "water seeks its level." If water is poured into the left side of a U-tube as in Fig. 6-18a, it will flow around and up to the same level in the other side. It does so because the water pressure at point *A* is transmitted equally in all directions. Being confined, the water transmits the pressure toward point *B*, where it is still the same as at *A*. At *B*, the pressure is again transmitted in all directions, but the water can only move upward, which it does until its height matches that on the left side.

If we now repeat the experiment, filling the U-tube with fine sand as in Fig. 6-18b, the water will not quite reach the same level on the right side. Some of the pressure is used in overcoming the friction as the water moves through the sand. Hence the pressure at point *B*, for example, no longer equals that at *A*. In fact, pressure (or head) is being lost continuously as the water moves through the interconnecting pores.

Although the groundwater below the water table (Fig. 6-17) is not confined as in the U-tube experiment, it is under the influence of the water pressure in the overlying pore spaces minus the frictional loss as the water seeps through the soil. A combination of factors, such as bedrock slope beneath a hill, the difference in water pressure beneath hills and valleys, and the difference in the resulting rates of motion cause the water table to become a subdued replica of the overlying topography (Fig. 6-17). Because of the slope of the water table, water pressure is greater beneath hills than beneath the lowland, causing a flow from higher to lower water-table regions as shown by the broken arrows.

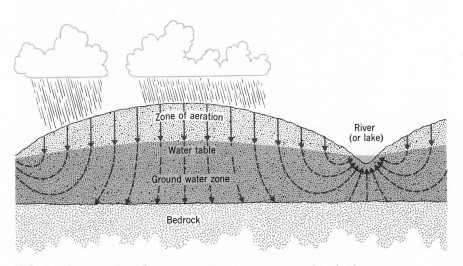

FIG. 6-17. Profile showing water table above saturated soil. Flow lines of ground water in the saturated zone are shown by broken arrows. Where natural depressions cut the water table, bodies of water—lakes or streams occur.

Where the water table cuts the earth's surface, as in the valley in Fig. 6-17, groundwater will seep out. If the depression is an enclosed basin, a lake will accumulate, with the lake surface marking the level of the water table. If it is a river valley, the groundwater provides the nourishment that maintains streamflow. If an artificial hole is cut below the water table, accumulation of groundwater produces the water well.

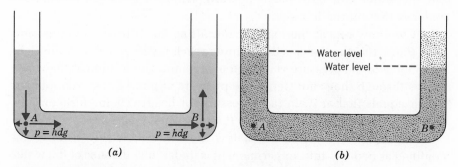

(a) *(b)*

FIG. 6-18. (*a*). In a clear U-tube, if water is poured into the left side, it will rise to the same height in the right side. The upward pressure at B will equal the downward pressure at A. (*b*) In a U-tube filled with sand, if water is poured in the left side, as water moves from A to B, pressure will be lost due to friction and the water will not rise as high in the right side.

Solution and Precipitation

The geological effects of groundwater are most prominent in humid regions where a large proportion of the bedrock is limestone. Although limestone is hardly affected by freshwater, it becomes more soluble in contact with water containing carbonic acid (H_2CO_3). Rain and snow dissolve some small amounts of CO_2 from the atmosphere. After entering the ground the water picks up more CO_2 from the soil, where it forms from bacterial decomposition of organic matter. As we noted earlier in this chapter, CO_2 dissolves in water to form carbonic acid:

$$H_2O + CO_2 \rightleftharpoons H_2CO_3$$

Limestone ($CaCO_3$) reacts with carbonic acid to form calcium bicarbonate—$Ca(HCO_3)_2$—which is soluble in water:

$$CaCO_3 + H_2CO_3 \rightleftharpoons Ca(HCO_3)_2$$

The soluble calcium bicarbonate is thus slowly removed by groundwater. It ultimately is transported into streams and then to the ocean, where it reprecipitates to form limestone beds.

Typical solution forms remain as relics of groundwater action. Limestone caves are perhaps the best-known. These vary from quite small rock cavities to the huge cavernous openings exemplified by Carlsbad Caverns in New Mexico, Mammoth Caves in Kentucky, or Howe Caverns in New York, to cite a few familiar cases (Figs. 6-19, 6-20).

The presence of limestone caverns are of a geological significance beyond the direct solution of rock material. The floors of the caverns are at present mostly above the water-table level, although one may occasionally get wet feet while sightseeing. But the entire cavern must have been in the groundwater zone (below the water table) when it formed. The presence of caverns thus indicates a former higher groundwater level than exists now. We will see later that much of the temperate and subtropical belts were considerably more moist during the great Ice Age that ended some 15,000 to 20,000 years ago. The caverns may well be relics of the more humid ice-age climate when groundwater levels were probably much higher in regions of present caverns. In fact, Carlsbad Caverns are now in an area of desert-type climate, but we know from other lines of evidence that the region was well watered during the most recent stage of glaciation.

FIG. 6-19. "Temple of the Sun" in Carlsbad Caverns, New Mexico. Icicles of stone (stalactites) hang from the roof. Stalagmites grow upwards from the floor. Columns form where the two meet. (Grant Kennicott, National Park Service.)

In many areas limestone caverns are so close to the surface that their thin roof layers have collapsed or been washed away to form features known as *sink holes*.

Although known caverns are naturally above the present water table, many caverns are being formed by solution along underground water courses below the water table. Such openings are of course not now accessible to us. Their presence is indicated in limestone regions where surface streams abruptly plunge underground to disappear from view. During their subterranean flow, subsurface streams slowly dissolve a considerable quantity of limestone. Such streams may return to the surface unnoticed when reaching a larger, deeper surface stream valley, or may reappear as a gushing spring that then flows into some surface waterway.

FIG. 6-20. Waterfall of stone, Carlsbad Caverns. (Grant Kennicott, National Park Service.)

Very striking and often beautiful precipitation forms contribute greatly to the beauty of limestone caverns. Water emerging from ceiling fissures may seep through slowly enough to evaporate before falling. In evaporating, calcium carbonate is precipitated. Continued seepage gradually extends the deposit into a pendantlike icicle of stone known as a *stalactite*. If the water falls to the cavern floor and then evaporates, the more blunt *stalagmite* may be gradually built up, and if this reaches the ceiling or merges with a stalactite a continuous *column* develops.

Mount Fairweather, Alaska a region undergoing active erosion by stream, and glaciers. (Photo by Austin Post, University of Washington.)

6B THE ACTIVE AGENTS OF EROSION

STREAMS AND STREAM EROSION

". . . All the rivers run into the sea; yet the sea is not full;
unto the place from whence the rivers come, thither they return again."

ECCLESIASTES

There are streams to fit all moods from the nearly calm, placid broad rivers of the green lowlands to the turbulent, frothy brooks that hasten down the slopes of mountains. To a watcher at the seacoast there seems little doubt that the fury of breaking waves must have a most destructive effect on the lands. But it may come as a surprise to realize that ribbons of running water have from time immemorial exerted a far greater effect on the denudation of the lands.

Weathering softens the bedrock. Mass-wasting transfers weathered material from higher to lower levels. Groundwater dissolves bedrock, principally limestone, and carries this load to streams whose valleys are cut below the water table. But it is the streams that ultimately transport all of this continental rock material to the sea and by so doing, they slowly and inexorably lower the continents. Of all the agents of erosion, streams are by far the most widespread and effective. They are found at all elevations and in all climates and constitute the most powerful erosional agent at work on the earth's surface and crust. As with groundwater, streams are a part of the hydrologic cycle, and their erosional effect is incidental to their flow to the ocean. In the course of this incidental erosion, many landforms are produced. Although most of these forms are quite temporary geologically, they constitute some of the most common elements of our scenery. The nature and behavior of streams thus merit particular attention and study.

Sources of Stream Water

Although stream water is atmospheric in origin, most of this water first enters the ground to become groundwater as described in the previous section. Some small fraction of the total precipitation from the atmosphere seeps into stream valleys whose channels are below the water table. A lesser amount flows directly down slopes as surface runoff.

Many of us, on seeing the steady flow of streams, have often wondered what maintains them even after fairly long periods without local rainfall and visible runoff. Groundwater is the great invisible reservoir that nourishes the flow of permanent streams (those cut below the water table). Some streams are also fed by large water bodies, such as the Great Lakes, whose volume is adequate to keep rivers that drain them in good supply. But the Great Lakes in turn are fed to a large extent by groundwater.

Streams consist of water of meteorologic origin flowing downslope in channels that lie in the bottom portion of larger, more or less V-shaped troughs known as *valleys*. Depending on the stage of development of a stream valley, the channel may occupy all or only a portion of the valley bottom, as in Fig. 6-21.

We now know that stream valleys do not predate their streams, as

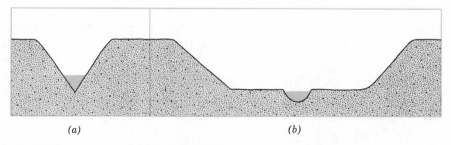

(a)   (b)

FIG. 6-21. Stream channels at the bottom of (a) a narrow V-shaped valley and (b) a broad, flat-floored valley.

was believed as recently as the early nineteenth century. Rather, streams carve their own valleys. This once revolutionary doctrine was first clearly enunciated by Playfair in 1802 with the words:

"Every river appears to consist of a main trunk, fed from a variety of branches, each running in a valley proportioned to its size, and all of them together forming a system of valleys, communicating with one another, and having such a nice adjustment of their declivities, that none of them join the principal valley, either on too high or too low a level; a circumstance which would be infinitely improbable, if each of these valleys were not the work of the stream that flows in it."

Energy and Velocity

In the course of eroding their valleys, streams consume the lands. Remember that as far as the stream is concerned, the valley is simply the waterway serving to return water to the oceans. When a particular volume of water first starts downslope, it has a large amount of energy to expend. This energy of position (potential energy) it acquired long before, when it absorbed the energy of sunlight and evaporated into the atmosphere. The pull of gravity quickly converts the potential energy to energy of motion—kinetic energy. This kinetic energy of moving water causes most of the erosional work of streams.

Kinetic energy depends on velocity according to the simple formula:

$$\text{Kinetic energy} = \frac{1}{2}MV^2$$

where M is the mass and V the velocity of the water (or any moving medium). Because velocity is the important variable in this formula, we must understand how streams acquire their velocity to best appreciate their erosional work.

Stream velocity depends primarily on the steepness of slope (gradient) of the valley, friction with the channel sides and bottom, and the volume of water being transported. The effect of the gradient is quite familiar. A ball will roll more rapidly down a steep hill or slope than down a gentle one, and will of course fall most quickly if dropped vertically. These cases are summarized in Fig. 6-22, where the pulling force down the slopes is equal to g sine θ where θ is the angle of slope above the horizontal.

Friction between the running water and the sides and bottom of the channel, as well as with the overlying air, may slow the stream considerably. Also, the rougher the channel surfaces, the greater is the friction and the lower the velocity. Added to this is the factor of turbulence or eddy action within the stream. In Fig. 6-23, which shows the flow in a hypothetical channel, the water in the central region has a linear movement (known as *laminar flow*) while that at the margins is thrown into both vertical and horizontal eddies, known as *turbulent flow*. The latter motion is most important in the erosional and transportational behavior of streams. As a result of friction and turbulence, stream velocity varies roughly as the square root of the slope ($v \propto \sqrt{\text{slope}}$).

The fact that volume has a stronger effect on velocity than the gradient was the somewhat surprising result of a study (by L. Leopold and T. Maddock) of the records of stream-gaging stations for many rivers over much of this country. It is again a familiar observation that a heavy ball will roll faster than a light one because its greater momentum overcomes friction more easily. This is also true of streams. Also important with streams is that the larger the volume, the larger must be the channel. Because friction and turbulence decrease away from the channel sides and bottom, streams with large channels and volumes tend to have a large central zone where the flow is laminar.

Stream Discharge

Streamflow is usually described most usefully by the *discharge*, which is the volume of water that passes a given point each second (commonly expressed in cubic feet per second—cfs). In Fig. 6-24 if all of the water

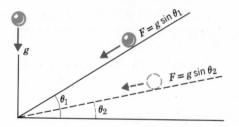

FIG. 6-22. As a valley slope decreases, the force of gravity pulling the water downhill decreases in proportion to the sine of the angle of slope.

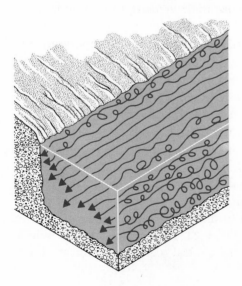

FIG. 6-23. Friction causes turbulent flow between a flowing stream and the channel bottom and sides; in the central part of the channel the water flow is more "laminar-flow" with fairly linear parallel streamlines.

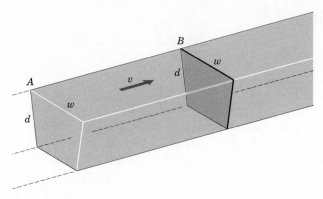

FIG. 6-24. Stream *discharge,* or volume of water passing a given point each second is equal to the area of the channel (*w* x *d*) times the velocity of flow.

between points *A* and *B* were to flow past *B* in one second, the discharge of this stream would simply be the volume of water between *A* and *B*, or the area of the channel (equal to width *w* times depth *d*) times the velocity *v*. At different points along the length of a stream the discharge usually varies because of differences in channel size, velocity, and differences in actual water volume. Many streams show a great variation in discharge at the same point during the year as the influx of water increases or decreases from seasonal changes in rainfall or snowmelt. An increase in discharge by ten times or more is not uncommon. And, as is all too familiar, a sufficient increase in supply may result in rampaging floods as streams break out of their channels to inundate adjacent lowlands.

Since discharge is equal to the channel area times the velocity (discharge = $A \times V$) a stream in flood stage may experience a great velocity increase, as the channel fills with water, a factor of great importance in the erosional work of streams. Table 6-1 gives discharges of some major rivers of the world and indicates the huge volumes of water transported by them. Note how the mighty Amazon overshadows all of the others.

The Erosional Process

We noted earlier that streams, more than any other agent of erosion, offset the effects of continental uplift. Through the expenditure of energy as they travel downslope, streams consume the uplands by direct erosion of their valleys and by transportation of material washed into them. Erosion is accomplished by three general processes: *solution, abrasion,* and *hydraulic action.* Each of these erosional mechanisms must be coupled to *transportation*—the carrying away of material eroded directly from the land and carried in by mass-wasting.

Material in solution includes the carbonates and products of chemical weathering, most of which are carried in by groundwater. Dissolved matter is not visually apparent in stream water, but its presence is visible

TABLE 6-1 Average Daily River Flow

River	Location	Flow (billions of gallons)
Jordan	Israel-Jordan	0.9
Clinch	Virginia and Tennessee	2.9
Delaware	New York and Pennsylvania	12
Hudson	New York	14
Tigris	Turkey, Syria, Iraq	14
Colorado	Southwest United States	15
Euphrates	Turkey, Syria, Iraq	18
Rhine	Germany	50
Nile	Northeast Africa	65
Hwang Ho (Yellow)	China	75
Yukon	Canada and Alaska	120
Indus	Pakistan	130
Niger	West Africa	139
Danube	Central Europe	141
Columbia	Northwest United States	170
Mekong	Southeast Asia	252
Ob	Soviet Union	290
Mississippi	Central United States	400
Yenisei	Soviet Union	400
Ganges	India	426
Brahmaputra	Southern Asia	452
Yangtze	China	497
Congo	Central Africa	900
Amazon	Brazil	4800

indirectly in the thousands of feet of chemical an organochemical sediments deposited in the seas. All of the material composing these deposits was transported to the sea by streams.

Clear water can flow over solid insoluble bedrock indefinitely and cause little erosion; but streams transporting a sediment load, particularly of sand size or larger, can scour the bedrock bottom of the channel by abrasion as the load is dragged along. When eddies are present, as described for the turbulence illustrated in Fig. 6-23, abrasion becomes concentrated in small circular areas. Once a depression forms below the normal channel floor, the increased eddy action promotes an even more rapid circular grinding "mill," resulting in stream *potholes* (Fig. 6-25). As a valley bottom becomes covered with numerous merging potholes, the entire bed of a river is lowered. The irregular bed of a stream undergoing active erosion of its bedrock floor leads to a high degree of turbulence that produces the familiar "white water" of *rapids*. If a vertical ledge develops from the presence of a uniformly resistant rock layer, the water plunges over this ledge to form a *waterfall*.

Hydraulic action of streams involves the removal and transportation

FIG. 6-25. Brook flowing in resistant granite in Medicine Bow Mountains, Wyoming. The channel is deepened and widened by the coalescence of potholes eroded by eddies in the stream. Potholes now exposed formed during higher flood stage.

of loose material—perhaps the single most important erosional process. Material loosened previously by weathering processes is first plucked from the channel sides and bottom and is then carried away by the current. The swifter the current, the larger is the size of the particles that can be transported. Studies have shown that the *competency* of a stream, or volume of a particle that it can carry, varies approximately with the sixth power of the stream velocity. This means that if the velocity is doubled, the particle size carried can increase by 64 times. Streams in flood may experience a tenfold increase in velocity, leading to an increased competency of one million times (10^6). This is why streams in flood stage are often observed to shift almost impossibly large objects such as huge boulders, automobiles, heavy machinery, and so forth.

The total load that can be removed and transported depends on the discharge of the stream. Obviously, the more water, the larger the possible transportation. The transportation process really involves three quite separate forms of movement: gravel and sand roll along the bottom by *traction:* some sand is bounced along (*saltation*); and clays and silts are carried by *suspension.* In the latter case the fine materials are buoyed up by vertical eddies.

Deposition

A decrease in the transportation ability of a stream causes deposition. As the stream slows, due to a decrease in gradient or discharge or both, streams deposit *alluvium.* Coarsest (and densest) particles are dropped first, followed by a progression of finer materials, until a balance once again exists between stream load, discharge, and velocity. This progression, from coarser particles on the bottom to finer ones on top, gives alluvium a characteristic graded appearance, as shown in Fig. 6-26.

FIG. 6-26. Stream deposits are commonly graded with coarse particles on bottom grading upward to finer sizes.

Thus, when streams in the flood stage break out of their channels, the relatively coarse material is dropped along the banks to form *natural levees,* whereas the less violent water carries the finer muds out over the valley floor to form the floodplain. This flat fertile alluvium makes ideal farm soil. Unfortunately, the flood—the cause of the rich and inviting lowland—is often the cause of much devastation in the very act of building the floodplain.

Where streams enter larger, quieter, bodies of water such as lakes, bays, or even larger streams, more sediment is carried in than can be transported by the quieter water. The resulting sediment deposits, known as *deltas,* have a characteristic structure and form as in Fig. 6-27a. A slow progression of shorelines occurs from deltaic deposition. Deltas of all sizes are built, depending on the stream size and load. The Mississippi delta on which the city of New Orleans is built is one of the largest and best known (Fig. 6-27b). Deposition of a delta increases the length of a stream beyond the initial mouth as the stream flows over its own deposits. After flood stages the delta portion of the channel may become choked with sediment, causing the waters to break out into numerous small *distributory channels* as shown at the extremity of the Mississippi River.

Alluvial fans are forms of importance in arid and semiarid lands.

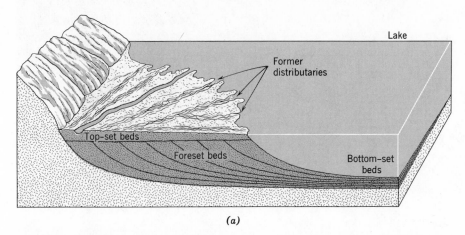

(a)

FIG. 6-27a. Typical subsurface structure of a delta showing topset, foreset, and bottomset beds which are deposited as a stream terminates in a quieter body of water. The stream channel is actually extended over the delta sediments as shown in the photograph of the Kander River delta Switzerland. (Courtesy A. Strahler, *Physical Geography,* John Wiley & Sons.)

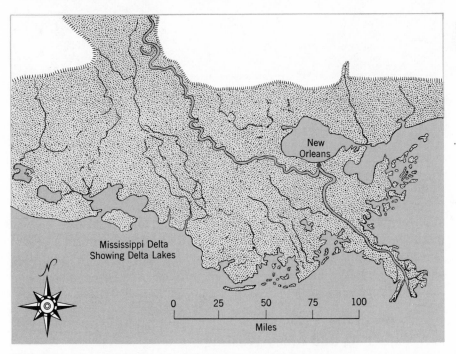

FIG. 6-27b. Sketch of the end of the Mississippi Delta whose area is about 12,000 square miles. New Orleans built on the delta is now 100 miles north of the river mouth. (A. K. Lobeck, *Geomorphology,* McGraw-Hill Book Co.)

These semicircular forms are deposited along the break in slope between a relatively flat valley floor and a steeper upland surface (Fig. 6-28). In regions of low rainfall, upland areas are often more moist. As a stream emerges from the deeper upland valley, a twofold change takes place; its gradient decreases with a consequent loss of velocity, and an actual loss

FIG. 6-28. Series of alluvial fans in arid country at the foot of a California mountain slope.

of volume occurs from greatly increased evaporation and percolation of water into the dry silt and sand of the valley floor. The loss of volume and velocity causes a fairly rapid decrease of total discharge and an inability to continue the transport of most of its load. The normal graded nature of stream deposits may be absent in fans because of the rapid changes in velocity and volume. Instead, a rather heterogeneous particle-size assortment can occur.

Relation to Weathering and Mass-Wasting

Were it not for the rock destruction of weathering and consequent downslope movement under the direct influence of gravity, streams would require an almost infinitely longer time to erode away the solid bedrock of the earth's crust. Actually, much of the erosional work of streams involves the plucking and transportation of weathered material and material that has mass-wasted down the valley sides.

Without the aid of prior weathering, a stream cutting down through resistant bedrock would maintain a valley width about equal to the channel width. The sequence of stages in valley development would be somewhat as shown in the series of transverse valley profiles in Fig. 6-29a. A narrow precipitous gorge would be developed, as shown, for example, by the well known "flume" (Fig. 6-30), a deep boxlike valley eroded down through resistant granite in the region of Franconia Notch, New Hampshire.

But most stream valleys have a flared V-shape resembling the stages in Fig. 6-29b. The direct erosional scouring by the stream would only cut the gorge shown by the broken line. Once the rock on the valley sides becomes exposed to the atmosphere, the decomposition and disintegration effects of weathering may result in mass-wasting of the material into the stream. In this manner, the shaded region of the valley is gradually removed, with the valley ever widening as the stream cuts down. The volume of rock material removed in this manner clearly far exceeds that removed by more direct erosion. In arid regions, the steep, angular, and relatively narrow valleys that are a consequence of slower rates of weathering contrast strikingly with views of broad, soft-looking valleys of humid areas.

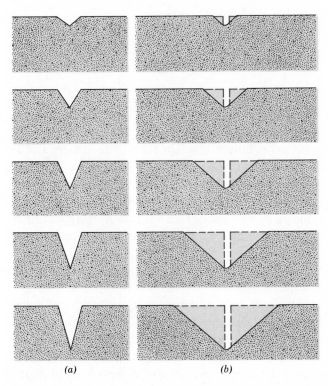

(a)   (b)

FIG. 6-29. (a) Stream downcutting with little weathering and mass-wasting produces narrow valleys. (b) Weathering and mass-wasting of valley sides above the channel causes a broad, V-shaped valley to form.

FIG. 6-30. The "Flume," a narrow gorge in the White Mountains indicating little weathering of the resistant granite walls.

Meanders and the Valley Floor

Most stream valleys have a rather remarkably level, flat bottom, as indicated in Fig. 6-21b. Nothing in the simple stream erosion processes described in Fig. 6-29 indicates how this might form. The answer lies in stream meanders, the swinging, curvelike pattern developed by nearly all streams after an early stage of growth. Meanders develop as a consequence of the physical principle that in most cases a fluid (not confined to a tube) moving in a straight line is in an unstable condition. By this we mean that any disturbance from this linear motion results in an even further departure from such motion.

Three stages of a stream are shown in Fig. 6-31. In a, the stream is flowing directly down slope. Imagine the stream to develop a single curve from the effect of an obstacle or rock inhomogeneity or turbulent water motion or the like. It is evident that the banks of the channel at points 1, 2, and 3 receive a greater impact than the intervening stretches of the banks. The water striking point 3 tends to erode the bank a little more toward the side producing the new curve shown in stage c. And as the stream comes around the new curve and strikes at point 4, a third curve will begin.

As the stream flows around the meander curve, its velocity must be greater on the outer than on the inner side of the curve. Hence, erosion

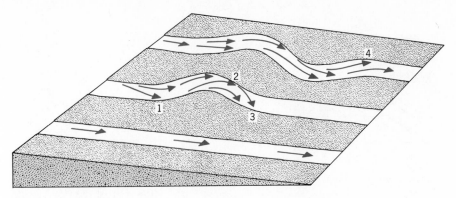

FIG. 6-31. Stages in the development of stream meanders.

is greater on the outer part of the bank, which becomes oversteepened as in Fig. 6-32. Excess material eroded from the outer banks is deposited along the bank on the inner part of the meander curve, where the velocity is less; it thus develops a gentle slope, such as the spur at A in Fig. 6-32. Water impact is concentrated particularly on the downstream part of the outer portion of the meander, as in Figs. 6-31 and 6-32. Deposition on the inner bank and erosion of the outer, particularly the lower part, causes enlargement and a downstream migration of all of the meander curves. In the course of this downstream migration, rock spurs such as A and B of Fig. 6-32 are eroded away, with the production of a fairly flat floor of rather uniform width. A summary of this planation process is in Fig. 6-33, which illustrates the gradual widening of a narrow valley floor as meanders sweep their way downstream. Once this flattening occurs, the bedrock bottom becomes covered with the fertile alluvial floodplain.

Base Level and Grade

If one could view a long reach of a stream valley from its very head or point of beginning to some point well down in its lower region, it would appear somewhat as in the schematic diagram in Fig. 6-34. From points A to C the stream and its discharge would grow in size from continuous influx of water from the ground and from side tributaries (not shown in the diagram). From points A to B the valley sides would converge directly at the water line. This is the youngest part of the stream and valley, which extend themselves upstream by *headward erosion;* in time the stream in the diagram would slowly wear its valley upslope to point O. The further downstream one goes, the older is the valley. Between points B and C, the meanders that are not shown in this schematic diagram, have developed the flat floor. Note that between B and C the stream is flowing on its own flood plain deposits rather than on the bedrock floor beneath.

From point B to its mouth, the stream is in the condition of *grade.* This is the condition in which, on the average, the discharge just permits the stream to carry its load. A decrease in velocity causes temporary

FIG. 6-32a. Streams flow faster on the outer portion of a meander curve where the water travels a greater distance in the same time taken by the water flowing along the inner portion of the curve. The length of the arrows is proportional to flow speed.

FIG. 32b. A narrow V-shaped valley with a meandering stream. Note how the valley is steeper on the outer sides of the meander curves where velocity is highest, and gentlest on the inner banks where deposition has occurred from the slower velocity.

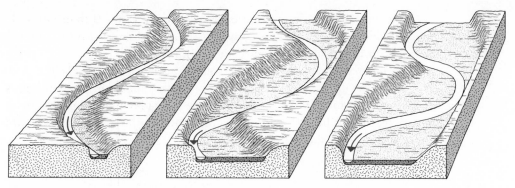

FIG. 6-33. The asymmetry of stream flow causes the valley to be eroded strongly where flow is greatest. Meanders thus migrate sideways and downstream thereby developing a flat-floored valley as shown in the series of diagrams. (After Longwell, Knopf, and Flint, John Wiley & Sons.)

deposition. An increase causes erosion of the bed. Normal seasonal variations in discharge cause continual small adjustments in velocity, volume, and load, but on the average, a well-balanced relationship exists. If discharge decreases, deposition raises the stream bed locally, thus increasing the slope or gradient. The slight increase in velocity increases the discharge again. A decreased load or increased discharge causes erosion of the bed, tending to decrease the gradient and consequently the velocity, again returning the stream to a graded condition.

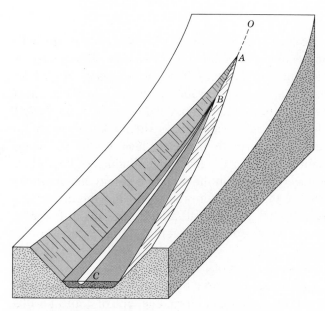

FIG. 6-34. Valley development from headwaters to lower portions of the valley. (Meanders are omitted for simplicity.)

Clearly, stream downcutting is greater above point B than below it. Hence, the lowering of a valley proceeds more and more slowly from the upstream to the graded downstream portion.

The ultimate level to which a stream can cut is sea level—the *base level* of stream erosion. Most of the great rivers that drain the continents have much of their valleys at or very close to sea level. Further upstream the valleys rise to higher levels, thus maintaining the flow, which would otherwise become stagnant.

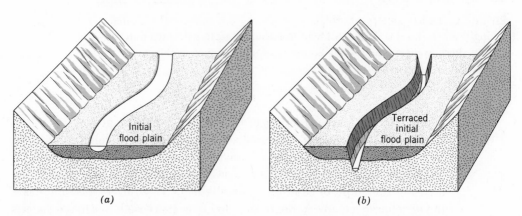

(a) *(b)*

FIG. 6-35. Stages in the rejuvenation of a stream. The stream in (*a*) is flowing on its own deposits. After rejuvenation, increased downcutting occurs and the stream channel becomes entrenched in the bedrock beneath its deposits. All of the meander curves are preserved.

In the course of erosion toward base level, a stream can do what people cannot—it can become rejuvenated. For example, at any stage of a stream's development, its gradient may become increased, often from crustal uplift. A stream that was in a graded state will no longer have its discharge and load in equilibrium, and the increased velocity will stimulate vigorous downcutting. Vast changes then take place in the appearance of the meanders. Normally the water level within the meanders of a graded stream is close to the level of the adjacent floodplain, as in Fig. 6-34. But renewed erosion in the stream bed may deeply incise the river bend well below the floodplain and often into the underlying bedrock as summarized by the "before and after" diagrams in Fig. 6-35. Some of our most scenic rivers—for example, that in Fig. 6-36*a*—are the products of rejuvenation.

In a late stage of erosion, spurs of incised meanders may be eroded through as a stream straightens its course by cutting off a tight meander curve. The relict spur and stream channel form one type of natural bridge (Fig. 6-36*a,b*).

Peneplains and Cycles of Erosion

A great river with its many tributaries and subtributaries compose a *drainage basin* or *watershed*. The Mississippi River, for example, drains all of the interior portion of United States from the crest of the Rocky Mountains on the west to the crest of the Appalachian Mountains to the east. Given enough time, stream erosion of a drainage basin, in combination with weathering and mass-wasting, would lower the region to sea level. But the process becomes progressively slower from decreasing energy as the lands become lower, so that a true erosional plane at base level is never quite reached. Instead, gently sloping surfaces of low relief appear to have formed many times in the course of earth history. Such landforms are known as peneplains (almost a plane).

Just as streams can be regenerated, so can entire peneplained mountain regions or drainage basins, on a much grander scale. The evidence of peneplain uplift lies in the uniform broad summit areas of large parts of present-day mountains. For example, the crests of many of the elongated mountain ridges of the Ridge and Valley Province of the Appalachian Mountains (particularly in Pennsylvania) lie at a rather remarkably similar elevation. This surface, which could easily be imagined to have once extended across the broad intervening valleys, has been interpreted as an uplifted and then deeply reeroded peneplain. In the same way, many of the accordant upland surfaces of the Rockies are similarly inferred to be elevated peneplains.

All such elevated and dissected surfaces imply to the geologist that the great mountain systems involved must have been broadly eroded to near base level—a process requiring tens of millions of years. From such observations and deductions we may also conclude that the present relief and irregularity of most of the world's mountain systems is not the direct result of the initial deformation and uplift. It is rather the result of at least one later cycle of erosion after a denuded mountain system was reelevated. In many cases present mountain relief, such as that in Fig. 6-36, is a consequence of the differential erosion of weak rocks into valleys leaving the protrusion of resistant rocks as ridges.

The presence of elevated peneplains provides a large part of the answer to an old enigma of erosion. From measurements of stream loads and of sediment accumulation in the oceans, it is possible to make an estimate of the total rock removal from the continents—an amount approximately equal to 10 billion tons per year. This is equivalent to an average lowering of the continents of about 1 inch per 1000 years. Such a rate of erosion would reduce the continents (whose average elevation is about one-half mile high) to sea level in 30 million years.

This rate of erosion might apply to the present geologic era, but we will see in Chapter 7, where we examine the criteria for estimating the age of the earth, that present erosional rates are rather high. Even if we double or triple or multiply by 10 the time necessary to level the conti-

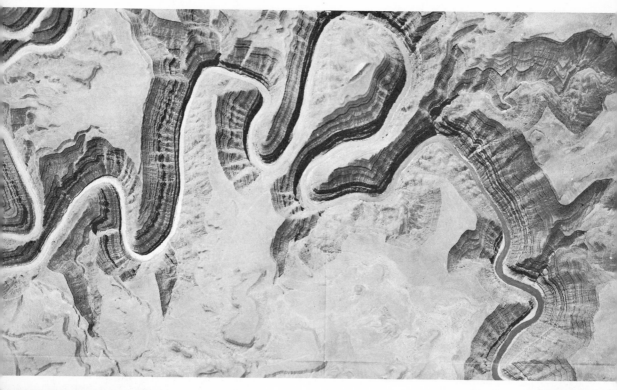

FIG. 6-36*a*. Deeply incised meanders along rejuvenated stream, San Juan County, Utah. (U. S. Geological Survey.)

FIG. 6-36*b*. Landscapes Arch, the longest stone span (291 feet) in the world, is an example of the erosional relict in which an incised meandering stream has cut through one of its spurs. Arches National Monument, New Mexico. (Photo by J. E. Boucher, courtesy National Park Service.)

FIG. 6-36*c*. Series of stone bridges marking the channel of an ancient meandering incised stream valley. Arches National Monument, New Mexico. (Photo by K. W. Williams, courtesy National Park Service.)

nents, it is clear that such leveling could have occurred many times over during 4 to 5 billion years of crustal history.

Despite this relentless wearing down of the lands, the continents are larger and higher than at any time in the known geologic past. The presence of raised peneplains tells us that erosion has in fact lowered most regions to some base level many times in the past and that such surfaces were then reelevated by profound movements in the earth's crust. Where does all this material come from and why haven't the oceans filled up with all of the erosional debris of the continents? The answer to these questions will be found in Chapters 8 and 9, where we will examine the dynamic processes that tend to maintain and renew the continents and ocean basins. Just as the hydrologic cycle involves a never-ending circuit of water from sea to air to the soils and rivers of the continents and thence back to the sea, an even grander cycle of rock material seems to exist, involving erosion and lowering of the continents, transportation to the seas, and return of material to the continents.

GLACIERS AND GLACIATION

To those of us who live north of the 40th parallel of latitude in North America, it may come as a surprise to learn that if we could have looked from a window about 20,000 years ago, we would have seen a fairly solid and continuous layer of ice thousands of feet thick. Major changes in the landscape were wrought by the slow but inexorable grinding movement of this ice as it spread southward. We can observe directly the action of streams and the processes of weathering and mass-wasting. But how do we know that vast sheets of ice, no longer present, once carved their way across much of the continental lands?

Evidences of Former Ice Sheets

The story starts with rocks called *erratics* that are very much out of place (Fig. 6-37). Normally, boulders and rock particles in the soil match the local bedrock from which they have been derived. Wind, gravity (mass-wasting), ocean waves, and even streams can move rocks a short distance. But none of these forces can transport huge boulders great distances from their source and leave them mixed with a most hetero-geneous mass of soil, gravel, and sand called *till,* all foreign to the bedrock on which they lie.

In the early 1800s, two Swiss scientists, J. Venetz and Jean de Char-pentier, who studied existing glaciers in the Alps, became convinced that only the movement of solid ice could account for the erratics observed along the Alpine valleys. Being solid, ice could transport rock of all sizes, both on its surface and throughout its interior.

But erratics are found thousands of miles from mountain glaciers and, in fact, are found in places where no mountains exist to support glaciers. The suggestion that ice transported all of the erratics was hard to believe and was not accepted until the imaginative field work and interpretations of Louis Agassiz, another Swiss scientist.

Agassiz knew he had to study the present to understand the past. He carefully observed the behavior of ice in the glacial valleys of the Alps and proved that these glaciers really moved, even if slowly, and that they transported huge amounts of rock debris in the process. This material is

FIG. 6-37. Large glacial erratic that lowered onto the bedrock surface as the ice which transported it melted away. Most erratics are of course much smaller. (Geological Survey of Canada.)

piled into ridges, known as *moraines*, at the side margins and end of valley glaciers (Fig. 6-38). After studying the glaciers of the Alps up close, Agassiz traveled widely in North America, where more than half of the continent is covered by moraines, erratics, and other features formed by moving ice. Erosional furrows and striae carved into bedrock surfaces (Fig. 6-39) by sharp-edged rock fragments beneath the moving ice are also unexplainable by any other erosional agent. The scientific world could not escape the arguments of Agassiz that not only were many mountain valleys of the past once filled with ice, but also, most of North America must have been covered by a thick sheet of ice that extended from the Atlantic Ocean to the Pacific Ocean and from about the 40th parallel to the Arctic Ocean.

Once the existence of former ice sheets was established, it became quite simple to find all kinds of other features, both depositional and erosional, that are quite characteristic of glaciers, as described below under "Glacial Erosion and Deposition."

Perhaps the most striking evidence of the presence of former glaciers is a feature not formed from the direct effects of glacial erosion or deposi-

FIG. 6-38*a.* Successive views of an Alpine (or Valley) glacier in the Alaskan Range. Debris slumping down the valley walls form *lateral moraines* along the sides of the glacier; along both sides the moraines are carried along as streaks of earthy materials (called *medial moraines*). The medial moraine, in the center of the main glacier, is from the joining of two equally sized but smaller glaciers in the distant background. Between the times of the two pictures a large avalanche and earth flow spread out as a giant apron across the lower portion of the valley glacier. The arrow shows the source of the slide, and in the lower photo indicates the scar that remained. (U. S. Geological Survey.)

FIG. 6-38*b.* A prominent valley glacier formed by the confluence of two glaciers. Many medial moraines extend from prior mergers visible in the right background. The sources of the glaciers are snowfields, some being particularly noticeable in the summit regions in the left middle ground. (Photo by Austin Post, U. S. Geological Survey.)

FIG. 6-39. Elongated glacial striae in the central Sahara desert
which was once covered by a continental ice sheet—about 450 million years ago. The ice moved toward the direction of the hammer-
head (south to north). (Photo by R. Fairbridge.)

tion. The earth's crust is quite elastic. The weight of the huge masses of
ice once present on the land depressed large elevated areas down to and
below sea level. When the ice melted, the lands slowly returned to their
original levels. But there was a lag between the time of melting and the
rebound of the land. During the time of this lag typical beach and
shallow-water features developed on the areas at, or just below, sea level.
Then uplift began, and these beach features, together with the shells of
clams and other shellfish characteristic of shallow-water conditions, were
raised from tens to hundreds of feet above the level to which the ice had
depressed them. (See, for example, Fig. 6-11.)

Coastal regions of northern North America and Europe abound with
such features. Typical beaches with fragments of shell fish are found up
to 900 feet above sea level along the eastern margin of Hudson Bay, and
the Bay itself is a relic of ice depression growing ever shallower until

its disappearance in the not-distant geologic future. To the south of Hudson Bay is a freshwater lake containing seals—typical, cold saltwater mammals. Once a small inlet of the ocean, this water body was cut off from the sea when uplift began. The seals adapted to life in the freshening water after separation from the ocean.

How Glaciers Develop

Over much of the temperate and arctic latitudes of the world, winter snowfalls often accumulate to several feet in thickness. This is true for both lowlands and mountain regions. With the warming of spring and summer, the snow normally disappears by melting and evaporation, except for very high mountain areas that may preserve snow throughout the summer. But if the snow accumulations are unusually thick or if the air temperature remains low enough, not all of the winter snow will melt during the warm season before renewed winter accumulation begins. A permanent *snowfield* thus forms, and if conditions unfavorable for melting persist, a small net thickening will take place each year. The underlying snow is soon converted to ice by the pressure of the snow above. When enough ice has accumulated in the original snowfield, it begins to move. It is this moving mass of ice that is called a *glacier*. The beginning of motion depends on both the thickness of ice and the ground slope on which the ice mass rests.

Glaciers that form in mountainous country actually accumulate in stream valleys that lead down from the highland. Glaciers of this type (Fig. 6-38) are known as *valley, mountain,* or *alpine glaciers*. Whether a mountain can support a valley glacier depends on the elevation. As we will see later (Chapter 14), the average temperature of the air decreases about 3.5°F per 1000 feet of elevation. Consider a warm equatorial land where the average temperature may be close to 90°F near sea level. At the rate of decrease of 3.5°F/1000′ the temperature will be about 32°F or at the freezing point at 18,000 feet. Hence mountain peaks above this elevation might easily have snowfields and glaciers throughout the year—a matter of well-known observation (Fig. 6-40) for the equatorial regions of Africa and South America.

Although Alpine glaciers still exist on many of the high mountains of the world and were even more numerous during the recent great Ice Age, by far the largest expanses of glacial ice occurred in the lower lands of the continents. These widespread glaciers once covered hundreds of thousands to millions of square miles of continental areas. *Continental glaciers* or *ice sheets* originated in the high-latitude lands of North America, Europe, and Asia and reached their greatest size and southern-most extent in North America (see Fig. 6-41).

The movement of ice in continental glaciers did not require any initial ground slope, as in the case of alpine glaciers. The pressure of the thick

FIG. 6-40. Glaciers on Nevado Huascaran in the Andes Mountains in low latitudes of the southern Hemisphere.

ice (which in continental glaciers reached 8000 to 10,000 feet) in the interior of the glacier caused a slow but continuous movement of ice toward the margins, as indicated in Fig. 6-42.

Glacial Erosion and Deposition

Where glaciers have overrun the earth's surface, they have strongly modified, and at times obliterated, the effects of prior agents of erosion. As with streams, a glacier erodes by abrasion as moving ice scours the underlying rock surfaces with the waste embedded in its bottom. Individual sharp rock fragments cut *grooves* and finer fragments cut striae into the rock (Fig. 6-39), while the fine rock flour actually polishes the rock beneath much as a jeweler's rouge polishes a gem.

Although bedrock with polished and striated or ground surfaces is still common in the glaciated areas of the world, such features are really only "skindeep" and are fast disappearing from the effects of weathering.

Glaciation in
Northern Hemisphere
during ice ages

FIG. 6-41. Distribution of glaciers in the Northern Hemisphere during the last glacial stage.

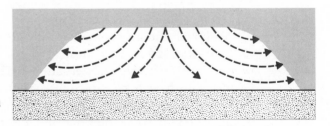

FIG. 6-42. Glacial flow within an ice sheet.

Again, as with streams, the more important mode of glaciation involves a process somewhat like that of the hydraulic action in which loosened rock is removed and transported elsewhere. But the load carried by streams consists mostly of the products of decomposition and disintegration, whereas much of the load carried by glaciers was *plucked* directly

FIG. 6-43. U-shaped glacial trough in Glacier National Park, Montana. A field party is crossing the foot of Grinnell Glacier. (J. J. Palmer, National Park Service.)

from relatively unweathered masses of bedrock. In the plucking process, rock is first loosened by freezing and thawing action from glacial meltwater. When sufficiently loosened, the moving ice, which is frozen onto the rock, simply tears it away from the main mass. By this process, a glacier may literally devour huge masses of the bedrock that surround and underlie it. A glacier can carry rock material of all sizes from fine clay to huge house-sized boulders.

Although the geologic effects of the vast continental ice sheets are far more widespread than those of alpine glaciers, the erosional features of the latter are much more prominent. These features are so distinctive because they have been carved into spectacular mountainous and upland regions. Alpine, or valley, glaciers transform the originally V-shaped stream valleys in which they accumulate to straight, steep-sided U-shaped *glacial troughs* such as that pictured in Fig. 6-43. The heads of glacial valleys are commonly widened by strong ice plucking into large amphitheater-shaped depressions called *cirques*—a characteristic feature of any glaciated mountain region (Fig. 6-44, 45). Stages in the transformation of an initial stream-covered mountainside into a glaciated mountainside, together with other features typical of such regions, are summarized in Fig. 6-46.

FIG. 6-44. Saddlebag Glacier, a valley glacier in Alaska, still present in its trough. A lake of melt water with floating ice is present ahead of the glacier terminal. A small stream can be seen transporting the meltwater down the valley. Fields of snow are present in the small bowl-shaped "cirques" from which subsidiary glaciers originate. Jagged peaks, known as "horns" marking the highest summits are erosional relics of larger mountain summits since eroded away by ice plucking and weathering-mass wasting effects at the heads of cirques. (Photo by Austin Post, University of Washington.)

All regions that have undergone glaciation exhibit smooth, rounded rocky knobs that formed first from the plucking action of glaciers followed by the smoothing and polishing effects of the overriding ice. Such *roches mountonnées* (sheep rock) are always asymmetric in the direction of glacial motion as in the Fig. 6-47; the gentle face slopes opposite to, and the steeper, plucked face slopes in the direction of, motion. Surface grooves, which may be present, also show the direction of motion.

Glacial deposition may occur directly from the ice or from glacial meltwater in the form of streams or lakes. Glacial *till*, which is deposited directly by ice, is extremely heterogeneous in size distribution, since ice

FIG. 6-45. Typical bowl-shaped cirques covered by individual valley glaciers in a Greenland ridge. (U. S. Coast Guard photo.)

is not limited to a particular load size. In an active glacier, ice is always moving toward the margins, where wastage by melting, evaporation, or calving (breaking into the sea) takes place. Hence till is carried continuously toward the margins, where it is dumped as ridgelike accumulations of mixed material called *end moraines* (Fig. 6-48). The end moraine marking the furthest extent of the ice is the *terminal moraine,* a prominent example of which is the low ridge that extends continuously from the eastern tip of Long Island westward through the central part of the island until interrupted by the water of New York Bay. The terminal moraine resumes its course across the Bay through Staten Island and on into New Jersey.

The regions that lay beneath the great ice sheets of the past are covered by a rather thin veneer of glacial till known as *ground moraine.* These surfaces bear the effects of both glacial erosion and deposition (Fig. 6-49*a*). For the most part, thick moraine accumulations were deposited only at the extremities of continental ice sheets, whereas valley glaciers accumulated thick deposits along the glacier sides from material that slumped down the valley walls in addition to deposits at the ends.

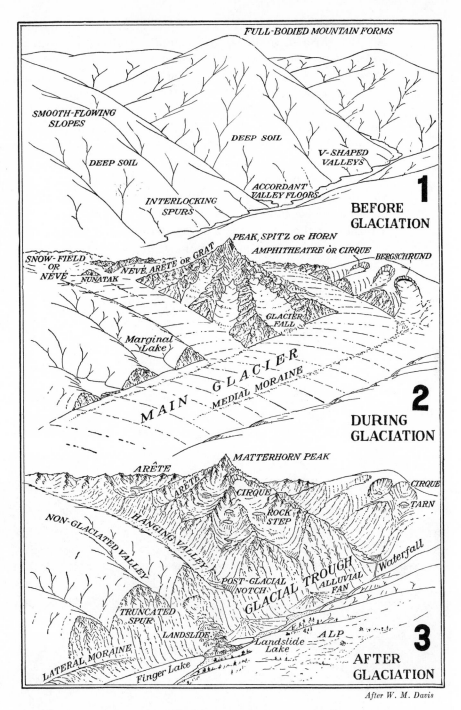

FIG. 6-46. Schematic views of a mountain region before, during, and after glaciation. (A. K. Lobeck, McGraw-Hill Book Co., after W. M. Davis.)

Beyond the margins of any type of glacier, meltwater often spreads out as a sheet of sediment-laden fluid. Resulting deposits are stratified and sorted according to size, as with typical stream sediments. The large apron of layered sands and muds that forms beyond the terminal moraine is known as the *outwash plain*. To the south of the well-known terminal moraine ridge crossing Long Island is the widespread outwash plain that covers most of the southern half of the island. The famous white sand beaches of Long Island are residual sediment left after ocean-wave action removed the finer muds and transported them further out to sea.

Of great importance in working out the sequence of glacial events are the finely stratified deposits that accumulated in meltwater lakes between an ice lobe and an adjacent hill. Thick seasonal layers were deposited in these lakes from the relatively rapid melting and deposition in summer because glacial ice often is loaded with loose rock material. These usually light-colored summer deposits contrast strongly with thin,

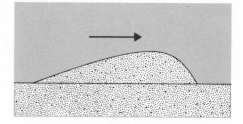

FIG. 6-47a. Profile of a roche moutonnée relative to glacial motion.

FIG. 6-47b. Roche moutonnée in St. Nicholas Park, New York City. The ice moved from left to right over the original rough surface.

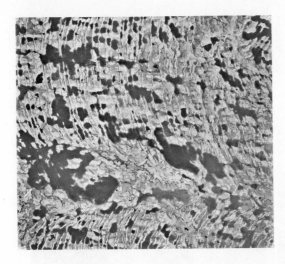

FIG. 6-48. Series of morainal ridges left behind the ice margin as a glacier retreated. (Royal Canadian Air Force Photo.)

dark winter deposits (Fig. 6-49*b*). The number of annual layers or *varves* can be counted readily, and the sum may indicate the duration of glacial activity in a particular region. Conclusions from varve observations play an important part in working out ice-age history as discussed in Chapter 15.

SEA AGAINST LAND

Despite the grandeur of mountains carved by glaciers, or the pastoral beauty of stream valleys cut into hillsides, the ever-changing, often dramatic quality of coastal landscapes composes, beyond doubt, the most interesting, varied, and beautiful of all of the sculptured aspects of the earth's face. Nowhere is the combined cause and effect of erosion so immediately obvious as along a steep coast under direct attack of waves. Fortunately, the building processes in the earth's crust seem to be more than a match for the destructive forces of erosion. Streams, as we have seen, are and have been continuously lowering the earth's surface; yet the lands were never higher. Sea waves have for billions of years pounded away at the shores, seemingly in an attempt to obliterate them; yet the continents were never more vast.

Energy of Shore Processes

Once again we must look to the sun for the energy that drives the seas against the land. Sea waves are the means by which the sea attacks the land. And sea waves are generated by wind that blows over the water. Wind, in turn, is air set in motion by the energy of the sun, as explained in Chapters 14 and 16. As long as the sun shines, the seas will drive relentlessly against the shores.

FIG. 6-49a. Stereo-pair of region in Hamilton County, New York, showing the "Black Cat," an interestingly shaped lake of glacial origin. The north-south lineation in the topography resulted from glacial modification of the original ground surface. A portion of one of the larger Finger Lakes of Central New York appears on the right (east). Lakes of the type visible result from a combination of glacial erosion along initial stream valleys followed by deposition of moraine material that dams the stream to form lakes. (U. S. Geological Survey and University of Illinois.)

FIG. 6-49b. Glacial varves. Seasonal deposits in a former glacial lake. Their number tells the number of years the lake existed. Variations in thickness indicate variations in climate from year to year. (Photo by R. F. Flint; courtesy Longwell, Flint, and Sanders, John Wiley & Sons.)

When waves roll in toward the shore, they increase in height as the sea bottom becomes shallow. If the bottom slopes gently away from shore, waves steepen into breakers that begin to spill over while still far out. But on steeply sloping seacoasts, waves travel in quite close before they grow to great size and plunge over as giant breakers (Fig. 6-50).

The energy expected from sea waves can be calculated from the formula:

$$E = 8LH^2 \text{ foot-pounds}$$

FIG. 6-50. Breaking Wave. The destructive power of a breaking wave can easily be sensed from a view like this and has been experienced by bathers in the surf zone. (Photo by R. W. Linfield.)

where 64 is the weight of a cubic foot of water, L is the length of the wave (distance between successive wave crests), and H is the height of the wave (vertical distance from trough to crest). From this formula we can calculate that a fairly common winter sea wave with a length of 500 feet and a height of 10 feet will have 400,000 foot-pounds of energy for *each one foot along the wave*. Another way to consider wave energies is to convert their energies to the familiar units of electric power. Ten-foot waves striking a coastline such as the south shore of Long Island (about 125 miles long) may release continuous energy of a half million kilowatts compared to an output of about 1.5 million kilowatts by the Hoover Dam as it converts water power to electric power. Unfortunately, we have not yet learned how to harness the immense sources of power available in ocean waves.

Waves have been known to make toys of some of the most impressive shore engineering structures. For example, the end of a breakwater at Wick Bay in Scotland, which contained concrete blocks totalling 1350 tons, was shifted about into a useless position during a storm. After being rebuilt into a unit that weighed 2600 tons, storm waves again moved it, seemingly with ease. It is little wonder that coastal storms create such havoc.

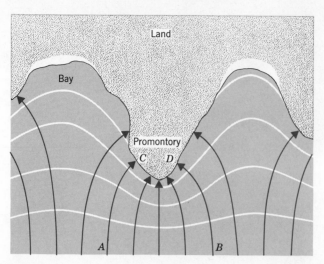

FIG. 6-51. Changes in a wave crest as it approaches an irregular coastline. Wave energy becomes concentrated on promontories, eroding them back. Deposition in the bays tends to make the shoreline more regular.

Along irregular coastlines, wave impact on the shore is actually increased on the headland (the land projecting into the water). Figure 6-51 illustrates a wave crest advancing from the open ocean. The incoming wave "feels" the shoaling sea bottom in front of the headland sooner than in front of the bay and begins to slow down. The wave crest becomes bent as the part in deeper water moves more rapidly and begins to resemble the outline of the shore. As the crest approaches the headland, the crest length, initially between *A-B,* becomes condensed into the length *C-D.* As water is crowded into this shorter distance, it must go somewhere—thus the wave height increases to accommodate the additional water. The wave energy and pressure thus increases greatly, causing rapid erosion of the headland (Fig. 6-52*a* and *b*).

The eroded sediment drifts into and across the bay. The effect of this process is to straighten irregular coastlines; headlands are worn back and bays become filled with sediment.

Despite the tremendous energy of breaking sea waves, their destructive force is confined to a narrow coastal zone, compared to the other erosional processes that attack and lower continents over their length and breadth. Also, coastlines are very dynamic regions. The relative position of land and sea have changed so frequently in the past that waves have rarely had an opportunity to complete a cycle of coastline erosion. For example, the sea level has risen continuously from glacial melting that began about 18,000 years ago. Thus, the point of wave attack shifted continuously as the sea level rose. Crustal uplift of the western United States has elevated the lands so that the level of wave attack has been lowered.

FIG. 6-52a. Wave erosion of a promontory on Santa Catalina Island, California. Storm waves in the placid Pacific Ocean of this view have isolated portions of a once-continuous headland into individual "sea stacks." These too will be consumed by continued wave action. (Courtesy American Museum of Natural History.)

FIG. 6-52b. Roche Percé, a wave eroded and isolated promontory off the eastern end of the Gaspé Peninsula, Quebec. The larger block has been further eroded through completely to form both a sea tunnel and a sea stack. The rock itself, visible in the foreground is a coarse conglomerate of Devonian age.

FIG. 6-52c. Sea stack on the northern coast of the Gaspé Penninsula, Quebec.

Characteristics of present shorelines depend primarily on the height and steepness of the lands along the coast plus the secondary effect of the resistance of the rocks to wave erosion.

Features of Steep Coasts

After waves break on a newly formed steep coast for what is a short time, geologically, a characteristic shore profile develops (summarized in Fig. 6-53 and 6.54). The original sloping rock face becomes connected to a steep cliff as wave erosion cuts back the lower part of the original slope. A nearly flat benchlike surface is left behind by the receding cliff. Sediment (usually sand) eroded from the slope may carpet the bench to form the *beach*. A large amount of the sand is transported below tide level to form a *wave-built* or *shore-face* terrace as shown in Fig. 6-53. During the year, sand is continuously drifting under wave action from the beach to the terrace (usually in winter) and back to the beach again (in summer).

Longshore currents resulting from wave action also drift sand laterally along the shore. Man-made jetties, seen so commonly on beaches, are constructed to decrease the loss of sand through this migration. It is a familiar sight to see sand piled high against one side of a jetty while the other side is quite exposed. Periodically, shore engineers scoop up the sand on the high side with giant cranes and return it to the other side of the beach.

Features of Coasts of Gentle Slope

Should the sea level fall of the coastal land be raised, emergence of the very gently sloping sea floor will occur. Soon after such emergence, transportation of sand either from a cliff source somewhere along the coast, or from the shallow offshore sea bottom, accumulates in a long narrow ridge just off and parallel to the shore. A shallow lagoon separates the ridge, known as a *barrier* or *bar*, from the shore as in Fig. 6-55. The barrier is usually kept open at one or more inlets by the daily sweeping of the tide as water periodically enters and leaves the lagoon.

Since sand barriers are deposited by wave action, one might ask why they project above sea level. Storm waves, particularly at high tide, heap sand above normal sea level. Winds may further enhance the development of the low sand ridge above normal sea level.

Along irregular coastlines, sand barriers may connect directly to the land on one side of a bay. Longshore currents drift sand, first removed by wave action, into a narrow sand ridge that slowly closes off the opening of the bay. Water eddies at the open end of the barrier often cause a curvature or hooking of the barrier into a *spit* as illustrated in Fig. 6-56.

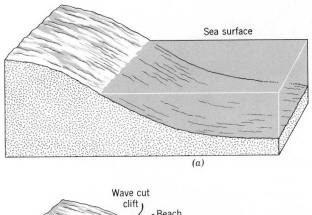

(a)

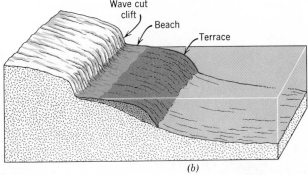

(b)

FIG. 6-53. Effects of wave action on an initially steeply sloping coastline.

Most of the eastern coast of United States south of New England is low and gently sloping. Barriers and spits characterize much of this coastline. The lagoons between them and the shore form the famous "inland coastal waterway" so well known to small-craft boatmen. It is possible for small craft to travel for much of the distance from New York City to northern Florida in the shelter of the calm lagoons, avoiding the rough waters of the western North Atlantic Ocean.

A variety of typical shore features is shown in the striking photograph in Fig. 6-56.

FIG. 6-54a. Sea cliff and marginal beach along the coast of Alaska. When the notch caused by wave action at high tide undercuts sufficiently, the overhanging portion will collapse. The cliff is maintained by this undercutting and collapse process. (Courtesy American Museum of Natural History.)

FIG. 6-54b. Portion of sea cliff being isolated by wave action from mainland, Pt. Reyes Natural Seashore California. (National Park Service Photo.)

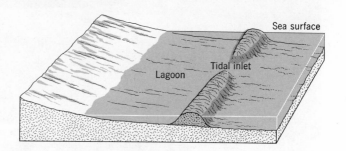

FIG. 6-55. Effect of wave action on a gently sloping shore as is common on or after emergence of a former shallow sea bottom.

WIND AND THE ARID LANDS

Wind Features

Wind is air in motion. Wind, unladen with sediment, can blow for eternity against solid rock and do little damage. But wind carrying sand will slowly abrade the exposed surfaces of bedrock. It is a familiar highway site in regions where loose sand is available to see the lower parts of telegraph poles protected by collars from the wearing effect of blowing sand.

The causes of wind motion and the source of its energy are described in Chapter 13 on the atmosphere. Here we are concerned with it only as an agent of erosion. Although abrasion by blown sand produces interestingly shaped rock exposures (Fig. 6-57), the long-time effect of such erosion is rather small.

Wind, like running water, does most of its geological work by *deflation,* the picking up and removing of material loosened by other processes. Depressions called *deflation hollows* (Fig. 6-58) form from wind eddies, much as potholes form from stream eddies. The finest particles, such as dust of all sizes, can be carried to great heights in the atmosphere. Coarser material such as sand is too heavy to remain airborne very long. It is transported by being bounced along, much as streams alternately pick up and then drop coarser sediment.

In desert regions deposits of sand take on characteristic *dune* shapes. The more familiar type is the *barchan,* illustrated in Fig. 6-59*a, b* in which sand is carried up the gentle windward (wind-facing) slopes. On reaching the dune crest, sand grains then tend to roll or slide down the steeper lee slope. By this process the dune actually "marches" downwind. Where a large supply of sand is present, barchans may be so numerous and closely spaced as to appear like irregular waves of sand. Other dune forms exist, but the crescent-shaped barchen is probably the most common and picturesque.

FIG. 6-56. Shore features on the eastern end of Martha's Vineyard. Cape Cod lies along the northern (upper) third of the photograph. Several spits with typical curved or hooked patterns are visible along the coast of Martha's Vineyard. A long sand bar has closed off the bay on the southern coast. (Photo by Lowry Aerial Photography.)

FIG. 6-57. Desert pebbles from Mongolia showing typical examples of facets produced by the sandblasting action of wind. (Courtesy American Museum of Natural History.)

FIG. 6-58. Deflation hollow, about 25 feet deep, caused by wind erosion in delicately-colored sandstone, southeastern Wyoming.

(a)

FIG. 6-59a. Crescent-shaped barchan dunes in the California desert. (U. S. Geological Survey.)

(b)

FIG. 6-59b. Large, interconnected barchan dunes in Great Sand Dunes National monument. These are some of the largest dunes in the country. (Photo by G. A. Grant, courtesy National Park Service.)

Tough dune grass that can grow in the very inhospitable sand environment is often planted to stabilize the sand, greatly restricting or even preventing further migration.

Stream Erosion in Deserts

Deserts are characterized by an excess of evaporation over precipitation. Precipitation, when it does occur, is often in the form of heavy showers and thundershowers of short duration but high intensity. Despite their infrequent occurrence, desert rains do produce strong erosional effects because the weathered bedrock material is not protected by a cover of vegetation. Consequently, numerous stream channels form. Only at times of rainfall do the channels contain water—at other times they are quite barren.

FIG. 6-60. Inselbergen in the Imperial (desert) Valley of California. The mountains shown in this air-photo project through sand formed from their erosion.

When desert showers do occur, a flood of water washes down mountain slopes into these previously dry channels. Loosened, weathered rock is carried along by this *sheet wash* into the channels, where it is transported further downslope to the floor of the lower land that surrounds desert mountains. Unlike streams in humid regions, those in the high deserts seldom reach the sea. Upon descending from the mountains, the water sinks into the dry desert floor and disappears, leaving a thick alluvial deposit that spreads and thickens after each heavy rain.

Slowly the mountains are consumed by sheet wash and channel erosion. The products of this erosion do not travel very far, but slowly build up and engulf the very mountains from which they were derived. Ultimately, the mountains project as islands (*inselbergs*) above the surrounding apron of loose sand and rock. Features of this kind are common in the arid southwestern portion of United States. (See Fig. 6-60.)

STUDY QUESTIONS

6.1. Describe in detail the processes by which a granite may be converted to a beach sand. Explain in particular what happens to the quartz and feldspars present in the original granite.

6.2. Explain briefly the energy process involved in chemical weathering.

6.3. What is the real energy source in mechanical weathering?

6.4. Why does chemical weathering involve a release of heat and result in the formation of low density minerals?

6.5. How does chemical weathering differ from rock metamorphism in regard to the process in question 4?

6.6. Explain chemical weathering in terms of the Bowen reaction series.

6.7. Explain the natural process by which limestone is dissolved.

6.8. How does hydrolysis cause mineral decomposition? Give an example.

6.9. Distinguish between hydrolysis and hydration.

6.10. Explain briefly the chief processes of mechanical weathering.

6.11. Why does weathering proceed more rapidly in a humid than in an arid type of climate?

6.12. Describe the topographic differences between weathering effects in humid versus arid climates.

6.13. Draw a diagram and label, with brief descriptions, the units of a typical soil profile.

6.14. What is the overall importance of weathering in determining the nature of the earth's topography?

6.15. Describe the importance of water in each of the processes involved in mass-wasting.

6.16. What do we mean by the term "hydrologic cycle"?

6.17. Draw a labeled diagram showing the water table beneath a sloping hillside and indicate where water may issue forth as a spring. In a separate diagram show how the water table is related to lakes and swamps.

6.18. Describe the features related to underground caverns.

6.19. How is the flow of water maintained in permanent streams between periods of rainfall?

6.20. Describe and explain the origin of stream valleys.

6.21. Explain the factors that control the velocity and discharge of a stream.

6.22. Distinguish between laminar and turbulent flow.

6.23. Explain and compare the erosional importance of abrasion and hydraulic action.

6.24. What is the importance of "meandering" in the process of stream erosion?

6.25. Describe briefly several erosional and depositional features produced by streams.

6.26. What is meant by the term "base level" of stream erosion?

6.27. How can a stream become rejuvenated and what is the most obvious evidence of rejuvenation?

6.28. How do rivers increase their length?

6.29. Explain "erosion cycle" and the meaning of the term "peneplain."

6.30. How is the presence of former glaciers determined long after the ice is gone?

6.31. Explain how seashell-covered beaches in high latitudes came to lie at elevations of many hundreds of feet above sea level.

6.32. What conditions are necessary for the formation and maintenance of glaciers?

6.33. Compare the processes of erosion by streams and glaciers.

6.34. List and describe briefly some erosional and depositional features of glaciers.

6.35. How are glacial "varves" used in the study of glacial history and climate.

6.36. Explain the energy transitions that cause the development of waves and their ultimate erosion of a shoreline.

6.37. Draw labeled diagrams to show the shore profiles following wave erosion of both initially steep and gentle shorelines.

6.38. In what climate is wind erosion most effective? Why?

6.39. What is a major difference in the courses of stream valleys in arid and in humid regions?

6.40. What is an important difference in the stream erosional process in these regions?

6.41. Describe barchan dunes and explain how they migrate.

6.42. Distinguish between a peneplain and a pediment.

6.43. Distinguish among the primary energy sources that affect the earth's interior and surface processes.

chapter 7

The valley and gorge constituting the Grand Canyon of the Colorado River in northeast Arizona. This collosus of canyons contains one of the most complete and extensive records of geologic time found in a single place on earth. From the Precambrian floor to the Triassic crest, the canyon displays a discontinuous time span of about 2 billion years. See Fig. 7-1 for details.

Measuring the Past

The primary goal of geological science is the deduction of the natural history of the earth. How did it come to look the way it does today; how did the core, mantle and crust develop; whence came the continents and ocean basins and the atmosphere and oceans that surround them; how were mountains formed and what forces and processes produced the tremendous variety of the earth's scenery?

To solve these problems, which lie at the very heart of earth science, geologic events that can be interpreted from rocks must be set in order of occurrence. The history of the earth needs a time scale just as does the history of man. To this end we have had to establish a geologic calendar. This permits us to give an "age" to a particular rock or to a geologic event of the past.

THE GEOLOGIC CALENDAR: RELATIVE TIME

The determination of the geologic calendar is a somewhat circuitous process. The time units, analogous to years, months, and so on, are worked out from a study of certain rock sequences. Then, the rocks are fitted into the calendar, which was based on them in the first place. It is somewhat like determining our present time system from events in man's history, rather than from the independent criteria of motions of the earth and

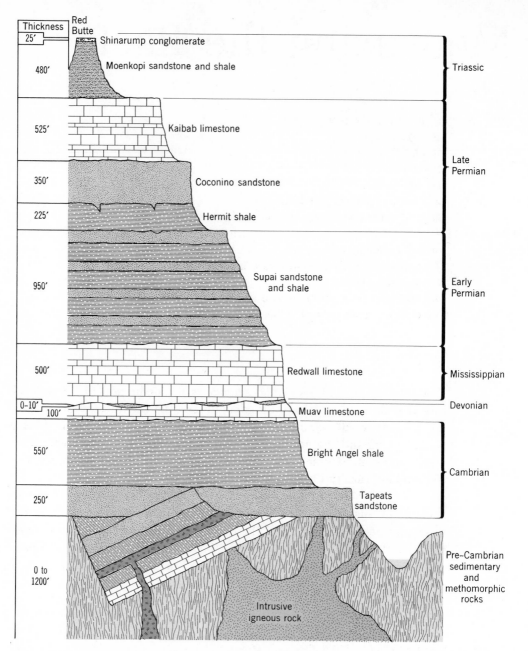

FIG. 7-1. Geologic section of the Grand Canyon of the Colorado River (Illustrated in Fig. 3-6). On the walls of the Grand Canyon are found rocks with an age spread of billions of years from bottom to top. The time range is from early Precambrian to Triassic. See Table 7-1.

FIG. 7-2. (Opposite) Conformable beds in normal (*a*) and overturned (*b*) positions as seen in two east-west vertical sections. Formation 1 is the oldest in each case. In (*b*) the broken lines show the reconstructed fold prior to erosion. After erosion the fold may not be obvious and if only the eastern half of the section were seen by the geologist, the reverse sequence of deposition might be inferred.

moon, and then assigning historical events to this man-based calendar. A very few basic principles and criteria have been used to erect the geologic time scale.

Superposition

The *law of superposition,* so fundamental to the history of the earth, is so simple a concept that one might almost pass over its importance. According to this law, sedimentary rock layers (strata) have been deposited on top of each other in order of decreasing age. As a consequence, the relative time relationship within a sequence of sedimentary strata is simply that each layer is younger than the one beneath. In the grand exposure of sedimentary rock in Figs. 7-1 and 7.2a, it seems obvious that the age of each stratum decreases upward. Rock layers are said to be *conformable* when they have been deposited without interruption.

In normal geologic practice, however, this simple law often becomes quite complicated. Rock deformation involved in mountain building may cause the rock layers to be overturned beyond 90° (Fig. 7-2b). In such cases, older rocks will lie above younger layers, and if much of the original structure has been eroded away, the overturning may not be very apparent. This situation must be realized and corrected by the geologist before any interpretation can be made.

Fossils

Fossils are the remains of former living animals and plants (fauna and flora). The study of ancient animal life has been the more important of the two in working out details of geologic time because animal fossils are more numerous and better preserved. And of the large and diversified animal kingdom, it is primarily the animals without backbones, the *invertebrates,* that are of particular importance. Examples of the invertebrate groups, both living and fossil, are illustrated in Fig. 7-3. Some of these forms will be quite familiar; others that lived in more remote times will be less so.

The importance of fossils in geology was first remarked by the English surveyor, William Smith, about the year 1800. Smith's classical observa-

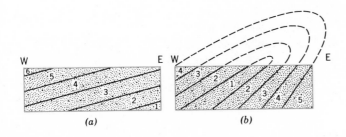

(a)　　　　(b)

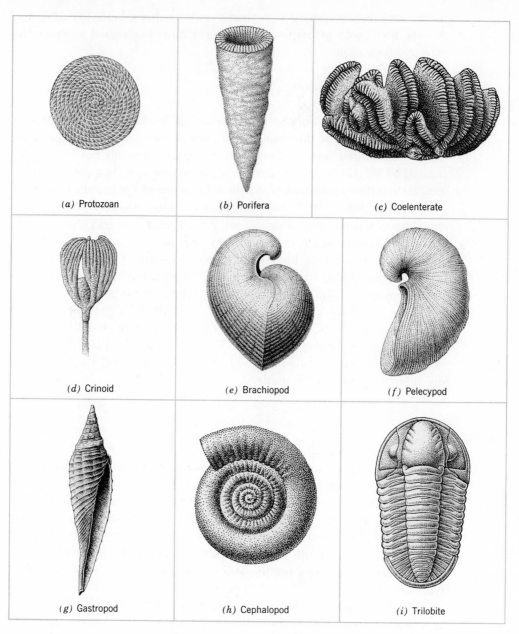

FIG. 7-3. Fossil invertebrates representative of the forms used in sedimentary rock correlation (a) protozoan or one-celled animal—about 10 x; (b) Porifera or sponge—the simplest of the multi-celled animals; (c) a coelenterate—or coral; (d) a crinoid or sea lily; (e) a brachiopod—one of the most important types of paleozoic index fossils; (f) a pelecypod, the group to which clams and oysters belong; (g) a gastropod—the snail group; (h) a cephalopod—the group that includes the octopus and squid; (i) a trilobite—an important paleozoic form that became extinct by the end of the Paleozoic Era.

tions are one of the great milestones that lifted geology into the realm of science. Smith was a surveyor responsible for the engineering of a canal system in England. He quickly observed the characteristic layering of sedimentary rocks in the canal cuts. He also recognized the presence of invertebrate fossils in many of the layers.

Smith soon became more interested in fossils and rocks than in surveying and realized that particular rock formations always contained the same types of fossils regardless of where they occurred. Soon he could use the fossils to match rock layers quite remote from each other and could also tell where particular formations lay in a series of other layers on the basis of fossil content, even if all of the units were not visible.

Correlation

In order to develop a geologic time scale applicable to the entire earth, it is necessary to establish which rocks were formed at the same time. *Correlation*, matching rocks of similar age that are geographically far apart, was made possible by the pioneering work of William Smith. Certain fossils, called *index* or *guide* fossils, have a widespread geographic occurrence but a very limited vertical or geologic range. Their presence provides a very positive way to match rocks of the same age. Rock formations may not always contain a specific index fossil, but often contain a particular group of fossils called a *faunal assemblage* that is distinctive for the formation. It is not possible to overstress the importance of fossils in unraveling the history of the earth.

Unconformities

Before we go on to the snythesis of the geologic time scale, we must examine this last of the major criteria used. Uninterrupted deposition of sediments leads, as we have seen, to a conformable sequence of rock layers that are structurally parallel to each other as in Figs. 7-1 and 7-2. But if normal marine deposition is interrupted by emergence of the sea floor followed by subsequent resubmergence, an omission of some vertical thickness of rocks will occur, compared to a place where continuous deposition took place. This is shown in Fig. 7-4, where the layers are numbered in accordance with age; bed 1 on the bottom is, of course, the oldest unit.

In Fig. 7-4a, a sequence with continuous deposition of six formations is shown. In b, part of bed 3 has been eroded away during a short interval of emergence. In c all of beds 4 and 5 are missing as the result of a longer period of emergence. Despite these gaps in the geologic record, the beds above and below the *disconformity*, as the surface showing the time break is called, are parallel to each other. Disconformities thus indicate crustal movements of minor severity.

Angular unconformities between rock units indicate severe crustal

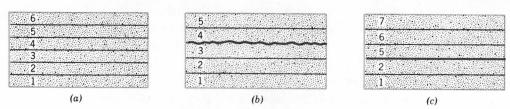

FIG. 7-4. Conformity and disconformity. In (a) six undisturbed, conformable beds are shown. In (b) part of bed 3 is missing as the result of temporary emergence and erosion, but beds are parallel above and below the disconformity (heavy line between beds 3 and 4). In (c) a stronger disconformity is present between beds 2 and 5 since beds 3 and 4 are missing from more prolonged emergence.

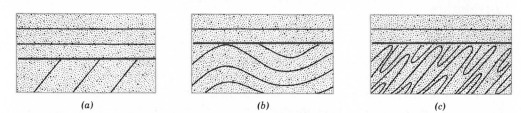

FIG. 7-5. Angular unconformities showing increasing degrees of deformation from (a) to (c), in the formations beneath the unconformity.

deformation as occurs in the mountain-building process. In Fig. 7-5a–c, the surfaces separating the deformed beds from the overlying horizontal beds are angular unconformities involving different intensities of deformation.

The relatively simple surfaces of unconformity really represent prolonged time gaps in the geologic record and permit a history such as the following to be deduced: The deformed rocks below the unconformities were originally horizontal marine sediments that were raised well above sea level in the deformation process. Perhaps huge mountains formed; subsequent erosion wore the region down to a near level surface. Submergence of this erosion surface beneath the sea then occurred, followed by renewed deposition of sedimentary layers.

Note that within both the undeformed and the deformed rock groups above and below the surface of angular unconformity, the layers are quite conformable to each other.

Creating the Geologic Calendar

Unconformities constitute the basic framework for the division of geologic time. Correlation has been the tool with which the calendar is made universally applicable.

Geologic time is divided into major and minor units. *Eras* are the grandest divisions of the time scale. They are divided into *periods,* which are in turn subdivided into *epochs.* These are broken into *stages.*

FIG. 7-6. Photograph of the classic angular unconformity south of Hudson, N.Y. between strongly folded and crumpled Ordovician shales and gently folded Devonian limestones. Silurian rocks are absent. This unconformity established the Taconian Disturbance that resulted from a major orogeny in late Ordovician time.

Eras are bounded in time by the most profound of crustal deformations, which often involved very widespread continental uplift or emergence. Vast mountain-forming movements (*revolutions*) are used to help delineate the ending of an era. All of this is recognized by widespread angular unconformities above strongly deformed, often metamorphosed rocks. Great evolutionary changes in living forms occurred at times of revolutions, so that eras are also characterized by certain dominant biologic groups. Periods are separated from each other by intervals of intense but less widespread activity known as *disturbances,* which also result in angular unconformities. The angular unconformity that represents the *Taconian disturbance* at the end of the Ordovician period (Table 7-1) is illustrated in Fig. 7-6. Time breaks involving less intense and less widespread crustal activity often of the type producing disconformities form the boundaries of epochs. Stages, one of the smallest of the time units, are often recognized by environmental changes not involving obvious crustal movements.

Before we turn to the actual time chart, it might be helpful to see how the scheme works by means of a simple example. In Fig. 7-7*a*, a single prominent angular unconformity is present, while in Fig. 7-7*b*, in a different part of the continent, two such unconformities exist. Assume that by a careful correlation with index fossils or faunal assemblages or both, it has been possible to match formations deposited at the same time. These have similar numbers, again with bed 1 being the oldest.

Although bed 5 is missing, presumably from erosion, in Fig. 7-7*a*, the

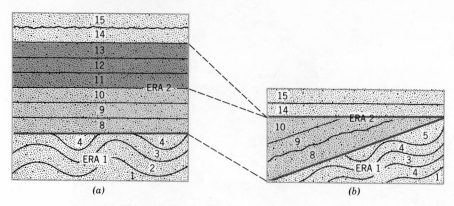

FIG. 7-7. The use of unconformities in the architecture of the geo-logic time scale. (See explanation in text.)

early groups in (a) and (b) have certainly been strongly folded in a crustal deformation that must have occurred at nearly the same time. We can tell this because the same bed, 8, lies on both angular unconformities. The widely distributed nature of this deformation would prompt us to call it a revolution and assign the formations below the unconformity to an earlier era (arbitrarily here, Era I) and those above, to the next era.

The angular unconformity below bed 14 in Fig. 7-8b is not found in 7-7a. Since the deformation responsible was not too widespread, we will call it a disturbance and can thus recognize two periods in Era II. Rock formations from 8 through 13 would belong to the earlier period and beds 14 through 16 would belong to the later one. Within each of the large time units, irregular contact surfaces of otherwise parallel beds would represent disconformities, providing the basis for epoch divisions. By painstaking application of this procedure to the rock record, the chart of earth history has been worked out in considerable but not yet complete detail.

Chart of Earth History: The Geologic Calendar

Table 7-1 shows the principle divisions of geologic time. The names of the eras are based on the general level of life present: Paleozoic means ancient life, Mesozoic, middle life, and Cenozoic, modern life. With this time scale is shown also some of the major geologic and biologic events of the different time divisions.

Two very important observations related to Table 7-1 must be made:

1. The time chart established by the procedure described is simply the relative scale of time. If a rock is identified by fossil content as belonging to the Ordovician period, its relative position in time is really stated—it is younger than Cambrian and older than Silurian rocks. (The absolute chronology in the last column was developed after 1960 by the methods described in the next main section. Prior to this, only relative ages were implied by the time chart.)

TABLE 7-1 The Geologic Calendar[a]

Era	Period		Geologic Events	Biologic Events	Age in Millions of Years
Cenozoic	Quaternary	Recent	Time since last glacial stage		0.0015
		Pleistocene	Time of glacial-interglacial stages	First man — Age	2
	Tertiary	Pliocene		of	10
		Miocene	Cascadian revolution (West Coast Ranges)	Mammals	25
		Oligocene			38
		Eocene			55
		Paleocene			65
Mesozoic	Cretaceous		Laramide revolution (Rocky Mountains)	Age — First mammals and Great flowering plants	135
	Jurassic		Nevadian disturbance (Sierra Nevadas)	of dinosaurs — First birds	185
	Triassic			Reptiles	225
Paleozoic	Permian		Appalachian revolution (Appalachian Mountains)	Extinction of enormous numbers of Paleozoic species	280
	Pennsylvanian		Glaciation	First reptiles	320
	Mississippian			Great coal-forming swamps	340
	Devonian		Acadian disturbance (New England)	Age of fish (First amphibians	405
	Silurian			Earliest insects and land plants	430
	Ordovician		Taconian disturbance (New England)	First vertebrates (fish)	500
	Cambrian			Beginning of good fossil record of plants and animals	600
Eras of precambrian time					
				Oldest primitive plant fossils	3,300
			Oldest dated rocks		3,600
			Formation of earth		5,000

[a]This is a table of earth history. Epoch dimensions are given only for the last two periods of geologic time. The method of determining the absolute ages in the last column is explained later in the chapter.

2. The vertical spacing on this time chart is in no way proportional to absolute age. This is especially important for the lowest major divisions referred to as Pre-Paleozoic (more commonly as Precambrian). We now know that this division corresponds to about $\frac{7}{8}$ of geologic time and the portion from Paleozoic on, to about $\frac{1}{8}$. The present state of completion of the geologic calendar is simply proportional to our degree of knowledge

of the past. Rocks of Precambrian age have been subjected to vast erosion in the formation of later sediments and when present are often covered by younger sedimentary rocks. Also, when exposed they are often so metamorphosed by later rock deformation as to be very difficult to interpret. And perhaps worst of all, they are essentially unfossiliferous, making their correlation very difficult.

One of the many overriding problems in earth history is the almost explosive rise of numerous and advanced forms of life in the very beginning of the Cambrian period. The "written record" of the rocks is said to begin at this time. What was happening in evolution prior to this time? Very slowly the Precambrian is giving up its secrets as more lands are explored by more geologists.

ABSOLUTE GEOLOGIC TIME

The assignment of absolute time to the events of earth history awaited the discovery of radioactivity by Henri Becquerel in the nineteenth century and the development of measurement techniques in the twentieth century.

Natural Radioactivity

In Chapter 4 we reviewed the characteristics of the normal or stable atomic nucleus, which basically contains a number of uncombined protons (positively charged) and a number of neutrons (no charge). Recall that the atomic number of an element is given by the number of protons present in the nucleus and the atomic weight by the combined protons and neutrons. A neutron is composed of a tight combination of a proton and an electron that neutralize each other. Since the electron is so light, it contributes little to the atomic weight. Some elements have nuclei with different numbers of neutrons, so that different atoms of the same element may have different atomic weights although their atomic number is constant. Such atoms are called *isotopes* of a given element.

The protons and neutrons of the nucleus are held together by very strong forces. Although the magnitude of the energy involved is known, the actual nature of the force is still not really understood. In the case of the atomic nuclei of some of the elements, the available energy cannot keep the nuclear particles adequately "cemented" together and a natural disintegration, called *radioactivity*, occurs. The disintegration process varies somewhat among the different radioactive elements.

In some forms of radioactivity, the nucleus ejects a very-high-velocity *alpha particle*, which consists of two protons and two neutrons bound tightly together (Fig. 7-8a). The loss of two protons lowers the atomic

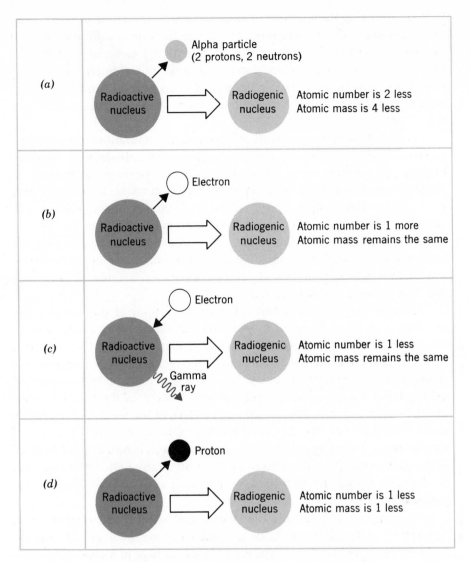

FIG. 7-8. Four of the common nuclear reactions in radioactivity showing the relationship of the original radioactive nucleus and the resulting radiogenic nucleus.

number by 2 and in this way actually leaves the nucleus of a different element behind. The atomic weight of the original nucleus (element) is decreased by 4 from the loss of the four particles. In another form of decay, a *beta particle,* or electron, is emitted (Fig. 7-8*b*). Since the nucleus has no free electrons, this particle must come from a neutron. Loss of the electron leaves an additional proton in the nucleus, thereby raising the atomic number of the newly formed element nucleus by 1. Instead of emitting an electron, some nuclei exhibit radioactivity by capturing an electron from their own innermost electron shell (Fig. 7-8*c*). The captured electron then combines with a nuclear proton to form a neutron, thereby reducing the number of free protons in the nucleus by one and hence reducing the atomic number by one to form a different element nucleus. To rid itself of the extra energy of the electron capture, the nucleus emits a very penetrating short-wave radiation known as a *gamma ray*. A decrease in the atomic number to form a new element also occurs by a fourth decay process in which a proton is ejected from the nucleus (Fig. 7-8d).

Radioactivity always results in a natural transmutation of the original radioactive or *parent* element into a newly formed radiogenic or *daughter* element. In the process of either gamma-ray or particle ejection at extremely high speed from radioactive materials in rocks, the surrounding mineral material is often altered as shown by the radioactive *halos* surrounding small crystals of the mineral *zircon* in Fig. 7-9. It is the radioactive element uranium in the zircon that has decayed to damage the enclosing mica grain. It was once hoped that the diameter of a halo might indicate the age of the altered crystal, but too many variables are involved and this attempt at dating failed.

Radioactivity has provided a beautiful method of determining the absolute ages of rock, mineral, and fossil materials. It does so because the *fraction* of radioactive nuclei that will decay in a unit of time is a constant, known as the *decay constant*. Thus, for a particular radioactive substance, the amount of decay is greater for a large initial mass of material than for a small mass, but the same *fraction* of the initial number of atoms will decay in a given time regardless of the mass. For example, if we start with one pound of radioactive material, half will decay in a certain time; if we began with only an ounce, a half ounce would decay in the same time. The time it takes for one half of any radioactive material to decay is called the *half-life* and is also a constant for a particular substance.

It is apparent from this discussion that careful measurements of radioactive materials can indicate the length of time involved in the decay because the same fraction of material always decays in a given time. The particular application to different radioactive elements varies with the material. Let us therefore examine some examples of important "nuclear clocks" and see how the process works.

The Radiocarbon (Carbon-14) Clock:
A Decay Clock

Radiocarbon or C^{14} (carbon-14) was the first radioactive element capable of use in dating events of the historic past and the recent geological past back to about 35,000 years ago. So important has this use become to scientists and historians that the American chemist W. F. Libby received the Nobel prize for demonstrating this application of the element in 1947. Most of the radioactive elements used for age determinations before this decayed so slowly (with half-lives from millions to billions of years) that at least a million years had to elapse before a detectable change in the

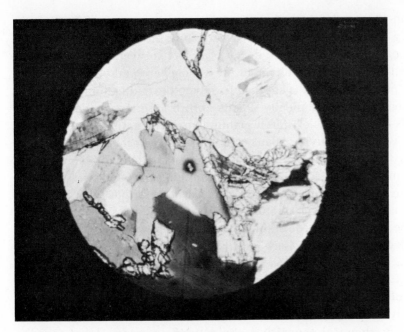

FIG. 7-9. Photomicrograph x 80, of an alteration halo about a small crystal of the mineral *zircon* embedded in a larger grain of mica in center of field. Uranium within the zircon is the source of the radioactivity. The rock, the Manhattan Schist of Ordovician age, is one of the three foundation rocks of New York City.

radioactive material occurred. But carbon-14 has a half-life of 5730 years permitting age determinations to be made readily on objects about a thousand years old. Materials older than 35,000 years have lost so much of the carbon-14 by normal decay that too little remains for accurate age measurements.

Here's how the carbon clock works. Cosmic rays, which are high-velocity nuclear particles, bombard the earth's upper atmosphere continuously. Nuclei of gases in the atmosphere are shattered from collisions with the very-high-energy cosmic particles and release a large number of neutrons into the upper atmosphere. Most of these neutrons are absorbed by nitrogen, which has an atomic weight of 14 (N^{14}). In the absorption process a free proton is ejected by the nitrogen nucleus, which is thus transformed into a new element—carbon-14. Carbon-14 (C^{14}) has an unstable nucleus that disintegrates by emitting a beta ray (electron) from one of the neutrons in its nucleus. The latter thus regains the proton it lost when the neutron slammed into it and forms a nitrogen-14 nucleus again (Fig. 7-10).

The common variety of carbon has an atomic weight of 12. Carbon-12 (C^{12}), the stable isotope of carbon, is the principal carbon component of the carbon dioxide in the oceans and atmosphere and of the carbon in all of the existing organic world. Prior to the period of nuclear testing in the atmosphere in the 1950s and 1960s, the ratio of atoms of carbon-14 to carbon-12 was about 1 to 1 million. Although it increased significantly following the test explosions, for a very long time earlier, the rates of production and of radioactive decay of carbon-14 were equal, resulting in a constant ratio of C^{14} to C^{12} in the atmosphere and oceans.

The carbon-14 clock is applicable only to the dating of materials that have been derived organically. Plants absorb carbon dioxide in photosynthesis—the process of starch manufacture. The carbon in the carbon dioxide within the living plant will have the constant C^{14}/C^{12} ratio. Animals that eat living vegetation or that eat other animals that ate vegetation will all have the same ratio in their bones and tissues. Shells of sea life are often composed of calcium carbonate ($CaCO_3$). The carbon in the seashells of living creatures will have the constant C^{14}/C^{12} ratio.

Once any organism dies, it stops acquiring new carbon dioxide and the clock goes into action. Carbon-14, decaying radioactively, is no longer replaced, so that the ratio C^{14}/C^{12} gets smaller and smaller. After the elapse of about seven half-lives (7×5730 years,) only a fraction of a percent of the original carbon-14 is left—an amount too small to measure accurately. In practice, after the carbon content of a substance has been isolated by chemical separation, the rate of emission of beta rays is measured in a radiation counter. The ratio, C^{14}/C^{12}, then depends on the number of emissions counted for the particular quantity of carbon used. Because carbon-14 is applicable to historic and archeological samples as well as to the very recent geological past, it has been used in almost spectacular age determinations. The Dead Sea Scrolls as well as ancient pottery and antiques have been dated with considerable success. Important details of the advance and retreat of glaciers during the last great Ice Age, which ended only 18,000 years ago, have been revealed with great clarity. Without this chronology of glacial and late glacial events, critical knowledge on which to base ideas of Ice Age origin would be seriously lacking.

In the case of radiocarbon, the age is determined by measuring the amount of decay of the original radioactive element so that it is often known as a *decay clock*.

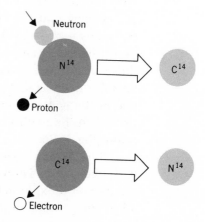

FIG. 7-10. Reactions in the formation and radioactive decay of carbon-14.

Accumulation Clocks

For the determination of ages from about a million to billions of years, a number of radioactive elements are used that work differently from the carbon-14 decay type of clock. In the accumulation procedure we must determine the ratio of original radioactive parent to the radiogenic daughter material that has accumulated since a particular mineral crystallized. Suppose, for example, that a certain amount of radioactive material becomes incorporated within a mineral crystal at the time of its formation. It will then begin to decay into a daughter product, and both parent and daughter may remain trapped in the crystal. If neither is lost, we can determine the time of crystallization by measuring the amounts of the two elements and by determining the rate at which the reaction occurs (the decay constant).

Let us see how the accumulation clock works in the case of the radioactive isotope of uranium, uranium-238 (U^{238}), which decays to form lead-206 (Pb^{206}). The decay constant for uranium-238 is $1/7.5$ billion, which means that one unit mass of uranium-238, say a gram, will decay to form $1/7.5$ billion ($1/7.5 \times 10^9$) grams of lead-206 in one year. Of course we cannot now measure such small quantities of material, so that a considerable time lapse is necessary to build up a measurable amount of lead in this procedure. To calculate the age of a mineral and of the rock containing it, let us review the following logic based on the above decay constant:

In one year, $1/7.5 \times 10^9$ g of Pb form from 1 g of U, or $Pb = 1/7.5 \times 10^9$ U, giving the ratio:

$$\frac{Pb}{U} = \frac{1}{7.5 \times 10^9}$$

In 2 years, $2 \times 1/7.5 \times 10^9$ g of Pb form from 1 g of U, or

$$Pb = \frac{2}{7.5 \times 10^9} U$$

giving the ratio of

$$\frac{Pb}{U} = \frac{2}{7.5 \times 10^9}$$

In 3 years, $3 \times 1/7.5 \times 10^9$ g of Pb form from 1 g of U, or

$$Pb = \frac{3}{7.5 \times 10^9} U$$

giving the ratio:

$$\frac{Pb}{U} = \frac{3}{7.5 \times 10^9}$$

In X years, $X \times 1/7.5 \times 10^9$ of Pb form from 1 g of U, or

$$Pb = \frac{x}{7.5 \times 10^9} U$$

giving the ratio:

$$\frac{Pb}{U} = \frac{x}{7.5 \times 10^9}$$

From the last ratio of Pb to U after X years, we can solve for X, the mineral age, as: $X_{years} = Pb/U \times 7.5 \times 10^9$.

To apply this to a hypothetical case, assume that in chemical analysis the ratio of Pb^{206} to U^{238} in a mineral crystal was observed to be 1:5, or

$$\frac{Pb^{206}}{U^{238}} = \frac{1}{150}$$

Then the time of formation of this crystal is found as

$$X_{years} = \frac{1}{150} \times 7.5 \times 10^9$$

$$= 50 \times 10^7 \text{ or } 50 \text{ million years}$$

This simple formula is good for U^{238}-Pb-206 decay during an interval of about 100 million years. Beyond this age, a more accurate and general formula must be used. So far, we overlooked the fact that a little less uranium is left each year as $1/7.5 + 10^9$ parts decay. Although unimportant for short time, this must be accounted for in the long run. The general formula for any accumulation type clock is:

$$Age = \frac{1}{C} \times 2.3 \log\left(1 + \frac{D}{P}\right) \text{ where}$$

C is the decay constant ($1/7.5 + 10^9$ for U^{238}), D the radiogenic daughter nucleus and P the radioactive parent nucleus.

For U^{238}, the formula becomes:

$$Age = 7.5 \times 10^9 \times 2.3 \log\left(1 + \frac{Pb^{206}}{U^{238}}\right).$$

If the ratio, Pb^{206}/U^{238} is 1/5 or 0.2 then the age is found by

$$Age = 7.5 + 10^9 \times 2.3 \log(1 + 0.2).$$

Standard tables of logarithms show that the log of 1.2 is 0.0792, hence

$$Age = 7.5 \times 10^9 \times 2.3 \times 0.0792$$

$$= 1.35 \times 10^9 \text{ years.}$$

This kind of calculation is also applicable to other accumulation types of radioactivity clocks. Of course, this procedure requires that none of the parent or daughter elements have been lost from the crystal, or an improper ratio would be obtained. It also requires that the mineral not be so old that all of the parent have vanished or has decayed too much to be measurable.

So slow is the rate of decay of U^{238} and the accompanying accumulation of Pb^{206} that the half-life of the process is 4.5×10^9 years (4.5 billion years). Because of this very slow decay, the U^{238}-Pb^{206} ratio has been extremely useful in measuring the ages of rocks of very great antiquity. It should be noted that U^{238} does not decay directly into Pb^{206}. There are a number of intermediate stages in which the resulting daughters are also radioactive. Lead-206 is the end product beginning with uranium-238.

The three other accumulation type reactions of particular importance in rock dating are uranium-235 (U^{235}), potassium-40 (K^{40}), and rubidium-87 (Rb^{87}). U^{235}, another radioactive isotope of uranium, also has a very long half-life and decays finally to Pb^{207} through a long series involving other radioactive daughter products.

Potassium-40 is a radioactive isotope of potassium (half-life, 91.3 billion years) that decays into the "noble" gas argon-40 (Ar^{40}). Since argon-40 is a gas, particular care must be taken to determine whether any has been lost from the crystal containing both the K^{40} and the Ar^{40}. Despite their well-developed cleavage planes along which gas molecules may be expected to escape, the micas have been found to be the best source of K-Ar for age determinations. K^{40}/Ar^{40} has made possible measurements on rocks whose ages vary from one million to billions of years for igneous and metamorphic rocks. The K-Ar method has become quite important because potassium is much more common than uranium and also because the method can be used over a greater time range. The K-Ar method has been used to date one of the oldest crustal rocks so far discovered. A K-Ar age of 3.6 billion years has been determined for a granitic rock near Lake Victoria in east-central Africa.

K-Ar ages are always considered to be minimum possible ages because of the possible loss of gaseous argon from ancient minerals. Remember, the less of the daughter element present, the younger the rock.

Rubidium-87, which decays into strontium-87, is one of the most recent of the accumulation clocks to be used with success in dating rocks of great age. This is particularly true because of its long half-life of 50 billion years. Like the K-Ar method, this has also become important because rubidium is very widely distributed in feldspars and micas, although in small amounts, so that a greater variety of rocks have become available for age determination than by methods involving lead. As of this writing, the oldest known rock is a granitic gneiss from the southwest coast of Greenland that has been dated by the Rb-Sr method at 3.98 billion years. This is close to the age of the oldest lunar rock of 4.15 billion years (returned by the crew of Apollo 15) although lunar soils, as referred to in Chap. 1, have ages to 4.6 billion years.

The Lead-Lead (Lead Ratio) Method
and the Age of the Earth

The question, How old is the earth?, has interested and fascinated scholars for ages. Ideas of age have ranged from an estimate of 6000 years on the basis of interpretations of the Old Testament to nearly five billion years on the basis of modern measurements of radioactivity. Between these two time limits, the earth's age increased progressively the more the subject was studied. With the emergence of the new science of geology in the early nineteenth century, the belief arose that a great age must be assigned to the earth to allow time for the slow accumulation of sedimentary rocks. This belief was reinforced by Darwin's *Theory of Evolution,* which clearly involved a long time span for the development of life as it now exists on the earth.

Three different numerical calculations of the earth's age were made just before and just after the turn of this century. Because they independently indicated an age of tens to a hundred million years, the latter value became difficult to dislodge from the records of science. The first of those three calculations was the work of the eminent British physicist, Lord Kelvin. On the assumption that the earth was initially a hot offspring of the sun, Kelvin calculated that cooling to the present conditions would require about 20 to 40 million years.

Shortly after this, Joly, an Irish geologist, determined an age of 90 million years. He arrived at this value as follows: The chemicals in the ocean are carried in by the rivers of the world after they have been dissolved from the rocks of the continents. The quantity of sodium (mostly as common salt or sodium chloride) in the oceans was very well known, as was the annual rate at which it is brought in by rivers. Hence the total quantity of sodium in the oceans divided by its rate of increase from streams gave 90 million years, taken as the total lifetime of the oceans and of the earth.

A very similar procedure was used with sedimentary rocks. The total thickness of sedimentary rocks deposited continuously was divided by the annual rate at which rivers are now transporting sediments to the sea. An age similar to the two previous values was obtained.

We know now that these three procedures had built-in errors. Kelvin's idea of the cooling of the earth was inaccurate and led to an earth much too young. Both of the other procedures erred in assuming that the rates of salt and sediment accumulation in the oceans were constant. These rates depend on rates of erosion of the lands, which are probably higher today than they were during most of the geologic past. The reason for this is that there is probably more land higher above sea level at present than in the past, and rates of erosion are clearly higher when there is more land to erode and when elevations are higher. And as regards sediments, it is difficult to include in the total thickness the vast quantities of Precambrian rocks that have been eroded, metamorphized, or covered by later rocks. Finally, the age of the oceans may be less than the age of the earth.

Until about 1930 the age of 100 million years was hard to shake. But radioactivity measurements showing great rock antiquity finally did it. In the midtwentieth century came the first good calculation of the radiometric age of the earth. It was difficult to do this with the procedures described above because it is hardly likely that the oldest rock would ever be found. The procedure is thus somewhat subtle and involves rocks from beyond the earth (meteorites) as well as measurements of lead-lead rates.

This cryptic procedure involves the ratios of the different isotopes of lead, namely, lead-204 (Pb^{204}), lead-206 (Pb^{206}), lead-207 (Pb^{207}), and lead-208 (Pb^{208}). The last three are radiogenic isotopes of lead, that is, they have formed as daughter products in the reactions $U^{238} \rightarrow Pb^{206}$, $U^{235} \rightarrow Pb^{207}$, and $Th^{232} \rightarrow Pb^{208}$. (The last involves thorium-232, a radioactive element that we have not referred to previously). Thus, the lead in the world today, called *common lead*, is composed of four isotopes. Lead-204 is nonradiogenic: it has existed in the same amount since the earth was formed.

Observations show that the younger the lead the greater is the proportion of the radiogenic lead to lead-204 and the older the common lead the less is the proportion of radiogenic lead to lead-204. Presumably then, if we go back far enough in time to the formation of the earth's crust, we can reach an age when the ratio of radiogenic lead to lead-204 would have had a minimum value. This would have been the time that the earth's gross structure formed. We know the present ratios of leads-206, 207, 208 to lead-204 and we know the rates at which each of the radiogenic leads formed from the radioactive parent. If we could determine the initial amounts of radiogenic leads we would know the initial ratio of radiogenic to nonradiogenic lead (Pb^{204}) and we could use the rates of formation and work backward in time until the ratios of radiogenic lead to lead-204 reached the initial ratio. The time required would be the age of the earth. This is where space rocks come in, because they indicate this primeval ratio.

Meteorites are stony and iron-stony rocks that occasionally bombard the earth's surface from interplanetary space. They are believed to be fragments of a disrupted planet or planets. According to the theory of earth origin reviewed at the end of Chapter 1, all of the planets must have formed at about the same time. Certain types of meteorites were found on analysis to contain a very low ratio of uranium to lead isotopes present. Since the uranium content is so low, it could not have contributed a significant amount of radiogenic lead to the combination of lead isotopes in the meteorites. Hence the present meteoric lead-isotope ratios probably equal the ratios present when the original planet formed. Because the earth probably formed at about the same time and from much the same material, we can assume that it had initially the same lead-isotope composition as the meteorites still have. But the earth accumulated a large amount of lead after formation from the uranium not present in the meteoritic source.

By working backward from the present lead-isotope ratios to the time when the ratios were equivalent to that in meteorites, an age of from 4.55 to 4.75 billion years has been obtained. If we go further and assume that it took some small fraction of a billion years for the formation of the earth's crust in which the present lead isotopes are found, the age of the earth as an astronomic rather than a geologic body can be estimated in round numbers to be 5 billion years.

Surprisingly, the epic flight of Apollo 11 and its intrepid crew to the moon in July 1969 greatly increased the importance of these age determinations on earth rocks. The Apollo 11 expedition was able to carry back a very scant sample of lunar rocks and dust. Despite its small size, the rocks collected in the immediate vicinity of Tranquility Base were dated at 3.6 to 3.7 billion years! And lunar dust and one rock sample from the Apollo 12 landing turned out to be 4.6 billion years old! The similarity between lunar sample ages and those measured and deduced from the oldest rocks on earth are of great significance in interpreting the history of the earth and solar system.

Despite all of the available rocks over the earth, only a handful of samples are close to the age of the lunar-rock samples and no terrestrial material comes close to the age of lunar dust. Why is there this difference in the age of earth and moon material? The lunar samples apparently represent primeval material that has lain unaltered since the time of formation. But as we shall see in later chapters, present rocks on the earth's surface have suffered many cycles of erosional destruction and regeneration resulting from the presence of the earth's atmosphere and oceans and the unstable nature of the earth's crust and interior. We know the moon has no atmosphere and surface water at present. Its rock ages tell us that it probably had none at any time. Future Apollo landings on the moon will convey further information related to the activity within the moon and the history of the earth-moon system.

The Absolute Geological Calendar

The dating of rocks mostly by the methods just described has enabled us to give absolute ages in years to many divisions of the scale of earth history. This dating is shown in the last column of Table 7-1.

Further applications and uses of this and other age data will be considered in following chapters.

STUDY QUESTIONS

7.1. How would you apply the law of superposition to sedimentary rocks having a vertical attitude (dip of 90° from the horizontal) as the result of deformation?

7.2. Distinguish between "conformable" and "unconformal" beds.

7.3. Explain how index fossils and faunal assemblages are used in rock correlation.

7.4. Draw a vertical crustal section of a region that had the following geological history.
(a) Deposition of six conformable sedimentary layers.
(b) Strong folding.
(c) Erosion to a flat surface.
(d) Submergence beneath the sea and deposition of six additional sedimentary layers.
(e) Tilting of the entire region followed by erosion to a flat surface.
(f) Deposition of four new sedimentary layers.
(g) Emphasize with heavy lines the angular unconformities present.

7.5. Explain briefly how the relative geologic calendar was created.

7.6. What is meant by "absolute geologic time"?

7.7. Explain how the loss of an electron from a nucleus increases the atomic number of that element.

7.8. Distinguish between atomic number and atomic weight.

7.9. What is an isotope? Give several examples.

7.10. Describe the nuclear processes involved in radioactivity.

7.11. Explain the "decay constant."

7.12. To what materials can the radiocarbon clock be applied? How does the "clock" work?

7.13. What are the requirements for measuring age by accumulation clocks?

7.14. Explain the accumulation clock in the case of lead (Pb) and uranium (U); which isotopes of each are involved in your explanation?

7.15. If the ratio of Pb^{206} to U^{238} found in a mineral is $\frac{1}{10}$, what is the age of the mineral?

7.16. How have the Apollo flights to the moon contributed to our understanding of the age of the earth?

chapter 8

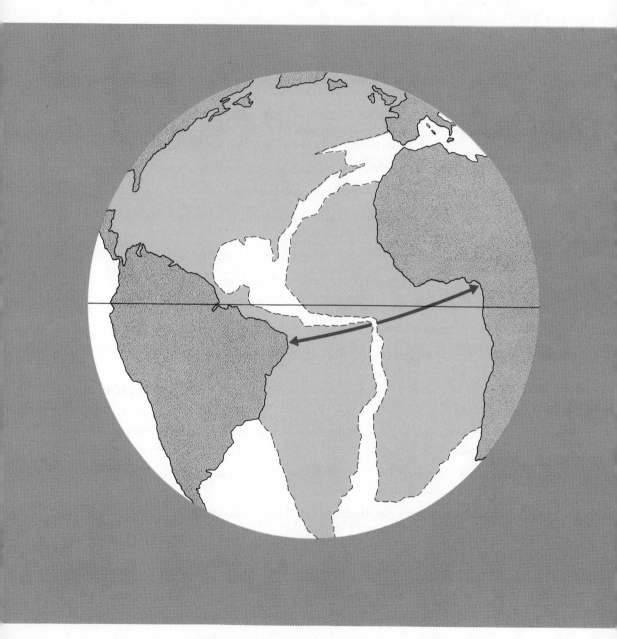

Separation of the continents across the Atlantic Ocean during the past
150,000,000 years (since the Jurassic Period).

Continental Drift and Sea-Floor Spreading

The study of the earth through geophysics and geochemistry expanded explosively in the middle of this century. Nothing less than a revolution in geologic thought occurred within a period of a few years in the attempt to explain some remarkable observations of the magnetism of the rocks composing the sea floor. In the explanation of these observations, very diverse information from other areas of geology, geophysics, and geochemistry was drawn together and synthesized into a single theory of earth behavior. The result not only explained the new marine magnetic observations but also provided insight into many other problems, both old and new. Chief among them is the classical concept of continental drift. Closely related are newer questions concerning the scarcity of sediment on the ocean floor, the apparent recency in age of the sea floor, the presence of the great marine ridge-rift systems with their unusual high rate of heat flow, and the origin of the deformational forces responsible for the great mountain ranges.

CONTINENTAL DRIFT

Wegener's Displacement Theory

Early in the twentieth century, Alfred Wegener, a German meteorologist, noticed a rather remarkable jigsaw-puzzle-like relationship between the west coast of Africa and the east coast of South America. It seemed to him that the two could have been a single continent that split apart and separated. Further thinking led Wegener to suppose that all of the continents were once joined as a supercontinent that he called *Pangaea*.* He imagined a sequence of events somewhat as shown in the series of map reconstructions (Fig. 8-1) taken from his monumental work, *The Origin of Continents and Oceans*, published first in German in 1915 and later (1924) in an English translation.

An immediate consequence of such reasoning is that the Atlantic Ocean basin can be no older than the time of separation of the Americas from Europe and Africa. Had Wegener known of the mid-Atlantic ridge (described in Chapter 3), whose trend matches so well the shapes of the coastlines of the continents on opposite sides of the Atlantic, he would have had further compelling arguments for his hypothesis. The fact that the basin of the North and South Atlantic Ocean system is broadest in the southern part led to a further conclusion that the split occurred between South America and Africa before North America and Europe separated.

Direct observation of changes in latitude and longitude that might prove or disprove drift are not easily made. Such changes would be so slight that more than a lifetime would be required to detect them. Because of this, the arguments in support of continental drift were necessarily indirect or "circumstantial" in character.

Evidences in Support of Drift

Many of the scientific observations supporting the drift hypothesis were well known in Wegener's time. Alternative explanations of these observations were offered that weakened the initial arguments of Wegener and delayed acceptance of the idea of drift. A review of some of the early arguments that have since regained much force is helpful in grasping this most dynamic geological concept.

1. A number of biologic forms found in rocks on opposite sides of the Atlantic Ocean once belonged to the same species; for example, certain forms of earthworms and land snails of the past were quite the same at the same latitudes of the old and new worlds. Once the split occurred, each pursued independent evolution thereafter. Also, the forms found on opposite sides of the South Atlantic began to differ at an earlier

---

*Gondwanaland is the name often applied to the initial continental mass in the southern hemisphere. It includes India, Australia, Antarctica, and parts of Africa and South America.

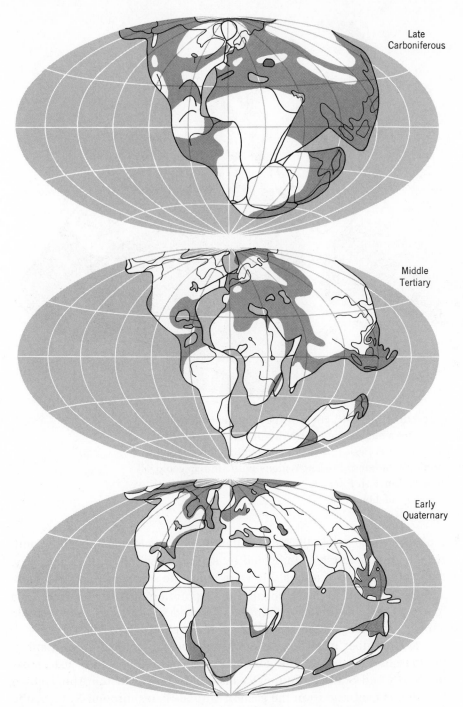

Late
Carboniferous

Middle
Tertiary

Early
Quaternary

FIG. 8-1. Wegener's description of continental drift by means of re-
constructions of the continents at three stages in the drift process.
The first stage shows the hypothetical predrift parent continent
(*Pangea*). Continental areas covered by shallow seas are shaded.
(A. Wegener, *The Origin of Continents and Oceans*, E. P. Dutton &
Co. N.Y. and Methuen & Co. London.)

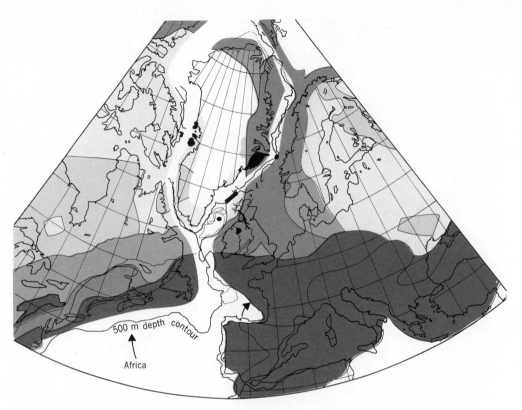

FIG. 8-2. This predrift reconstruction of the continents across the North Atlantic Ocean shows the continuity of similar geological provinces from North America to Europe, including Greenland. Areas with similar shading represent regions with similar geologic age and history. (After F. J. Fitch, *A Symposium on Continental Drift,* Royal Society of London.)

time than those across the North Atlantic. This could mean that the South Atlantic opened first, as indicated by its greater width. In the same way, many other forms of animal life were much the same for Australia, India and Ceylon prior to the Cenozoic Era. Separation of Australia from Asia in early Cenozoic could explain the observed similarity of pre-Cenozoic animal life. Many other examples could be cited.

2. A number of strikingly similar geological features are present in the marginal regions of the continents that face each other just across the North and South Atlantic Oceans. Folded rocks of the Appalachian Mountains continue from the eastern United States through Nova Scotia and southeast Newfoundland and apparently terminate at the ocean. Across the Atlantic, similarly folded rocks of the same age extend to the coasts of Ireland and Brittany and would apparently join those in North America if the continents were adjacent. Geologic similarities among the North Atlantic continents are shown in Fig. 8-2, with the continents reconstructed in assumed pre-drift positions.

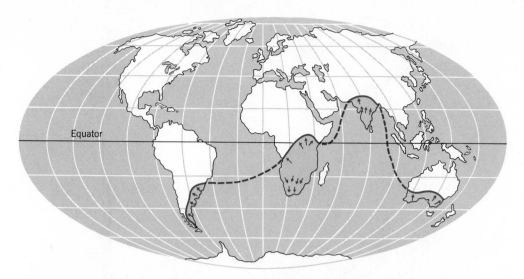

FIG. 8-3. The distribution of glaciation in Permo-Carboniferous time, showing its restriction to the Southern Hemisphere except for India. But the latter would have been joined originally to east Africa according to the Wegener reconstruction. Note that in South America and Australia the ice movement, shown by arrows, would have had to come from the oceans according to present geography. (A. Holmes, Principles of Physical Geology, Ronald Press.)

In the Southern Hemisphere folded rocks of Permian Age trend east-west and reach the coasts of both Argentina and South Africa. Other, more ancient metamorphic rocks of Brazil and South Africa also show great similarity. On the other hand, folded rocks of the lower Cenozoic (Oligocene) Age in the Atlas Mountains of North Africa have no counterpart in South America. This would be expected according to Wegener because the separation would have occurred long before, in the Cretaceous Period of the Mesozoic Era. Many other compelling geological arguments similar to these were offered by Wegener.

3. One of the strongest indirect evidences for continental drift, at least in the Southern Hemisphere, comes from the distribution of known glaciation that occurred in the middle and late part of the Paleozoic Era. Remains of this *Permo-Carboniferous* glaciation are found on Antarctica, South America, Australia, New Zealand, India, and Africa (from the southern tip to the equator), as shown in Fig. 8-3. Now a glance at a world map shows that most of these lands, except Antarctica, lie in or close to the tropics. Very pertinent to the argument is the strange fact that Paleozoic glaciation has never been found in the Northern Hemisphere, where so much of the land is in the high, and therefore cold, latitudes. Since all of the more recent large-scale glaciation is a feature of the higher latitudes, the pointed question arises, "How could the widely separated low-latitude lands of the Southern Hemisphere have possibly been glaciated in their present locations, without even greater glaciation occurring in the Northern Hemisphere?"

FIG. 8-4. Map of the Southern Hemisphere centered on the south pole showing the predrift reconstruction of the southern continents around the pole. The shaded region of Permo-Carboniferous glaciation can then be explained as a high latitude-polar-cooling effect. (After K. M. Creer and A. L. du Toit.)

Wegener concluded that not only were all of the southern continents joined rather closely at this time, but they were also at a higher latitude than at present, as in Fig. 8-4. And although Wegener didn't know it at the time, later observations further strengthened his argument related to the distribution of glaciers. It is possible to determine the direction of movement of glaciers by a careful observation of the grooves and striae they leave on the surfaces of bedrock scoured by the ice. Observations in South America, South Africa, and Antarctica indicate that the great Paleozoic ice sheets moved in from what is now the ocean, a situation quite impossible unless large land areas were present instead of the sea.

4. The rather remarkable fit of the continents that initially led Wegener to his controversial hypothesis came under question towards the middle of this century. It was argued that the match between coastlines on opposite sides of the Atlantic Ocean could be a coincidence of the present level of the seas. If sea level were to rise by a rather small amount, a vast change in the configuration of many continents would occur. Perhaps a similar change would follow a drop in sea level, so that the present fit is a fortuitous one arising from the present position of the sea level.

It became quite clear from the increased geophysical study of the sea floor and continental margins following World War II that the real margins of the continents are seaward from the present coastlines, and that the oceans overlap onto the borders of the continents. What would the fit look like if the real margins were used? Sir Edward Bullard, in England, developed a mathematical procedure to test the fit of the trans-Atlantic continents with the aid of modern high-speed computers and discovered that the best match occurred not for the present coastlines, but at the 500-fathom (3000-foot) depth line, the real margin of the continents. An example of his continent matching at this depth is shown in Fig. 8-5, which seemed to give renewed support to the Wegenerian hypothesis.

Then, to lend support to Bullard's reconstruction of the predrift continents, came the results of P. M. Hurley (of the Massachusetts Institute of Technology) and his colleagues; they showed a striking correlation of very ancient rocks on adjacent coasts of Africa and South America (Fig. 8-6). This matching of ages plus the similarity in the structural trends of the rocks on the opposing coasts is further indirect evidence that the two continents were once united.

The Objections from Geophysics

At about midcentury, geophysics was emerging as a very significant brance of earth science and geophysicists argued that it was simply impossible for the continents to drift through the strong, rigid rocks of the oceanic crust. This was a very powerful objection, especially since supporters of drift had never been able to offer a reasonable mechanism for the drift process. The absence of a known drift mechanism has certainly been one of the strongest arguments of opponents of the idea.

But barely had the arguments of geophysicists sent the supporters of drift reeling back, than new evidence from geophysics itself not only provided renewed support for drift but also led to a possible mechanism. Since the newest of the supporting data and arguments come from the recently emerged study of rock magnetism, let us digress a bit in order to examine the procedure and then look at some of the critical results.

ROCK MAGNETISM

The earth is a gigantic magnet. It is this fortunate property that controls the behavior of the compass needle, which has in turn made possible the global explorations and routine travel of the past few centuries. Navigation without a compass must have been difficult indeed.

The Earth's Magnetic Field

The earth's magnetic field emanates primarily from the earth's molten metallic core, as shown by the view of the magnetic field revealed through the lines of magnetic force (Fig. 8-7). The particular position and orientation of the magnetic field is a consequence of the earth's rotation and motions within the molten core and may be likened to the field that would develop from the spinning of a bar magnet in the center of the earth. Note

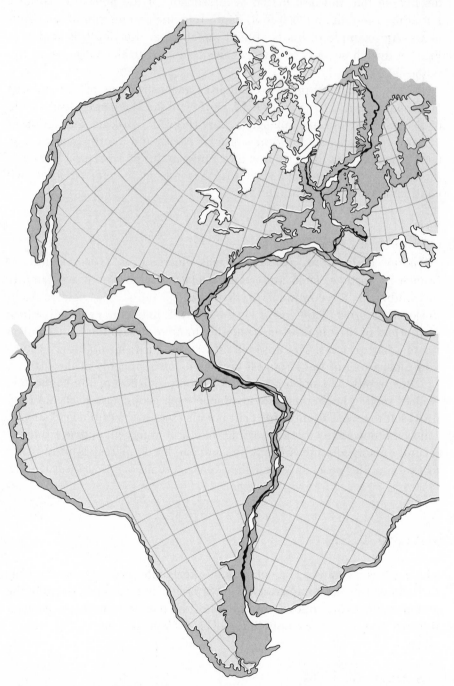

FIG. 8-5. The matching of the continental shapes bordering the Atlantic at the 500 fathom depth. (Redrawn after E. Bullard.)

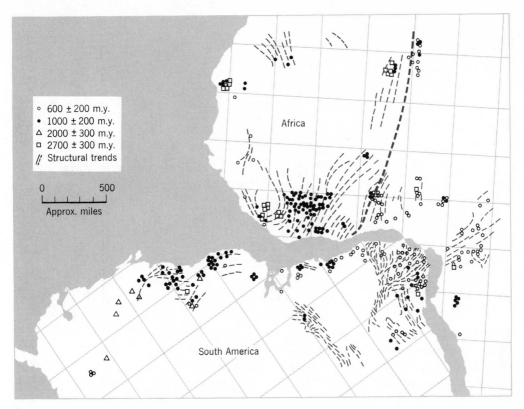

FIG. 8-6. The similarity in geologic age provinces in Africa and South America suggesting both continents were joined as shown. Similar trends in geologic structures can also be traced from one continent to the other. (Redrawn after P. Hurley, *et al* in *Science*.)

that the magnetic-field lines are horizontal at the magnetic equator and vertical at the magnetic poles. Between the poles and the equator the lines of force cut the earth's surface at an angle whose acuteness depends on the latitude.

Source of Rock Magnetism

The magnetic lines of force extending through the rocks of the earth's crust often become preserved within rocks that have iron-bearing minerals present. Lava flows are especially important for such magnetic-field preservation. It is well known that a magnetic substance loses its magnetism when heated above a certain temperature. Conversely, when the same magnetic substance is cooled below this critical temperature (the Curie point), it again acquires the magnetic field of the region. If a sediment contains magnetically sensitive mineral grains, these will rotate so as to align themselves parallel to the earth's magnetic field during deposition. Certain lavas or sediments can thus be considered to have the magnetic field present during their formation "frozen" into them. Once developed,

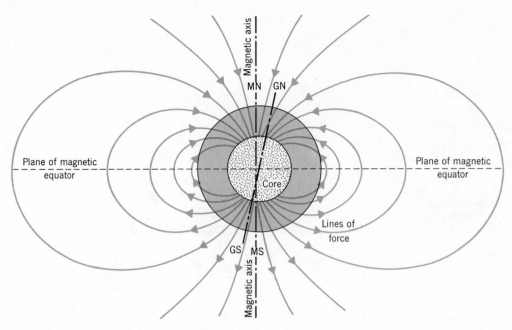

FIG. 8-7. (*a*) The earth's magnetic field shown by schematic lines
of force that emanate from the molten core. The present, or normal
field is conventionally taken to be directed from the magnetic
south (MS) to the magnetic north (MN) poles, as shown by the arrows. The lines of force show the orientation that is taken by the
north seeking end of a compass needle. (When the field is reversed,
as discussed in the text, magnetic north and south pole positions
would be reversed as would the direction of the arrows on the
lines of force). GN is geographic north and GS geographic south.
Note that the magnetic and geographic poles do not coincide at
present being separated by 11.5°.

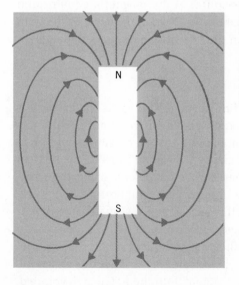

(*b*) The magnetic lines of force extending from the poles of a simple
bar magnet. The magnetic field of
the earth is very similar to that of a
bar magnet to which it is often
likened.

this *remanent magnetism* cannot be easily altered even by later changes in the earth's field. Hence the nature of the earth's field can be "read" from rocks hundreds of millions of years old.

Before going on, we must establish the axiom that on the average, the magnetic poles and the geographic poles coincide, or that the earth's rotational axis and the magnetic axis coincide. Convectional motions in the earth's molten core cause small displacements, as at present (see Fig. 8-7), between the geographic and magnetic poles; but over hundreds of years these differences are averaged out and the positions of both types of poles are believed to coincide.

If we were to heat a rock or iron-bearing substance above the Curie point and then allow it to cool, the magnetic field then developed within the rock would coincide with the present field. However, laboratory measurements of the magnetic-field orientation in geologically old specimens often show it to be different from the orientation of the present field. And the more ancient the rock, the greater is this difference!

Determining Rock Magnetism

The method of measuring rock magnetism is another example of how methods of science have triumphed in the measurement of minute quantities. Imagine that a small cube of rock, say a piece of lava flow, has been removed from its field location. Prior to removal, the appropriate orientation directions are marked on it. The problem now is how to determine the position of the magnetic field that was present in the rock at the time of formation—perhaps millions to hundreds of millions of years ago. Although the principle of this determination is based on one of the most elementary principles of science, the method is relatively complicated because of the small quantity to be measured.

When the rock cooled below its Curie point, some magnetic orientation of iron-bearing minerals developed. Thus the rock became a very weak magnet, with weak lines of force extending out from it, much as from the earth (Fig. 8-7). Now the scientific principle involved is this: When a magnetic field cuts an electrical conductor, a current is generated in the conductor. The cube of rock is placed within a coil of wire and spun rapidly on a rotating shaft (Fig. 8-8). If an ancient magnetic field is present, the weak current developed in the coil is amplified and measured. The cube of rock is spun on each of its three axes. From this information the true orientation of the magnetic field within the rock is determined by geometrically combining the three-component electric-current information.

Rock Magnetism and Continental Drift

Suppose the magnetic field within the rock does not match the present field. As long as the earth's magnetic axis is assumed to coincide always with the rotational axis, the explanation must lie in a change in the position of the rock relative to the poles. When the positions of the poles indicated by rock magnetism measurements are plotted on a map of the present surface, we find that they lie on a curve, with the older pole locations being progressively more distant from the present pole, as in Fig. 8-9a. This remarkable effect, known as polar wandering, is not yet fully understood. One explanation offered is that some outer layer or shell of the earth can slide around the earth's interior. Hence the polar curve Fig. 8-9a simply represents the points that once lay at the north pole and have since migrated away.

If only polar wandering occurred, regardless of the cause, past pole locations determined from all of the continents should coincide and the polar curves themselves should be identical. Now Fig. 8-9a shows the curve determined mostly in North America; it is quite similar to the polar wandering curve from rocks in Europe. But when the two curves are

FIG. 8-8. The Lamont spinner magnetometer. A rock specimen is visible in spinning position in the horizontal shaft in the center of the coil. When being tested for remanent magnetism, the shaft is rotated rapidly. Any magnetism in the specimen induces a weak electric current which is measured by the meter in the unit to the rear of the spinner.

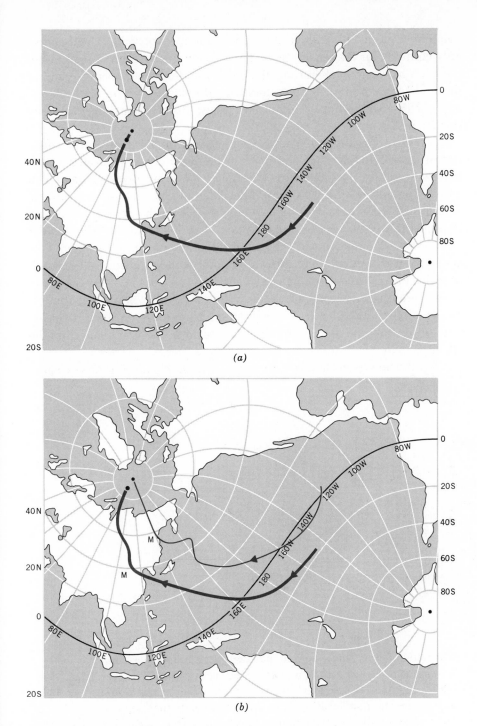

FIG. 8-9. (*a*) The apparent migration path of the north magnetic pole from Precambrian time to the present. (*b*) Comparison of the polar migration (wandering) paths determined from rocks from Europe (thin line) and North America (heavy line). Following the polar positions for Mesozoic time (M) the paths begin to converge to the present location. (After K. Runcorn.)

compared, as in Fig. 8-9*b*, a clear discrepancy is observed. The curves only coincide for the recent rocks. These show the position of the present pole, but as we go back in time, the polar wandering curves diverge until the mid-Mesozoic Era. Earlier than this, the curves are parallel to each other with a separation of about 30°. These observations are explained further in Fig. 8-9*c*.

These observations led to the renaissance of the concept of continental drift. Suppose that North America and Europe separated in the Mesozoic Era as postulated by Wegener from other evidence, and that movement continued until recent time. Then the magnetic-pole positions determined from these continents would match for the present but would be separated for older rocks as explained in Fig. 8-9*c*. For rocks older than Mesozoic, polar positions would have a constant separation. This is just what is shown by the calculated polar wandering curves for North America and Europe. If we imagine North America to move 30° eastward, it would rejoin Europe and close the Atlantic Ocean. When the continents are so moved, the polar-wandering curves from each continent overlap, thus supporting the theory that they separated in the first place from the effect of drift.

For the Southern Hemisphere, separate polar wandering curves, again

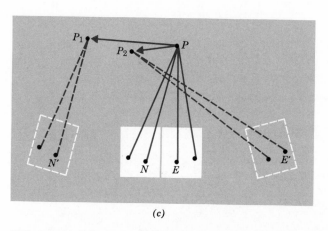

(c)

FIG. 8-9. (c) Schematic diagram showing how paleomagnetism indicates continental drift. If North America and Europe were joined as at *N* and *E* respectively in early Mesozoic time the magnetic fields formed in their rocks would indicate the single magnetic pole location at *P*. If the continents then separated by moving laterally with slight rotation as at *N'* and *E'*, the magnetic fields preserved in rocks formed in Mesozoic time would be displaced from their earlier orientation and would indicate the two different pole positions, P_1 and P_2, respectively. Rocks formed at progressively later geologic times would have magnetic fields that would show less and less polar separation for the two continents. If polar wandering is added to the drift effect, location of former locations lie on curves like those in Fig. 8-9*b*.

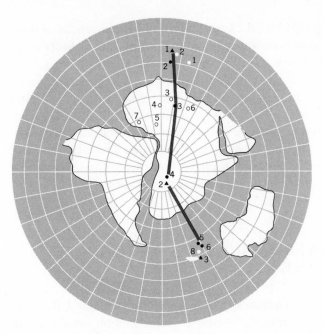

FIG. 8-10. If the continents of the Southern Hemisphere are displaced until each of their polar wandering paths merge into one (heavy line) the continents assume the positions shown about the south pole. This determination agrees with the glacial reconstruction in Figure 8-4. Numbered symbols show past pole locations from the different continents. (After K. Greer, The Royal Society.)

converging at the present south pole, have been determined for the individual continents. If the curves are brought together for south pole locations back to the Paleozoic Era, Australia, Africa, and South America appear as in Fig. 8-10, where they are seen to lie around the geographic south pole. (Other evidence indicates that Antarctica lay between Australia and South Africa). The *paleomagnetic* evidence now makes it easier to understand the wide distribution of Paleozoic glaciers in the Southern Hemisphere, referred to earlier. With this reconstruction, the Paleozoic ice sheet is distributed about the geographic south pole as in Fig. 8-4 and the entire problem of the origin of the Paleozoic Ice Age becomes simplified.

The conclusions from rock magnetism were given tremendous support by one of the great fossil finds of the century in December 1969. A group from Ohio State University working in the Queen Alexandria Range (Antarctica), one of the mountain ranges not fully ice-covered, discovered remains of a land-dwelling reptile in rocks of the Triassic period (about 200 million years old). Since the fossil found has been recognized as a key index fossil for rocks in the major southern landmasses, its presence in Antarctica seems to establish a former connection among all of the continents of the Southern Hemisphere, including Antarctica. And a month earlier, a different group from the same institution discovered elsewhere in Antarctica a more mixed assemblage of both freshwater amphibians and land reptiles all characteristic of Triassic rocks of other lands. Again these finds are strongly indicative of the connection of Antarctica to the other continents of the Southern Hemisphere.

A great variety of observational evidences have thus given increasing support to the drift hypothesis, but no explanation has been given of how drift occurred. The first real light was shed on this problem in the early 1960s and the new ideas were again sparked by observations of rock magnetism. It is these newest observations and their interpretation that have quite revolutionized geologic thought; these were the observations of magnetic reversals and the concept of sea-floor spreading.

MAGNETIC REVERSALS AND SEA-FLOOR SPREADING

The idea of sea-floor spreading has become well-established from the interweaving of a large number of seemingly unrelated observations that include rock magnetism, radioactive dating of lava flows, study of deep sea cores, accurate location of marine earthquakes, and measurements of heat flow in the ocean bottom.

Magnetic Reversals

Studies of rock magnetism by spinning small rock samples within a coil led to the discovery that the magnetic field preserved in successive layers of rocks often shows successive reversals. For example, in one layer the orientation of magnetic north may coincide roughly with that of the present fields; in an overlying layer the rock magnetism may be reversed from the present field; that is, magnetic north may point toward the Southern Hemisphere. Somewhat higher in the rock sequence, the rock magnetism may again be normal, or similar to the present, and so on.

This curious phenomenon was soon found in rocks of many types, including lava flows and sedimentary rocks. At first the magnetic reversals were regarded as anomalous occurrences and no ready explanation of them was at hand. But as the modern theory of the origin of the earth's magnetic field developed, it was shown that movements of the molten metallic core could provide for irregular alternations of the field every half a million to a million years. Also, when rocks tested for magnetism were dated by radioactive isotope measurements, particularly potassium-argon (K-Ar), it was found that rocks of the same age had the same magnetic orientation. And the times of change from one orientation to another could also be correlated on a worldwide basis. An absolute chronology of magnetic reversals has now (1971) been established back to 4.5 million years ago. Although reversals are common well back in the time scale, the present error in dating beyond 4.5 million years is too great

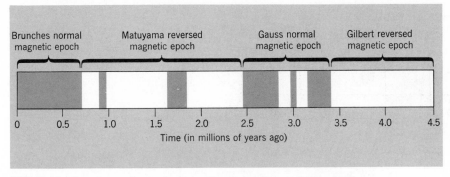

FIG. 8-11. The chronology of magnetic reversals as determined to 4.5 million years ago. (Shaded regions represent times of normal magnetic fields and unshaded sections reversed magnetic fields.)

to resolve the intervals of magnetic change. The broad features of the present magnetic chronology including absolute dating and the names of the magnetic epochs, are shown in Fig. 8-11.

Magnetic observations were also made on a large number of deep sea cores long enough to penetrate millions of years of marine sediments. Again, magnetic alternations in sediments were found whose ages, although not precisely datable, could be estimated to match similar magnetic events in rocks on land. Thus the important fact was established that simultaneous alternations in direction of the earth's magnetic field occurred over the entire earth that affected rocks both on land and at the sea bottom.

Magnetic Reversals and Ocean Ridge Systems

Since World War II it has been the practice to tow magnetometers astern of many oceanographic research vessels. These instruments make continuous records of changes in the strength of the earth's magnetic field as a vessel travels its course. Variations in the earth's magnetic field recorded in this way are due to variations in the magnetism of the igneous (basaltic) rocks that compose the floor of the ocean and whose effects become superimposed on the field of the earth. Whenever the tracks of research vessels crossed midocean ridge systems, rather remarkable records were obtained. A quite typical example is shown in Fig. 8-12, a magnetic profile across the East Pacific Rise in the South Pacific Ocean. To the casual observer this is just a wiggly line, but to a more practiced eye there is a striking symmetry of peaks and troughs about the broad central portion of the line. If one were to fold the profile in half about the indicated center line, the wiggles would almost exactly coincide! (Trace the line on a piece of thin paper and try this.)

This exciting discovery was duplicated by all magnetic profiles taken across all of the midocean ridge systems. It was quickly shown that the regions of magnetic peaks and troughs, which extended as belts parallel

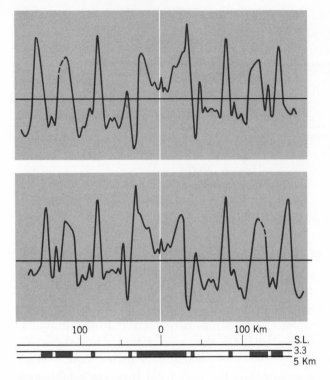

100   0   100 Km

S.L.
3.3
5 Km

FIG. 8-12. Variations of the earth's magnetic field (upper curve) observed along a ship's track that crossed the East Pacific Rise. The vertical line represents the crest line or axis of the ridge. Peaks are interpreted as indicating crustal rocks with normal magnetic fields that reinforce the earth's field; while troughs appear to indicate rocks with reversed magnetic fields that subtract from or weaken the earth's field. The crustal rock relationship is shown beneath the curves; dashed areas represent normal magnetization. The second curve is a mirror image of the first—or is drawn with right and left reversed. (Redrawn after W. Pitman and J. Heirtzler, Science.)

to the axes of the ridges, as in Fig. 8-13, could be explained on the basis of magnetic reversals. Belts of rock showing magnetic peaks have normal magnetic fields; belts overlain by magnetic troughs have a reverse magnetic orientation whose effect is to weaken the present magnetic field of the earth, thus decreasing observed magnetic intensity. This effect is indicated in Fig. 8-14. If the earth's crust were uniformly magnetized at a particular latitude, the magnetic lines of force would extend uniformly from the upper atmosphere down to the surface. But if alternating zones of reversed (R) and normal (N) magnetism were present as shown, the lines of force would converge toward the normally magnetized rock and diverge from that reversely magnetized, thereby giving the variations in magnetic intensities.

Interpretation of the Sea-Floor Magnetic Belts: Sea-Floor Spreading

The facts of magnetic reversals and related chronology have provided a most important and scientifically exciting means of explaining the belts of alternating magnetization of the sea floor. Sea-floor spreading seems to be the answer.

Let us imagine a slow but continuous process in the ocean crust and upper part of the mantle in which warm rock material rises vertically beneath much or all of a midocean ridge zone. On reaching the crust the rock cools and spreads laterally in the form of broad plates as indicated in Fig. 8-15. When the newly formed crust cools below the Curie point, it becomes magnetized by the earth's magnetic field existing at the time. The belt of magnetized rock splits continuously into two portions as the crust spreads sideways from the ridge. If the earth's magnetic field then reversed, a symmetrical pattern of magnetization would develop as in Fig. 8-15. The median rift valleys that mark the crests of the marine ridge systems are then explained as a natural consequence of the crust's being pulled apart in crustal spreading.

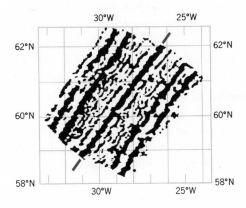

FIG. 8-13. Map showing the belts of alternating magnetic polarity on either side of the mid-Atlantic ridge southwest of Iceland. (Redrawn after J. Heirtzler, et al. Deep Sea Research.)

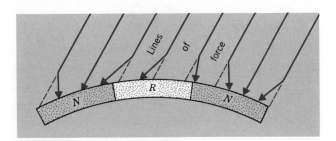

FIG. 8-14. Diagram showing the convergence of magnetic lines of force reinforcing the earth's field over rocks of normal magnetic polarity and the divergence over rock of reversed polarity.

The alternations of normal and reversed magnetic marine belts have been explained by the mechanism of sea-floor spreading during long periods of irregularly spaced alternations of the earth's magnetic field. It thus seems reasonable to correlate these magnetic belts with the dated latyers of normal and reversed magnetism shown in Fig. 8-11. The present belt of normal magnetism must correspond with the uppermost layer of normal magnetism (Brunhes epoch) and the two belts of reversed magnetism symmetrically adjacent to the ridges must then match the underlying Matuyama layer of reversed magnetism, and so on. Once this is accepted, the absolute dates determined for the vertical magnetic time scale can be applied to the equivalent marine magnetic belts. From this the rate of sea-floor spreading can be quickly calculated by dividing the distance from the ridge crest to the belt by the age of the belt.

The distribution and ages of magnetic belts determined for all oceans of the world as of 1968 are shown in Fig. 8-16. Lines of "0" age are at ridge centers. Displacements of groups of age bands are the result of submarine faults or fractures of the ocean floor that involved primarily horizontal slippage. From the ages of the magnetic belts, the shaded area showing the marine crust formed during the present (Cenozoic) era of the last 65 million years has been determined (Fig. 8-17). In other words, the sea floor has spread from the "0" age line to the margins of the shaded area during this era. Actual rates of spreading vary from about 1 to 15 centimeters ($\frac{1}{2}$ to 7 in) per year in different parts of the oceans.

We have now developed two parts of the rather simple model of sea-floor spreading; namely, rising of warm rock beneath ridges followed by horizontal spreading. A third aspect of the model that must be considered is—where does the spreading rock material go? Recall the presence of the profound ocean deeps or trenches (Chapter 3) and the occurrence of earthquakes along planes sloping downward toward adjacent continents beneath the ocean deeps or trenches. These observations can

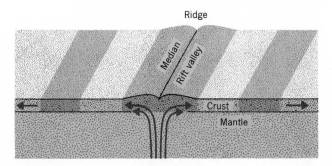

FIG. 8-15. The interpretation of the belts of alternating magnetic polarity by the rising of warm rock beneath midocean ridges followed by lateral spreading across the sea floor. The new crust at the ridge system would develop a magnetic polarity matching the earth's field at the time.

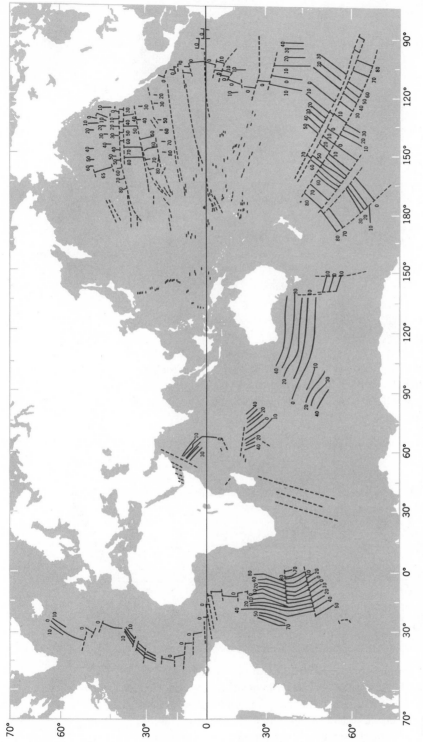

FIG. 8-16. From sea floor spreading rates determined as explained in the text, the ages of magnetic polarity belts can be interpreted as shown for the world ridge systems. Lines of 0 age are at ridge crests; the number of the lines of constant age represent millions of years of spreading from the crests. The ridges and related age belts are displaced by numerous faults. (Redrawn after J. Heirtzler, et al, *Journal of Geophysical Research.*)

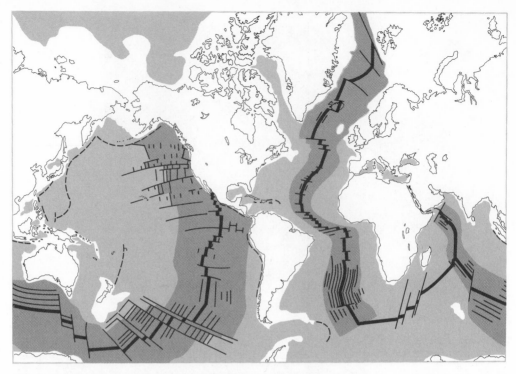

FIG. 8-17. The shaded bands show the area of new crustal rock formed during the Cenozoic Era according to the sea floor spreading theory. The width of this zone is based on the age belts in Figure 8-16, reproduced here. (After F. Vine.)

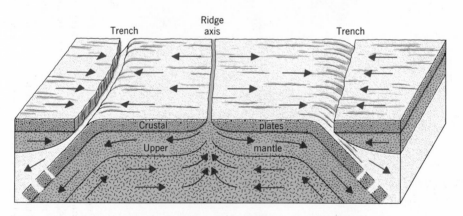

FIG. 8-18. Three-dimensional diagram illustrating the sea floor spreading model. Warm mantle rock rises beneath ridges and then spreads laterally toward ocean trenches. The solidified uppermost igneous layer is the ocean crustal plate which is carried by the underlying, spreading upper mantle. The combination of spreading crustal layer and upper mantle is now called *lithosphere*. (Modified from B. Isaacs, J. Oliver and L. Sykes in *Journal of Geophysical Research*.)

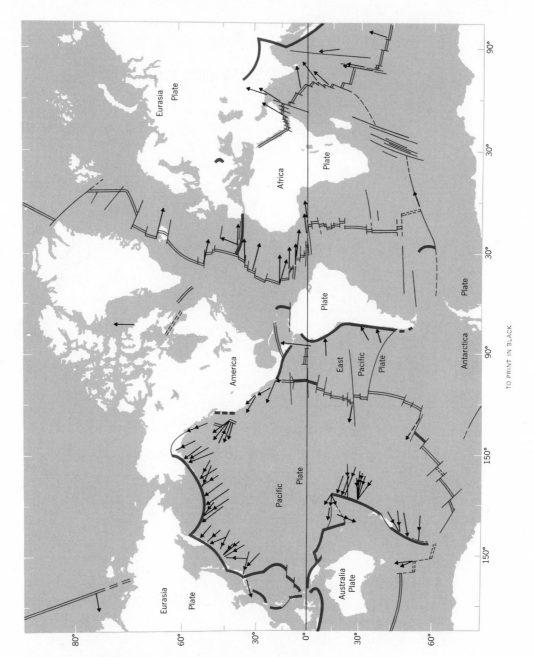

FIG. 8-19a. Map of the principle tectonic (crust and upper mantle) plates of the earth. Plates are bounded by ridges and trenches, the former shown by a double line and the latter by a solid line. Arrows indicate motion at both ridge axes and trenches. (After B. Isaacs, J. Oliver and L. Sykes in the *Journal of Geophysical Research*.)

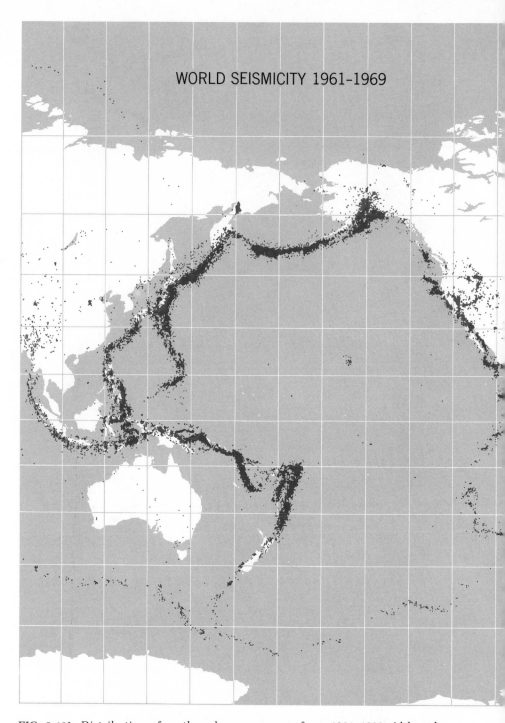

FIG. 8-19b. Distribution of earthquake occurrences from 1961–1969. Although each earthquake is marked as a single dot, the black areas represent places where the dots overlapped. Note the nearly perfect matching of the earthquake pattern and the tectonic plate borders. This distribution can now be understood in detail for the first time. (U. S. Coast and Geodetic survey, NOAA.)

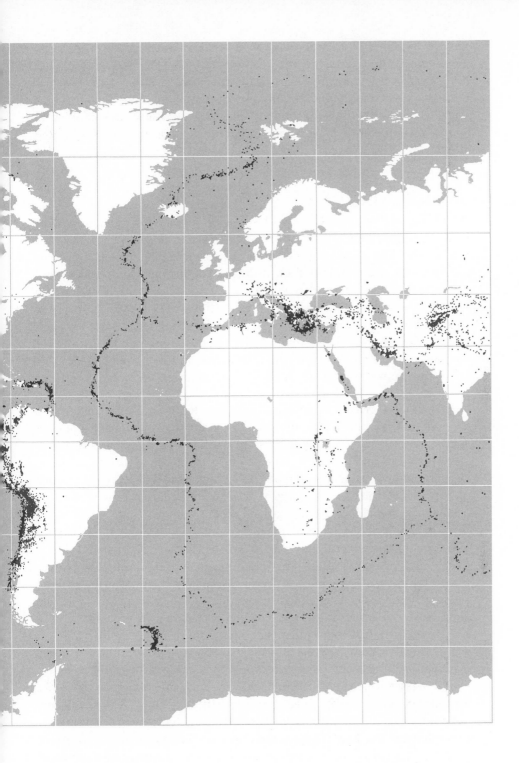

now be explained by the descent of the crustal material that has spread across the sea floor from the ridge systems, as in Fig. 8-18. We can consider that ocean crust is formed at ridge systems and destroyed at trenches. Because ocean crustal rock is manifestly rigid, spreading does not take place as simply as the convectional circulation of thick soup cooking in a pot. The explanation of sea-floor spreading involves the concept of rigid crustal plates.

Theory of Tectonic Plates

According to the present view of sea-floor spreading, warm rock rises beneath and then spreads away from active ocean ridge systems, as noted previously. The upper portion of the spreading layer, which preserves the magnetic field impressed on it during cooling, solidifies into a rigid crust. The plate is carried with the more mobile, less rigid underlying layer of the upper mantle some 60 miles (100 km) in thickness from the ridge (where it was created) to the trench, where it descends and becomes reincorporated into the mantle. This mobile 60-mile unit is called a tectonic plate. This theory of the generation and movement of oceanic crust and upper mantle (or subcrust) is summarized schematically in Fig. 8-18.

The global crust and subcrust appear to be naturally organized into six major active plates plus the smaller, less active east Pacific plate, as illustrated in Fig. 8-19a. The plates are bounded by ridges and trenches except in cases, such as the west coast of North America, where spreading has carried the continent over the ridge system. Note the close coincidence of the global earthquake pattern to plate boundaries (Fig. 8.19b), an occurrence that strongly supports the plate theory.

The complete model explains quite well the reason for the existence of the great midocean ridge system as well as the deep, elongated trenches that border island arcs and in some cases, continental margins. Ridges are present because warm rock is rising; trenches exist because the crust is dragged down as oceanic plates descend. A slightly more detailed scale view of plate motion at and near trenches is shown in Fig. 8-20.

It must, by this time, have become apparent to the reader that we now, through geophysics, have a mechanism for continental drift, thus overcoming the earlier objection of geophysicists. Instead of the continental masses having to push their way through the strong crustal rocks forming the ocean basins to accomplish drift, they can now be imagined to be simply carried along as continent plates on the moving upper mantle beneath them, much as a floating piece of wood is transported by a current of water. According to this mechanism the sea-floor spreading away from the mid-Atlantic ridge system was the cause of the separation of North and South America from Europe and Africa. These continents would once have been joined directly across the ridge system prior to separation in the Mesozoic Era.

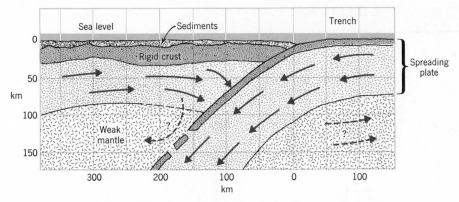

FIG. 8-20. Simplified schematic view of plate behavior at ocean trenches drawn to approximate scale in kilometers (multiply kilometers by 0.62 to convert to miles). The rigid upper crustal plate is thicker on the continent (left) side of the model. (Modified after B. Isaacs, J. Oliver, and L. Sykes.)

A summary of the mechanism of drift including the creation of a new ocean basin (like the Atlantic) is given in Fig. 8-21. Continents should always be older than the ocean between them. Geological age studies certainly support this conclusion.

Further support for the entire mechanism will also be given in Chapter 10, where the occurrence of oceanic earthquakes will be seen to match exactly the locations of ridges and trenches.

In order to probe more deeply into the ocean crust than was possible before, and also to test ideas advanced in this chapter, a new era of exploration was begun in 1968 by the research vessel *Glomar Challenger*. This vessel, which is operated jointly by a number of American oceanographic institutions under the support of the National Science Foundation, carried out a series of deep drilling ventures along the tracks shown in Fig. 8-22. The operators of the vessel developed procedures for drilling deeply into the earth's crust while holding a fixed position at sea, almost regardless of the water depth.

For the first time samples of bedrock from beneath the sea floor were brought up to be viewed and analyzed; for the most part the results give striking conformation to the model of sea-floor spreading. For example, the ages of rocks progressively farther from midocean ridges are in general progressively older than those closer to the ridges. Sediments of Jurassic Age (about 140 million years old) have been taken from the sides of the Atlantic Ocean basin. These locations for the oldest sediments so far discovered in the sea give further support to the spreading concept.

In summary, we have seen the following.

1. The earth's magnetic field has undergone alterations of normal and reversed direction during geologic time.

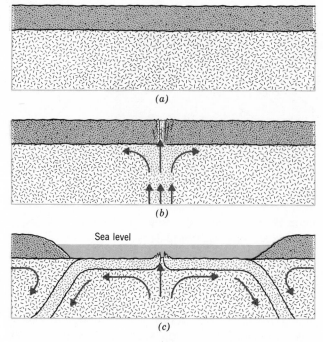

FIG. 8-21. Continental drift and the creation of a new ocean basin through the initiation and continuation of sea floor spreading. In (*a*) a single large continent is shown prior to rifting and separation. In (*b*) upper mantle spreading has begun tearing apart the continent with the intrusion of dense mantle rock to start a new ridge system. In (*c*) both segments of the continent have been transported far from the midocean ridge and trenches have been developed marginal to the continents where the upper rigid plate and spreading mantle beneath, descend.

2. The times of these variations have been dated quite accurately by radioactive measurements of lava flows on the continents.

3. The magnetic reversals have been shown to be worldwide in occurrence, affecting rocks on the continents and sediments in the oceans.

4. Belts of ocean crust showing normal and reversed magnetic fields lie symmetrically about midocean ridge systems.

5. By assuming that the crustal belt over the marine ridges is the youngest and correlates with the present field, then crustal belts progressively out from the ridges can be correlated respectively with the dated rock layers on land.

6. The result leads to a picture in which ocean crust is formed at midocean ridges and then moves away laterally as plates to descend at ocean deeps; the further the rock is from the ridge the older it is.

7. The concept of sea-floor spreading provides the first reasonable mechanism for continental drift.

FIG. 8-22. *Glomar Challenger*, the first of a new generation of ocean vessels, is capable of drilling into the sea bottom in the open ocean. The drilling derrick, visible amidships, stands 194 feet above the water line. Twenty-four thousand feet of drilling pipe lie in sections on the rack forward of the derrick. Tens of thousands of miles of drilling tracks have been made by the vessel as part of the Joint Oceanographic Institutions Deep Sea Drilling Project managed by the Scripps Institution of Oceanography.

RELATED PROBLEMS

The complete model of sea-floor spreading involving ascent of warm rock beneath ridges and descent at ocean deeps with lateral spreading between, provides an explanation for several other major problems of the sea floor.

Sediments of the Ocean Floor

With the expansion of seismic studies of the sea bottom following World War II, it quickly became apparent that a relatively thin veneer of sediments carpets the igneous crust of the oceans. The range in thickness of the unconsolidated deposits over all of the ocean basins is only 660 to 3300 ft. (200–1000 m) miles. An equivalent layer of possibly consolidated sedimentary rock lies between the loose sediments and the true crust.

If we use very conservative values for the rate of deposition in the oceans, the thickness of sediment that should have accumulated during geologic time is many times this observed sediment thickness. Why, then, is the existing sediment layer so thin? The answer may lie in sea-floor spreading. As the igneous crust moves away from the ridge systems, the sediment directly above may be carried along and may have thus prevented continuous thickening during the long years of geologic time.

Heat Flow from the Sea Floor

Rocks are very poor conductors of heat. Most of the heat generated within the earth by the contraction and radioactivity processes discussed in Chapter 1 is retained there—witness the molten core. But a very small heat flow does occur. That reaching the surface escapes into either the atmosphere or the oceans; the amount is very small and is about the same for the normal crust of both continents and oceans. On the continents the heat flow is measured in deep oil wells or test wells and in mines. In the oceans, the heat flow is now measured by very sensitive probes attached to the long coring tubes used to obtain sediment samples from the sea bottom. Electrical conductors run thousands of feet from the probes to the deck of the research vessel, where electrical current readings can be converted to heat-flow values in the ocean sediment.

Over much of both the ocean floor and continents, the heat flow outward is about 1.5 millionths of a calorie or 1.5 microcalories per square centimeter per second (1.5×10^{-6} calories/cm^2/sec). On the continents this value is much higher in volcanic regions like Yellowstone Park or Alaska. Such occurrences are expected and easily explained. But in the sea bottom, heat flow rises many times higher than the normal, reaching eight microcalories per square centimeter per second at the midocean ridge systems. These anomalous observations are also better understood from the sea-floor spreading model. The high heat flow over the ridges would be a consequence of the rising high-temperature rock beneath the submarine ridge systems and seems to be one more natural process embraced by the spreading theory.

All of the details of the combined drift-spreading model have not yet been worked out, and there are still some observations over which the theory stumbles a bit. More complete observations are being accumulated continuously, but the varied types of observations described here that can be synthesized into a scheme that seems at once to be so all embracing is certainly one of the greatest chapters in the history of earth science.

STUDY QUESTIONS

8.1. Describe briefly four observational evidences that support Wegner's theory of continental drift.

8.2. What was the principal objection to the theory of continental drift?

8.3. Draw a diagram indicating the relationship of the earth's magnetic field to the earth's surface and the core.

8.4. Explain "remanent magnetism." In what ways is it acquired by a rock?

8.5. How does the study of paleomagnetism suggest "polar wandering"?

8.6. In what way did the study of rock magnetism give renewed support to the theory of drift?

8.7. What is the evidence for magnetic reversals?

8.8. How is the chronology of magnetic reversals determined?

8.9. How are magnetic reversals related to ocean ridge systems?

8.10. Explain how the age of each strip of normally and reversely magnetized ocean crust is determined. What important assumption is made in this age determination?

8.11. Explain "sea floor spreading"—use a diagram(s).

8.12. How is sea floor spreading related to tectonic plates? Use appropriate diagrams.

8.13. How do the above concepts explain continental drift?

8.14. What problems are associated with the thickness of sediments in the oceans and rates of heat flow from the crust beneath the sea?

8.15. How are these problems explained by sea floor spreading?

chapter 9

Devil's Tower in Wyoming is believed to be the residual resistant igneous rock or *neck* that occupied the conduit through which the magma issued. The volcanic mountain surrounding the conduit has been removed by erosion. (Courtesy American Museum of Natural History.)

The Restless Crust:
Volcanoes and
Mountain Chains

The true scenic grandeur of the continents lies in their lofty mountain ranges. To an air traveler winging westward across the United States, the monotony of the interior farm and grazing lands is rather abruptly replaced by the dramatic topography of the Rocky Mountains. Rugged mountain scenery then prevails during the journey over the western third of the country. Some of the individual peaks and ranges are so high as to be above the snow line. Even in midsummer their snow-capped summits gleam brilliantly in the clear mountain air.

In the eastern United States, the Catskills of central New York State also show very rugged scenery, although at a much lower elevation. But the Catskills, according to our earlier discussion in Chapter 3, are really an eroded plateau and from a geologic point of view, the term Catskill *Mountains* is erroneous. Recall that plains and plateaus are underlain by undeformed, horizontal rock, whereas mountains are underlain by either some form of igneous rock or deformed (and often metamorphosed) sedimentary rocks. In popular usage, high regions of limited summit area are called mountains, whereas the geologic definition pays little attention to the relief or elevation involved. Thus, the low, rolling lands of eastern Canada and southern New England, including New York City, are mountains. These differ from the Rockies, Alps, Himalayas, and other prominent mountains in being geologically far older. Prolonged erosion has lowered the original highlands of the New York region. However, the highly deformed and metamorphased igneous and sedimentary rocks now found at the earth's surface bear witness to the prior existence of a chain of lofty mountains long since destroyed.

Why are some mountains old and others young? Why haven't the powerful erosional forces described in Chapter 6 reduced all but the most recently formed mountains to near sea level? Why, despite the long ages of erosion, are the continents higher and more expansive than ever before?

The answers to these questions lie in the nature, origin, and history of mountains. Many of these answers we now know with some certainty, but unsolved questions remain.

There are many kinds of mountains. Perhaps the simplest are the *volcanoes*, which are mountains that have accumulated from igneous material erupted at the earth's surface. Although volcanic action is responsible for much of the major relief of the sea floor (see Chapters 3 and 8), volcanoes are only locally important on the continents. Of more importance are the great mountain chains composed of folded, often metamorphosed, sedimentary rocks. Another type of mountain resulted from *faulting*, a process in which a portion of the crust is broken and displaced. Many fault structures are a direct result of the folding that formed mountain chains; others have had an independent history.

In this chapter we will examine some of the fundamental facts and ideas related to these types of mountains.

VOLCANOES

Great Eruptions

From the beginning of geologic time the earth's crust has been rent by eruptions of gaseous, liquid, and solid materials, often with great devastation to the surrounding environment. One of the most violent eruptions in historic times in the Dutch East Indies (between Sumatra and Java) in 1883 destroyed, rather than created, a mountain. After two centuries of inactivity, the old volcanic island of Krakatoa blew itself apart with an explosive force estimated to be equivalent to a 5000 megaton hydrogen bomb. The eruption hurled more than four cubic miles of igneous rock material into the atmosphere. Much of the material was so finely fragmented that it was borne high into the stratosphere, where strong winds

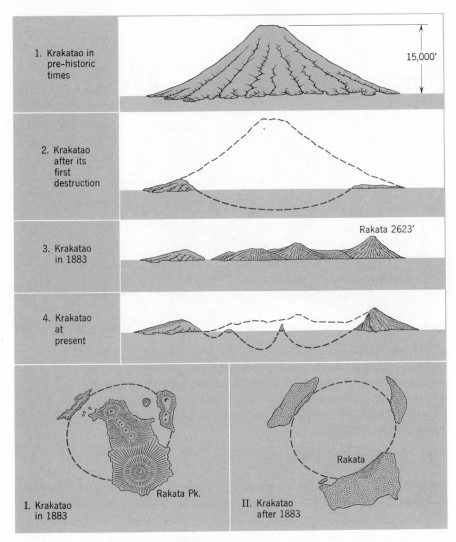

1. Krakatao in pre-historic times

15,000'

2. Krakatao after its first destruction

3. Krakatao in 1883

Rakata 2623'

4. Krakatao at present

I. Krakatao in 1883

Rakata Pk.

II. Krakatao after 1883

Rakata

FIG. 9-1. The sequence of events in which the present volcanic islands of Krakatao and Rakata formed in the Dutch East Indies. (After the Report of the British Royal Society.)

soon distributed the resulting dust around the globe. Spectacular red sunsets prevailed for several years over the earth from the scattering effect of the Krakatoa dust on sunlight.

A considerably less beautiful result was the drowning of some 36,000 inhabitants of Java and Sumatra by the giant sea waves (see tsunami in Chapter 10) more than 100 feet high that were generated by the explosion. Much of the original volcanic island of Krakatao subsided into the void left after the eruption of the huge volume of underlying magma, leaving in its place an expanse of water several miles in extent and about a thousand feet deep (Fig. 9-1). Great as was this volcanic explosion, it was

FIG. 9-2. Mt. Pelée on the Island of Martinique and the ruins of St. Pierre following the series of disastrous eruptions beginning May 8, 1902. The spine of rock, projecting from the peak of the volcano, rose slowly after the eruption began. After being elevated 1200 feet above the rim of the crater, the spine subsided into the supporting magma and disappeared from view. (Courtesy American Museum of Natural History.)

probably exceeded by the eruption of Tambora in 1813, also in Indonesia, in which nearly 40 cubic miles of lava and rock were blown into the air. In place of the original 12,000-foot volcanic mountain, a crater 4 miles across and 4000 feet below the original height remained.

One of the historically most famous and scientifically best-studied volcanoes is Vesuvius, which began its development in 79 A.D. The present Vesuvius occupies part of the large crater of Mt. Somma, which was an old volcano with no recorded activity prior to 79 A.D. In that year, violent eruptions destroyed the coastal towns of Pompeii and Herculaneum. Most of the ancient structures that remained were buried under the huge fall of volcanic ash. The present Vesuvius was built from a succession of eruptions of fragmented material and lava that created a new conical volcano entirely within the huge crater formed from the destruction of the original Mt. Somma.

Another famous volcano that can be used to justify the rephrased expression "crime may sometime pay" is Mt. Pelée on the island of Martinique in the French West Indies. One of the most devastating events of this or any century was the eruption of Mt. Pelée in the early spring of 1902 after its essential dormancy during historic time. A number of

FIG. 9-3. Burned buildings and shattered trees in St. Pierre caused by the eruptions of Mt. Pelée. (American Museum of Natural History.)

eruptions of seething lava and pyroclastic material occurred. Pyroclastics (fire broken) are airborne, often incandescent, fragments of solid rock material of all sizes often erupted by volcanoes. The lava repeatedly broke through the original crater and poured down the mountainside while huge black clouds of exploded material spread out from the peak. On May 8, in the most powerful eruption of the series, huge quantities of lava, fine volcanic ash, and coarser pyroclastic material were hurled down the volcano side that faced the port of St. Pierre. In a matter of minutes the hot noxious gases and axphyxiating volcanic ash annihilated all of the nearly 30,000 inhabitants of St. Pierre except for one prisoner who survived because of the less contaminated air in a deep dungeon (supporting our opening statement). Fires set by the heated gases and incandescent ash and associated earthquakes completed the devastation of the city Figs. 9.2 and 9.3.

The list and history of volcanic activity over the world is long and fascinating and fills far more space than is here available. But no reference to such activity would be complete without mention of the volcanoes of the Hawaiian Islands. These islands have been built by a series of overlapping, gently sloping volcanic cones. Despite their gentle slopes, the topographic relief of this mountain group is perhaps the greatest on the earth's surface. Mauna Loa, which reaches nearly 14,000 feet above the ocean, rises from the sea floor about 15,000 feet deep in the central Pacific Ocean, giving a total relief close to 29,000 feet! Because the volcanoes of

FIG. 9-4. A close group of three volcanoes on Java. The volcano
with the steaming crater is *Smeroe*. (Courtesy American Museum
of Natural History.)

FIG. 9-5. A model of Crater Lake in Oregon which occupies a huge depression, or *caldera*, believed to have formed from the subsidence of the entire
top of a former volcano into the underlying magma. Wizard Island is the
small cinder cone formed from an eruption within the crater sometime after
its collapse. (Courtesy American Museum of Natural History.)

this island chain were formed from relatively mobile lava, slopes of only
2° to 4° were produced and as a consequence the volcanoes now cover
large areas. The total amount of lava required to build the Hawaiian
Islands is 100,000 cubic miles, an enormous volume of material.

Examples of volcanoes and volcanic action are included in Figs. 9.4 to
9.10.

FIG. 9-6. Photograph of the perimeter of the Crater Lake caldera and Wizard Island in winter. (National Park Service.)

Distribution and Nature of Volcanoes

Figures 9-11 shows the global distribution of volcanic and earthquake activity. The common distribution of both activities is, of course, not accidental and implies a strong and significant geologic relationship between the origin of earthquakes and volcanoes. They both occur where the earth's crust is unstable, particularly along continental margins and midocean ridges. These are zones where crustal material descends or ascends in the course of sea-floor spreading.

In the building of a typical volcanic cone (mountain), the molten material (magma) from which a volcano is constructed rises through a relatively narrow conduit that extends deep into the crust or upper mantle.

FIG. 9-7. The "Castle" in Crater Lake is a result of recent extrusion. (National Park Service Photo.)

Upon reaching the surface, the liquid lava or aerially ejected (pyroclastic) materials are erupted from a relatively circular vent whose shape and size cause the ejected material to accumulate in a rather symmetrical cone through which the conduit becomes extended. The active vent at the summit of a volcano maintains a depressed hollow or crater. The steepness of the growing cone depends on the nature of the material ejected.

Cinder cones are the smallest of volcanic types (Figs. 9-12, 9-13). A well-formed cinder cone is a steep-sided conical feature that is composed largely of pyroclastic debris produced by explosive volcanic activity. The angular fragmental material tends to interlock as it is deposited and can thus maintain a high angle of repose (see p. 167). Shield volcanoes, however, contrast markedly with the relatively small conical cinder cones. *Shield volcanoes* are composed of congealed lava whose original very fluid nature permitted the lava to flow out great distances from the volcanic source, thus producing the huge volcanoes typified by those of the Hawaiian chain. Their name is derived from their topographic profile that closely resembles the gentle curvature of a shield. The grand, high-elevation volcanoes of greatest scenic beauty such as the 9000-foot Fuji-yama of Japan; Mt. Hood and Mt. Rainier in the American northwest; and Vesuvius and Stromboli in Italy, among so many others, are known as *composite volcanoes* (Fig. 9-14). In this case alternating layers of solidified lava and pyroclastic rock produce a stratified structure that indicates volcanic activity, including both explosive eruptions and lava flows. These volcanoes owe their steep slopes and great heights to protective layers of solidified lava that serve to cement and hold in place the very unstable but steep deposits of cinders.

FIG. 9-8. Shiprock projecting above the desert sands in New Mexico is another example of a volcanic neck whose surrounding volcanic mountain has been eroded, perhaps to form some of the loose sands that now flank the igneous projection. Several resistant dikes, one evident in the foreground projecting through the sand, radiate from the central neck. (Courtesy American Museum of Natural History.)

FIG. 9-9. Steaming Paracutin, a volcano that erupted in a farmyard in northern Mexico in 1943. A scorched forest is in the foreground. (Photo by Ivan Wilson, courtesy Heald and Robinson.)

FIG. 9-10. Lava spreading from Paracutin. A portion engulfed the neighboring village leaving the village church still uncovered. (Photo by Ivan Wilson courtesy Heald and Robinson.)

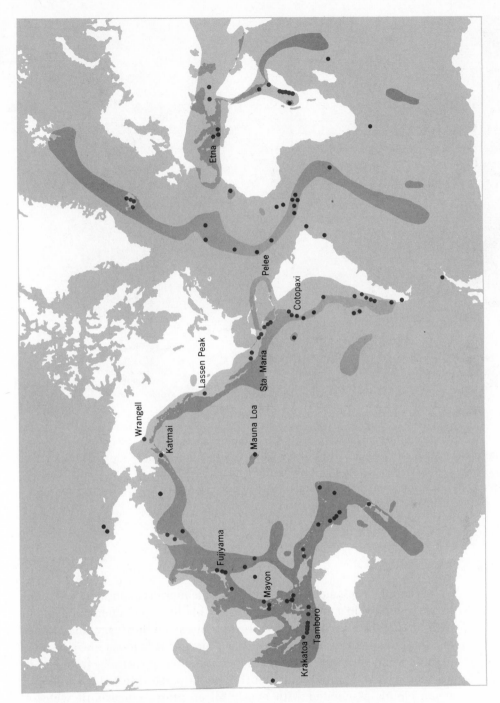

FIG. 9-11. Global distribution of principal volcanic activity. Some active and recently active volcanoes are located. Note how closely this distribution matches the belts of earthquake activity (Fig. 10-11) and the borders of crustal plates (Fig. 8-19).

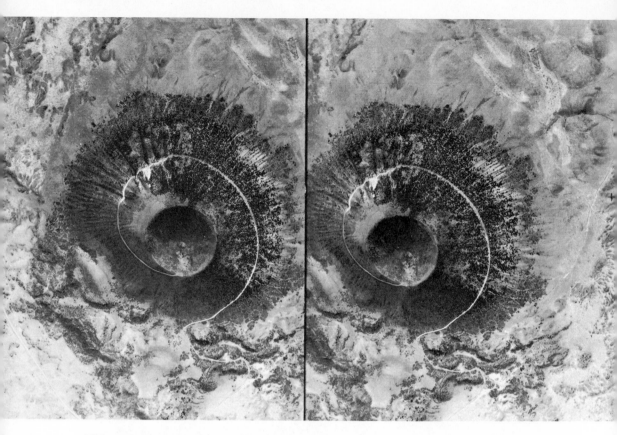

FIG. 9-12. Stereographic air-photo pair of Mt. Capulin, a cinder cone in New Mexico. The white line is a road that winds up to the rim of the crater. With a small stereoscope the volcano relief and crater hollow can be seen very strikingly. (U. S. Dep't. of Agriculture Photo, courtesy University of Illinois.)

When the rock material composing volcanoes is examined it is found to vary widely in composition. Because of the rapid quenching of the high-temperature lavas, the textures are always either very fine-grained aphanitic or glassy. For the most part, marine volcanoes eject lavas of basaltic composition (p. 127), whereas continental ejecta range from basalts to rhyolites (fine-grained equivalent of granite, p. 127). Most of the billowing steam clouds associated with nearly all eruptions consists of water of atmospheric origin that must have seeped through the ground into the volcanic conduit or the main magma chamber beneath. The high steam pressure generated by the intense heat of the rising magma ultimately blows out the solid rock of the volcanic conduit, and sometimes, as we have seen, the entire mountain itself.

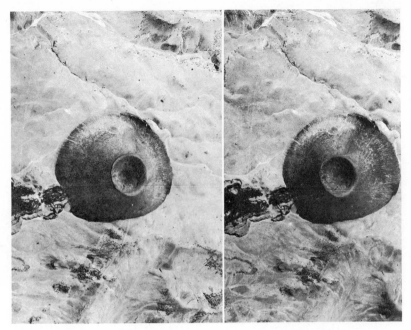

FIG. 9-13. Stereogram of cinder cone in Coconino County, Arizona. (U. S. Dept. of Agriculture photo, courtesy, University of Illinois.)

Origin of Volcanoes and Volcanic Magma

According to the almost identical distribution of volcanoes and earthquakes (Fig. 9-11), we must conclude that a closely related origin exists for both activities. Although the Hawaiian Islands are less beset with quakes than most of the Pacific volcanic zone, at least one of the volcanoes present there is perhaps the best studied because of its almost continuous activity and accessibility. About 20 miles down the slope from the crest of Mauna Loa lies another very active shield volcano, Kilauea, whose low elevation, 2000 feet, has made it an easy tourist site as well as an excellent natural laboratory. A striking photograph of its fiery activity is shown in Fig. 9-15.

For a long time the depth of the basaltic magma source of such volcanoes has been the subject of considerable speculation and study. Many scientists believed that chambers of magma originated wholly within the crust of the earth—which we now know to be only a few miles thick at the most, beneath ocean basins. The problem appears to have been settled following the establishment of a network of earthquake recorders (seismographs) around Kilauea by the U. S. Geological Survey's Hawaiian Volcano Observatory.

The seismographs have shown that earth tremors originate from 28 to 38 miles below the summit of Kilauea at the commencement of activity leading to an eruption. The small earthquakes are interpreted to mark

FIG. 9-14. Stromboli, a great composite volcano of which only one-third projects above the sea. (Courtesy American Museum of Natural History.)

the movement of molten rock into the conduits from a source some 25 to 35 miles deep in the mantle. As the magma nears the surface, the locus of earthquake activity shifts closer to the surface, indicating that an eruption is close at hand. A possible geological cross section through Mauna Loa and Kilauea, gleaned from the work of the Observatory, is shown in Fig. 9-16.

When viewed on a bathymetric chart of the Pacific Ocean (see Fig. 9-17), the main group of the Hawaiian Islands are seen to be part of a 1600 mile long submerged ridge system that, in places, forms volcanic islands reaching above sea level. Midway Island lies near the northwestern end of the chain and Hawaii near the southeastern end. It now seems that a fundamental weak zone extends through the crust into the mantle along this ridge-island system, providing localized conduits for the escape of magma from below.

Although the reason for the presence of the Hawaiian ridge and volcanic activity is not yet clearly understood, the volcanoes that border the Pacific Ocean and also lie on a roughly north-south belt in the Central Atlantic are now explainable on the basis of the sea floor spreading model described in Chapter 8. The Atlantic volcanoes lie on the Mid-Atlantic Ridge, which marks the zone along which new ocean crust is forming from the ascent of warm mantle rock. The island volcanoes of Iceland, the Azores, Ascension, Tristan Da Cunha, St. Helena, and the like that reach above the sea surface occur from concentrations of magma that emerge through local fissures on or marginal to the ridge axis.

Recall also from Chapter 8 that crust that is created in the ridges spreads laterally toward the continents as moving crustal plates, which then descend along continental margins. Often the zone of descent is at the ocean trench—island arc systems that lie just seaward of many continents. It is the extreme instability in these continent-ocean border lands that is associated with the very great volcanic activity common to the Pacific border zone as well as with islands of the Caribbean arc (West Indies). Most, but not all, recent volcanic mountains and volcanic activity seem to be causally related to the global tectonic pattern of sea-floor spreading between midocean ridges and the bordering island-arc systems.

Temperature Within the Earth

It is natural to ask what energy drives the process culminating in volcanic activity and what initiated and maintained the process at different places and times during the billions of years of the earth's life. Although we know the answer in general terms, namely that heat and gravity are the principal

FIG. 9-15. Night view of incandescent lava streams on Kilauea, Hawaii.

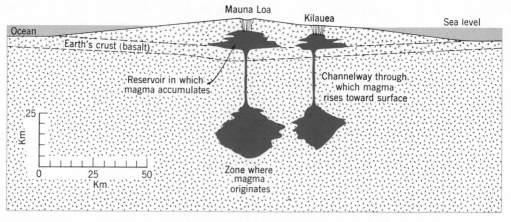

FIG. 9-16. Hypothetical cross section through Mauna Loa and Kilauea volcanoes, Hawaii. The magma, which is believed to originate within the mantle, rises through conduits to temporary reservoirs within the crust prior to eruption. (Redrawn after G. A. MacDonald in *Science*.)

energy sources, the exact method by which this energy is concentrated to produce localized volcanic action in the zones described is still elusive. But a number of important factors are already known.

If we measure temperatures in the crust of the earth—in mines, oil wells, drill holes, and the like—an average vertical temperature gradient of about 1.8°F per 100 feet is observed. (In some places it is much higher and in others much lower.) In other words it gets warmer the deeper you go at about 1.8°F/100 feet, or about 95°F/mile. Much of this gradient results from the effect of the hot interior, which we learned earlier is a result of the gravitational contraction of heavier materials toward the center, plus the effects of radioactivity. If we continue this gradient without change, the earth's center, 4000 miles deep, would have a temperature of about 100,000°F—an impossibly high value. The temperature at the center of the earth has been calculated from other technical considerations to be approximately 10,000°F, so that the gradient observed for the crust must be much higher than for most of the interior.

Since heat flows from high to low temperatures when a thermal gradient is present, the average heat flow out of the earth's crust can be calculated. The resulting heat flow rate from continents and ocean basins is surprisingly constant at about 1.5 millionths (1.5×10^{-6}) of a calorie per square centimeter per second or 40 calories per square centimeter per year. This heat flow from the earth's surface would raise the temperature of a layer of water 1 centimeter thick around the earth by 40°C (or a layer $\frac{4}{10}$ in, 72°F).

Although the temperature gradient—or heat-flow rate—is so uniform over most of the earth, over the midocean ridges these values rise considerably, reaching five times the world average. This was already touched on in the last chapter and is now explained on the basis of the ascent of warm mantle rock material beneath the ridge systems in the process of generating new ocean crust.

Two problems of importance remain: Why do the continent and ocean gradients have so nearly the same value (except for ocean ridges and regions of current volcanic action on land), and by what process is magma generated to form volcanoes in the unstable zones described?

In considering the first of these problems, we must recall that the continental crust is composed essentially of low-density rock, predominantly granite in composition, having an average thickness of some 20 miles, whereas the ocean crust is composed essentially of high density basalt with a thickness of about 3 miles. Now, lava reaching the earth's surface has a temperature of about 1800°F—approximately the melting point of the rock at the surface. Because this heat must be brought up

FIG. 9-17. The Hawaiian Islands represent a group of volcanoes that compose the southeastern end of a great midocean ridge of which Midway Island is the northwestern terminus. (Courtesy of National Geographic Society)

with the magma, we must conclude that the source rock presumably in the upper mantle should have at least this temperature. If the upper mantle is at 1800°F, how can the same thermal gradient prevail through 20 miles of continent as through 3 miles of ocean if the bottom of each of these units is at the same temperature?

We know that some of the answer lies in the high radioactive-mineral content of the continental rocks compared to the very low content of ocean basalts. The radioactivity of the basement rock of continents generates huge amounts of heat, which help maintain the high temperature gradients of the continents. But a problem still exists if we work back from the surface to calculate the temperature in the top of the mantle. For the continents, if we start with a mean temperature of 60°F, then at a depth of 20 miles with temperature increasing at 1.8°/1000 feet or 95°/mile, the temperature should be 1960°F, a reasonable value based on lava temperatures. But beneath the sea, if we start with a mean temperature of 40°F, at a mean ocean depth of 3 miles, the top of the mantle would only be $3 \times 95°$ or 295°, a value much too low. The explanation of this temperature problem still lies ahead and may await the long-sought—for drilling through the ocean crust.

The second question, regarding the cause of magma generation, still remains an unresolved problem. The erupted magma must carry up its high temperature from below. But the high pressures at great depths prevent a rock from melting at the same temperature as it melts at the surface unless the source rock is some low-melting-point fraction of the mantle. We know from the propagation of earthquake waves (Chapter 10) that the upper mantle is itself not molten. An adequate decrease of pressure will, of course, lower the melting point, but too great a pressure change would be required at the depths of known magma generation. Although radioactivity is now a well-known heat source, the great preponderance of lava is basaltic, and basalts and associated volcanic gases have a very low radioactive content, contrary to what would be expected. The details of magma generation remain as one of the frontiers of geology.

CRUSTAL UPLIFT AND ISOSTASY

Evidences of Crustal Uplift

Motions of the earth's solid crust may take place so slowly as to be imperceptible even from observations on the surface, or they may occur instantaneously and with a severity causing catastropic earthquakes as described in Chapter 10. Less severe motions may still be detectible by observed changes over time spans from several years to several centuries.

One of the classical examples of historical crustal motion is the behavior of the ruins of the ancient Roman temple of Jupiter Serapis on the seashore of the village of Pozzuoli, just north of Naples, Italy. Evidence

FIG. 9-18. Remains of the ancient Roman temple of Jupiter Serapis against the background of modern Pozzuoli, Italy. Stone-boring marine clams have pitted the lower 20 feet of the columns indicating a history of submergence of at least 20 feet, followed by subsequent emergence. (Photo by Maurice Rosalsky.)

FIG. 9-19. Elevated former beach on the coast of Alaska. Storm waves have cut back the cliff to form a modern tide-level beach. (Photo by George Plafker.)

of striking changes of sea level is preserved on the three columns that remain from the original structure (Fig. 9-18). About 20 feet above the ancient floor a distinct boundary separates the smooth upper portions from the water-worn, pitted lower portions of the columns. Some of the pits were formed by rock-boring marine clams; shells and shell fragments within the pits still bear evidence of their boring action. From these observations we can deduce that after construction by the Romans, the ancient temple was depressed through crustal downwarping to a depth of 20 feet. Following a period of uncertain duration, the shore zone was reelevated but did not quite reach its original level.

Many observations, usually of lesser historical interest, show similar crustal up- or downwarping. Other observations, which involve time intervals longer than historic time, are equally or even more striking. One of the best types of indicators of crustal movement during the rather recent geologic past is found along many coastlines of the world. As we noted in Chapter 6 on marine erosion, the sea carves characteristic beaches and sea cliffs through wave action. Along much of the western coast of North America, for example, many former beaches and sea cliffs now stand high above sea level (Fig. 9-19), and some of the elevated beaches still preserve remains of seashells of several types.

Other compelling evidences of crustal uplift are found in the series of raised coral reefs that surround many islands in the West and East Indies. Coral reefs grow up from the shallow sea bottom adjacent to tropical islands. Being formed of marine organisms, the living coral does not grow above sea level (Fig. 9-20), although wave action heaps coral

FIG. 9-20. Modern coral reef off New Guinea. The living coral (pale area) grows to mean sea level. Coral reefs above or below sea level imply either changes in sea level or crustal motion. (Photo by Rhodes Fairbridge.)

debris somewhat above sea level. But over much of the tropical Atlantic and Pacific Oceans coral reefs now lie up to several hundred to a thousand feet above wave action. Only crustal uplift can account for this.

Uplift of crustal rocks of much greater age and to much greater elevations are found widespread over the continents. In southwestern United States, the Colorado Plateau, which includes parts of Arizona, New Mexico, Colorado, and Utah, lies between 7000 and 8000 feet above sea level. But most of these flat-lying rocks were deposited in the sea. Approximately 100,000 square miles of sedimentary rock, mostly marine, must have been lifted several thousands of feet since their deposition at about sea level.

Possibly even more striking than the raised surfaces of ancient deposition are the raised surfaces of past erosion. For example, broad and fairly level areas of the Rocky Mountains lie at elevations reaching 14,000 feet. These summit levels are underlain by either crystalline igneous rocks or rocks that have undergone extreme crustal deformation.

Such surfaces are usually considered as former erosional surfaces carved on mountains of an earlier cycle. Their presence at high elevations must indicate a considerable amount of broad crustal uplift (Fig. 9-21).

All of the examples and references just given refer to essentially vertical movements of the earth's crust. In many of these cases, such as the elevating of the Colorado Plateau or of the surfaces of erosion, large segments of the earth's crust are involved. And in the case of an elevated erosion surface, a mountain system formed at an earlier stage by igneous action or crustal deformation may be reelevated after vast erosion.

Before we go on to consider the major process of mountain formation, namely, crustal deformation, it will be helpful to examine what is known about the vertical crustal movement, so opposite to the pull of gravity.

The Principle of Isostasy

Historically, the problem of isostasy (equal standing) really began in the middle of the nineteenth century when a precise survey of India was carried out under the direction of Sir George Everest (whose name is now borne by Mt. Everest). In the course of this survey, the distance between two stations, Kaliana and Kalianpur, separated some 375 miles along a north-south line, was measured with great precision by two different procedures. One method was the standard geodetic surveying procedure (known as triangulation) which involves the extending of a series of triangles from a carefully measured baseline. The other procedure was an astronomic one in which the angle of elevation of a particular star was measured at the two stations to determine their latitudes. As explained in Chapter 2, the difference in elevation angle of two stars measured when on the same meridian is a direct measure of the distance between them and is the way in which latitude is determined. Although both methods were carried out with extreme care, a discrepancy in

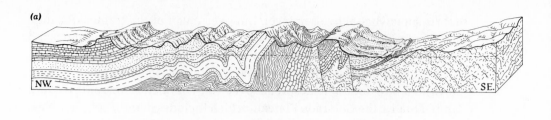

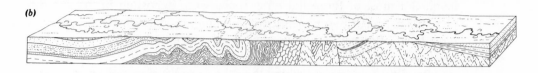

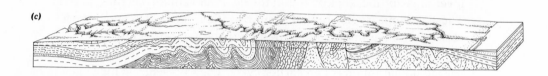

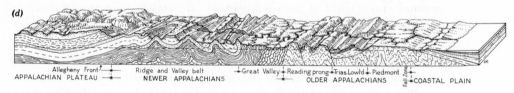

| Allegheny Front? | Ridge and Valley belt | Great Valley | Reading prong | Trias.Lowld | Piedmont | |
| APPALACHIAN PLATEAU | NEWER APPALACHIANS | | OLDER APPALACHIANS | | Fall Zone | COASTAL PLAIN |

FIG. 9-21. Example of an uplifted erosion surface formed over northeastern United States as exemplified by the simplified history in (a) through (d). The deformed rock structures in (a) are shown eroded to an essentially flat surface near sea level (peneplain) in (b). After up-arching as in (c), renewed erosion produced the topography of the regions labelled in (d). The summits of the ridges in (d) preserve remnants of the uplifted erosion surface. (Modified after E. Raisz, from A. N. Strahler, The *Earth Sciences*, John Wiley & Sons, Inc.)

distance of 500 feet occurred. This is equivalent to an error of only 5 seconds of arc in measuring the star elevations at the two stations. A resurvey led to repetition of the error.

The principle of isostasy emerged from these observations following a series of calculations by J. H. Pratt, a British amateur mathematician and archdeacon. Pratt's calculations were based on the following argument: Kaliana is in the foothills of the Himalaya Mountains while Kalianpur lies in the relative flatlands well away from the mountains. The astronomical latitude calculation assumed that the plumbline pointed to the earth's center (the broken line in Fig. 9-22) and referred the angle of elevation of the star to this broken line. (Recall the similar problem in

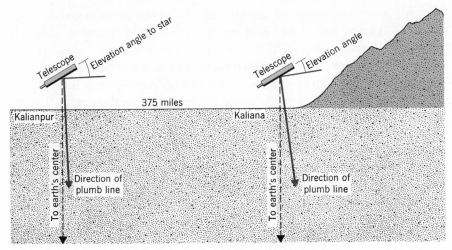

FIG. 9-22. In Everest's measurement of the distance between
Kalianpur and Kaliana an error of 5 seconds occurred in the star
elevation angles. Pratt tried to explain their error by assumming
the mountain deflected the plumbline more at Kaliana; but his calculations gave an even larger error of 16 seconds indicating he
overestimated the attraction of the mountain. The theories of isostasy were developed to explain the anomalous attraction of the
mountain.

connection with the geoid, p. 45.) Then, to explain the discrepancy, Pratt
assumed that the gravitational effect of the large close mountain mass
deflected the plumbline in the manner shown. He calculated that the
deflection should be 12 seconds at Kalianpur and 28 seconds at Kaliana,
much closer to the mountains. But the difference in deflection of the
plumbline from the assumed vertical is thus 16 seconds—not the 5 seconds
observed! In other words, the observed deflection of the plumbline was
one third of the value resulting from good calculations.

The observed discrepancy in observations of 5 seconds could not be
explained by errors in observational procedure, and the 16-second theoretical discrepancy could not be explained by error in the choice of the
Himalaya Mountain mass—the only variable involved in the calculated
deflection of the plumbline.

Airy's explanation. Within a few months of Pratt's surprising
calculations, a British astronomer, G. B. Airy, proposed the first solution
of the puzzle. Airy's explanation, which has come to be known as the
Mountain Root Hypothesis, proposed that the earth's crust is a rigid shell
of uniform rock density floating in a "liquid" subcrustal layer of greater
density than the crust. According to this concept of flotation, any elevated
portion of the crust must have a corresponding downward extension or
"root" as in the models in Fig. 9-23.

In Fig. 9-23a a series of wooden blocks of the same cross-sectional area and density but of different lengths are floating in water, normally denser than wood. Recall that a floating body displaces its weight (in water). The longer blocks thus extend deeper in order to displace larger amounts of water. Now consider a crust of varying thickness but uniform density floating in a denser *substratum*. As in the analogy with floating wooden blocks, the elevated regions extend deeper than the lowlands or ocean basins. But in this model we would find that the total mass under any given area down to the line labeled "compensation level" would be the same. The larger volume of rock under the mountain is balanced by the rock beneath the lowland because here, mantle rock of higher density comes closer to the surface. This model can explain the puzzle revealed by Pratt.

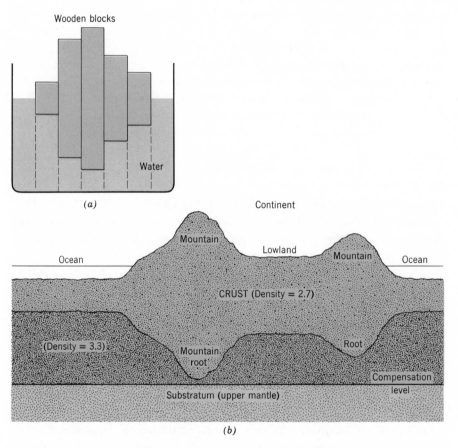

FIG. 9-23. The Airy model of isostasy. Airy assumed that mountains have roots that extend into the underlying denser rock. Because dense rock is closer to the surface beneath regions of low elevation, the same mass of rock underlies equivalent areas beneath mountains and lowlands. This is analogous to the simplified model in (a) where the total masses of adjacent columns of low density plus higher density water are equal.

If the bottom of the crust were smooth instead of mirroring surface topography, a plumbline close to the mountains would be deflected toward it more than one at a distance because of the excess mass projecting above the surface. But in the Airy model (Fig. 9-23a) the greater density of the rock a short distance below the surface tends to balance the effect of the high but less massive mountain.

Pratt's explanation. Several years after Airy's explanation of the problem raised by Pratt, Pratt himself came forth with an explanation, resembling but different from Airy's hypothesis. Pratt's model is summarized in the diagrams in Fig. 9-24. In *a,* five blocks of metal of the same cross-sectional area and the *same mass* but different densities are shown floating in the much denser liquid mercury. As each block has the same weight, it will sink to the same depth in the mercury. The mass of each metal is the same above this depth so that the least dense block—the aluminum (Al)—stands highest and lead (Pb), the densest metal, stands lowest.

Pratt considered that variations in density can account for variations in topographic elevation as in Fig. 9-23b. He explained that the rock beneath mountains must be lower in density than the rock beneath plains. Hence a thicker rock column would be necessary if the same mass of rock

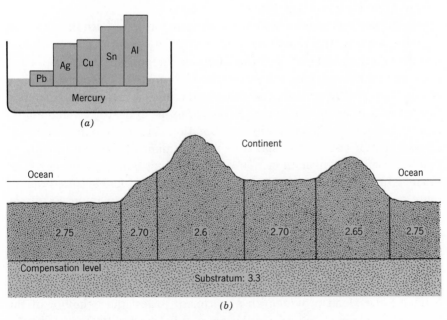

(a)

(b)

FIG. 9-24. In the Pratt model of isostasy regions of different elevation are considered to be underlain by rocks of different density. Low density rocks stand higher than higher density rocks so that the total mass down to a flat compensation level are equal. This is analogous to equal masses of different metals of the same area floating in denser mercury.

Crustal Uplift and Isostasy 311

were to be present beneath high- and low-elevation regions. Regions with density above average for the crust would stand low and regions with density lower than average would stand high; ocean crust, being below sea level, would be higher in density than continental crust.

This model also explains the original puzzle. The plumbline close to the mountains would not be deflected as much as originally calculated because of the deficiency of mass beneath the mountain compared with that beneath the lowland.

Both the Airy and Pratt theories involve conditions of balance or equilibrium among crustal units. In referring to this balanced state, the American geologist C. E. Dutton, in 1889, coined the term isostasy, meaning equal standing.

Isostasy and Crustal Uplift

Both the Airy and Pratt theories of isostasy accounted for the enigma for which they were proposed. It was not possible for at least a hundred years to discriminate between them. Both also accounted for many unexplained aspects of crustal movement. In either of the models shown in Figs. 9-23a and 9-24a suppose we slice a piece of a taller block and transfer the slice to the adjacent lower block. The weighted block would sink lower in the fluid and the shortened block would float up higher, although it would not quite reach the original level.

In the somewhat more realistic models in part b of both illustrations, suppose some of the mountain were eroded by natural processes to a lower elevation and the resulting sediment deposited on the adjacent lowland or the more remote ocean basin. The weighted lower crustal units would then sink lower into the substratum and the mountain, lightened through loss of mass, would rise. Repeated uplift of deeply eroded mountain regions can be explained by the "flotation" involved in both concepts of isostasy. Of course, the speed of readjustment of the real mountains would be much slower than that of the floating blocks in the artificial models. In Chapter 6, we noted that mountains that had suffered one or more periods of erosion to some base level seemed to undergo one or more periods of reelevation followed by renewed erosion. A good part of the explanation of this geologic cycling appears to lie in isostatic readjustment; uplift of eroded mountains is an expected consequence of isostasy.

The construction of the Hoover Dam on the Colorado River that formed Lake Mead provided an experiment to test the behavior of the crust when loaded. The accumulation of this large water body added a huge mass to the earth's surface. Within a matter of years crustal subsidence occurred in a roughly circular area centered on the lake (Fig. 9-25).

Nature performed an even larger experiment in the recent geologic past. As noted in Chapter 6, the last several million years was a time of

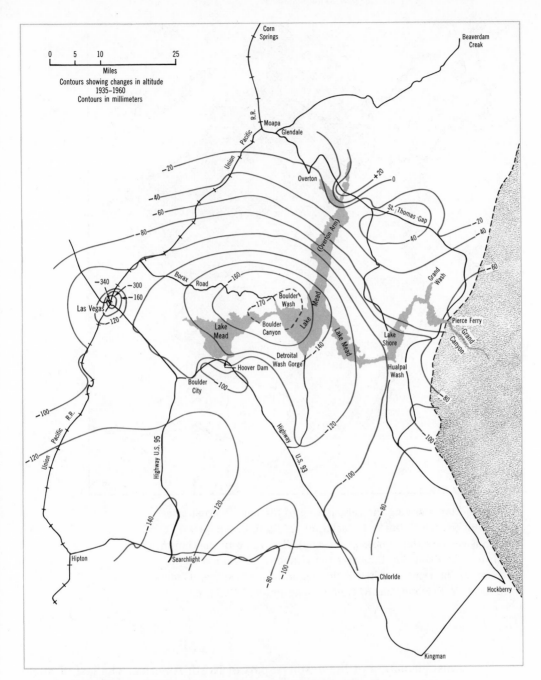

FIG. 9-25. Twenty-six billion tons of water plus an unknown amount of sediment accumulated in Lake Mead after completion of the Hoover Dam in March, 1935 to July, 1958. This increased mass depressed the earth's crust by amounts shown by the contours or lines of equal subsidence. The subsidence was greatest in a region centered on Lake Mead.

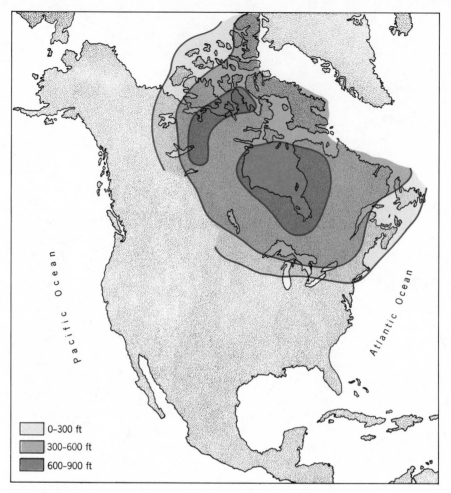

FIG. 9-26. Map showing the rebounding of the earth's crust after melting of the ice of the last glaciation. (Melting began 18,000 years ago and was completed 6000 years ago.) The region of maximum uplift over Hudson Bay indicates the region of maximum ice depression of the crust and hence the region of thickest ice. (Generalized after W. Farrand and R. Gajda, *Geographical Bulletin*.)

vast glaciation of the northern part of North America. The last of many glacial stages terminated about 18,000 years ago, at which time an ice sheet up to 2 miles thick covered much of eastern Canada. This excess crustal burden must have caused a vast subsidence of the crust beneath, because when the ice melted, most of eastern Canada was below sea level. Before isostatic reelevation of the ice-depressed land occurred, the sea left an indelible print over much of the region in the form of sea cliffs, beaches, and shell deposits. Today many areas of Canada have rebounded to at least 1000 feet above sea level. Hudson Bay seems to be a remaining relict of the once greater postglacial seaway (Fig. 9-26).

Airy or Pratt?

The concept of isostasy seems so basic to our understanding of many aspects of crustal behavior that much attention has been devoted to it. During the hundred years following the birth of the two theories, scientists sought for geologic tests in an effort to determine which of the two might be most applicable to the real earth. Both theories could explain all of the observations and processes described in the previous sections. With the almost dramatic "explosion" of geophysics into the realm of classical geology during the middle of the twentieth century, a more complete understanding of the problems involved became possible.

The application of seismology (Chapter 10) as a major tool in earth study revealed a number of important conclusions. It was shown from the study of earthquake surface waves that the continental crust has a thickness of some 20 miles and the ocean crust of only about 3 miles. It was also shown that continental mountains have "roots." The crust of the continents is thicker and extends deeper beneath most mountains and is thinner and more shallow beneath lowlands. These results resemble the Airy model in Fig. 9-23. But gravity and seismic as well as direct sampling showed that the continental crust of a density of 2.7 g/cc is ten percent lighter than ocean crust, whose density is 3 g/cc. Also, some lateral density differences do exist between highlands and lowlands on the continents.

We must conclude that both the Airy and Pratt models apply to isostatic balance within the crust. But the details of the balance are more dependent on the Airy model involving variable thickness than on the Pratt model of lateral density differences.

Modern Thoughts on Isostasy

Grand as the classical ideas of isostasy may be, they do not explain many of the important problems related to crustal uplift. The two models we discussed describe a reason for the existing balance between different segments of the earth's crust and indicate how uplift may occur in order to maintain the balance. But these models of isostasy do not indicate how the initial crustal arrangement a la Airy or Pratt came to be. For example, large plateaus, such as the Colorado Plateau, are now some 8000 feet above sea level over an area of about 100,000 square miles. But prior to uplift, the sedimentary rocks composing most of this 8000-foot thickness formed in an inland extension of the sea and thus lay at and below sea level.

In accordance with the Airy model, the uplifted plateau of rather low-density rock must contain a deep crustal root also of low density. As the plateau is a geologically young (middle to late Cenozoic) feature, its root must also be young. Since isostasy requires that the low-density root be present in order to support the uplift, the question arises, whence came the root. Either the low-density material was carried in from elsewhere in the crust, creating a huge problem involving subcrustal transfer, or it was developed in place.

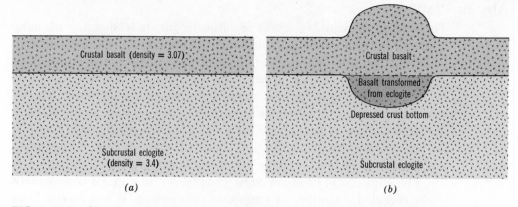

FIG. 9-27. Explanation of isostasy by a phase change in the upper mantle. If the upper mantle rock could be converted locally to a rock of the same composition but lower density (phase), a volume change must occur causing elevation of the surface. Eclogite is a dense rock which can change to basalt with a density 10 percent lower; a 10 percent increase in volume would occur from such a phase change.

The generation of low-density, deep crustal or subcrustal rock from previous high-density rock requires a fundamental change in the physical properties of the original rock. According to present speculations, an appropriate rise or fall of temperature, pressure or both may produce internal changes in the mineral composition analogous to metamorphism (Chapter 5), resulting in a rock type different in density from the original. For example, recall that we believe the crust to consist of a globe-encircling layer of basalt composing the ocean crust and extending beneath the granitic rock of continents. Basalt is composed essentially of plagioclase feldspar, pyroxene, and olivine and has a density of about 3. Under great pressures, the minerals of basalt can be transformed into the denser garnet and jade forms of pyroxene, yielding a rock known as *eclogite*, whose density is about 3.4. Basalt and eclogite are said to be different phases of the same rock composition—basalt being the low-density phase and eclogite the high-density phase.

Isostatic balance may be retained during crustal thickening through phase changes at great depth; in fact, crustal thickening may be caused by such changes. Suppose eclogite composes a large part of the upper mantle. A localized increase in temperature in the crust-mantle border zone might cause an expansion of eclogite, producing lower-density basalt. A ten percent decrease in density would occur in such a phase change, causing a ten percent increase in volume. Such an increase could only be brought about by an upward expansion, and if a considerable volume of rock is involved, a large crustal uplift would result, in the manner summarized in the schematic diagrams of Fig. 9-27.

In *a* of this diagram, a uniformly thick crustal layer lies on the denser subcrustal layer of eclogite. A phase change from eclogite to basalt might occur from a localized temperature increase or pressure decrease as in *b*, causing the basalt-eclogite contact to shift downward while the over-lying column expanded upward to accommodate the increased volume.

Although this presentation is simplified, it does indicate the way modern science is adding to the classical models of isostasy and indicates again one of the frontier regions of earth science.

We have now examined the development of mountains from igneous action and from isostatic uplift of older base-levelled mountains. Let us now combine much of the information gained so far and examine the major process of mountain building called *orogenesis*.

FOLD MOUNTAINS

Rocks are most commonly deformed by either folding or faulting but the great mountain chains of the continents are formed of rocks that have been strongly folded and are known as *fold mountains*. Such mountains are often complicated in structure through rock dislocations or *faults*. In the faulting process rock masses are displaced along a fracture or fault surface.

Fold Structures and Patterns

We have already learned that sedimentary rocks are formed in nearly horizontal strata. But examine the attitude of the strata in the photographs in Fig. 9-28. Upon observing these remarkable yet common contorted rock structures, one is tempted to remark, "How can solid rigid rock become so bent, twisted, and deformed without shattering into innumerable frag-ments?" Careful analysis of folded rocks plus laboratory experiments have shown that the deformation occurs from several internal adjustments on a microscopic level. As a result of these minute changes within rocks, seemingly rigid layers can behave as plastically as toothpaste squeezed from a tube and take on shapes like those in Figs. 9-28*a* and *b*.

Under conditions of less intense deformation, more regular folds develop like that in Fig. 9-29. Upfolds are known as anticlines and down-folds as synclines. The line along the crest of an anticline or trough of a syncline is the axis of the fold. Most regions of folded rock resemble conditions in Fig. 9-30, where erosion has modified the terrain into a series of parallel ridges and valleys. More resistant rocks form ridges, and weaker layers are eroded by streams into valleys between the resistant strata.

Note that the axes of the folds in Fig. 9-30 are horizontal. But the axes of most folds are inclined, or *plunge* beneath the horizontal, as in Fig. 9-31. In the erosion of folds with horizontal axes, the surface exposure (*outcrop patterns*) of the folded rocks takes on a parallel alignment of

FIG. 9-28a. Contorted rock, initially horizontal, of Ordovician age on the north shore of the Gaspé Peninsula, Quebec.

FIG. 9-28b. Simple deformational arch with extremely contorted rock to the left on the coast of Italy near Portofino. (Photo by K. E. Lowe.)

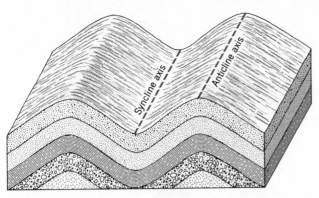

FIG. 9-29. Three-dimensional (block) diagram showing two uneroded upfolds or anticlines with an intervening downfold or syncline.

strata and of resulting ridges and valleys (Fig. 9-30). Where a plunging anticline is breached by erosion (Fig. 9-32), the surface view shows an outcrop pattern that converges in the direction of plunge. For a plunging syncline, the outcrop pattern converges opposite to the direction of plunge. The rock outcrop pattern overlying a series of plunging folds thus assumes a zigzag form as shown by the resistant rock pattern in Fig. 9-32, which includes one syncline between two anticlines.

The erosion of plunging folds composed of sedimentary beds of different resistance to erosion yields a distinctive topography characterized by converging and diverging ridges (and valleys) or zigzag ridges. A number of typical views of the erosional effects of plunging folds appear in the photographs of Fig. 9-33. Much of the rugged terrain of chains of fold mountains is simply a reflection of the differential erosion of beds of different resistance. The topography of a part of the Appalachian Mountains as depicted in the maps of Fig. 9-34 illustrates this effect quite well.

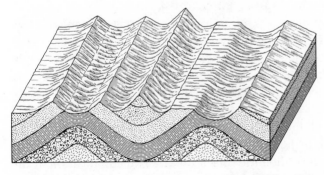

FIG. 9-30. Erosion of the geologic structure in Fig. 9-29 results in a series of parallel ridges and valleys that form over the resistant and weak rock strata respectively.

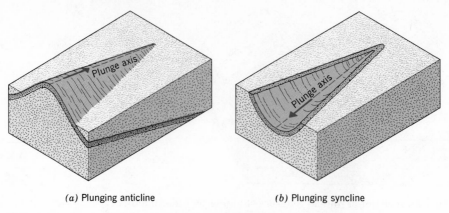

(a) Plunging anticline   (b) Plunging syncline

FIG. 9-31. Illustration of plunging folds where the fold axes dip below the horizontal in contrast to the folds in Fig. 9-29 where the axes of the folds are horizontal.

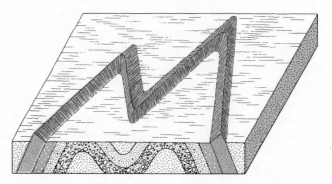

FIG. 9-32. Erosion of a pair of anticlines with an intervening syncline results in a zig-zag pattern of ridges in contrast to the parallel ridges of Fig. 9-30.

Origin of Fold Mountain Belts: The Geosyncline

Paradoxically, the real birth of a mountain system starts with subsidence, not with uplift. The fold mountains we see today are supported by tens of thousands of feet of sedimentary rock that were horizontal prior to the deformation and uplift. Since these rocks are mostly marine in origin, they must have been deposited in a downwarped crustal trough of some kind.

Our knowledge of the early history of mountains begins with the pioneering field work of James Hall, which has become a classic in the history of geology. Hall studied the rock structure and composition of the northern Appalachian Mountains, particularly in Pennsylvania and New York State, in great detail and made a number of profound observations that have since been discovered to apply to most other great mountain chains.

FIG. 9-33a. The end or "nose" of a plunging anticline in Arkansas. Note how the ridges (formed of resistant rock layers) and valleys (underlain by weak rock layers) alternate to reveal the details of the structure.

In an epic publication in 1859, Hall reported that the folded rocks of the Appalachians, where studied, reached a total thickness of 40,000 feet and were composed mostly of shales, sandstones, and limestones that were deposited initially in *shallow water*. He deduced the shallow-water marine nature of the sediment primarily from the presence of a vast number of fossils of shell life known to live only in the shallow parts of the seas (with depths no greater than the present continental shelves). He supported this with observations of other shallow-water features like mudcracks, raindrop impressions, and submerged plant remains. Rocks of the same age and often the same type, composing the plains to the west of the Appalachians, are at most only a few thousand feet thick.

FIG. 9-33b. Plunging anticline in Hot Springs County, Wyoming. The axis of the folds trends NE-SW. The southeast flank of the anticline dips gently to the southeast (toward the lower right corner). Where the stream crosses the resistant beds in the south-central part of the photograph, the narrow parts of the V-shaped notches point in the direction of dip of the beds. The northwest flank of the fold dips more steeply to the northwest. This is indicated by the asymmetry of the structure which is broader on the southeast flank where the dip of the beds is gentle and is fairly narrow on the northwest flank where the dip is steep.

When the thick folded rocks studied by James Hall are traced along their entire extent from northeast to southwest, the general model of origin becomes more complete. The basic fold structure extends from Newfoundland southwestward into Alabama and is some 150 to 200 miles across. The details of the rock structure and composition are somewhat complex. For example, in places the sedimentary rocks have become metamorphosed into slates and schists in response to very severe deforma-

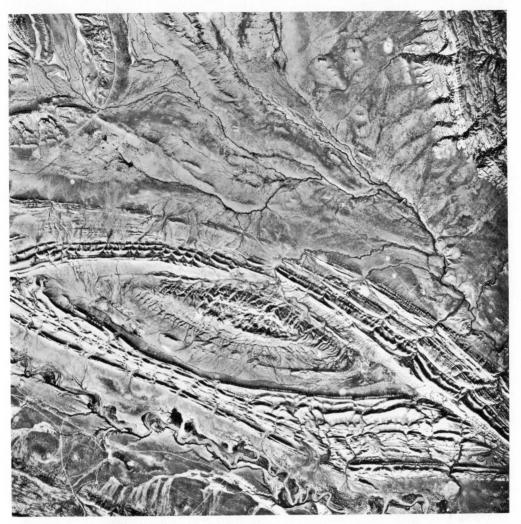

FIG. 9-33c. An anticline plunging at both ends as photographed in the Rocky Mountains of Wyoming. Again the structure is brought out by the differential erosion of resistant and weak beds. Ridges, such as those visible here, formed of steeply dipping resistant rocks are called hogbacks—a feature more common in the western part of the United States. (U. S. Geological Survey Photo.)

tion; in other places igneous rock is present in the form of both volcanics and intrusive (possibly granitized) crystalline masses. But the generalized picture of the Appalachians is one of a very elongated narrow belt of strongly folded, primarily shallow marine sedimentary rocks, tens of thousands of feet thick.

We have learned since the time of Hall, that deeper water sediments also occur in geosynclines, although shallow-water rocks usually predominate.

From this description there emerges the view of a long narrow trough filled with seawater of shallow depth that must have been present where now we find the Appalachian Mountain system. This downwarped crustal

Fold Mountains   323

FIG. 9-34a. Topographic map showing a zig-zag ridge near Hollidaysburg, Pennsylvania. The north-pointing nose represents an anticline plunging north; that pointing south is a syncline, also plunging north. Contour lines are drawn at vertical intervals of 20 feet.

zone was named *geosyncline* by J. D. Dana in 1873. But how could shallow water be present in a seaway that trapped up to 40,000 feet of sediments? The answer to this must be that the geosyncline subsided slowly as sediment accumulated, as suggested by the construction in Fig. 9-35.

Examination of all of the great mountain chains reveals that they, like the Appalachian, also arose from the deformation of sedimentary rocks of great thickness that were deposited initially in a slowly subsiding geosyncline. Today, some of the highest mountains are capped with sedimentary rock that bear fossil seashells of former shallow-water marine life.

Several other questions arise from these observations and conclusions: What caused the elongated geosyncline troughs to subside? Could

FIG. 9-34*b*. Geologic map of the region in (*a*). Each shaded band represents
the surface distribution of a particular rock formation. The map shows
clearly how the topographic ridge pattern in Fig. 9-34*a* is related to and controlled by the rock structure. The rock formation that follows the crest of the
ridge (the Shawangunk quartz-sandstone) is one of the most resistant rocks in
the Appalachian Mountains.

the simple weight of sediments accumulating in a very shallow initial
trough depress the crust tens of thousands of feet and if so, whence came
the initial trough? Or did some independent process generate geosynclines, with the sedimentation simply being an important byproduct?
Although these are some of the most cogent questions in the evolution
of the earth's crust, only some speculation based on present knowledge
can be offered now.

James Hall, to whom we owe the discovery of the geosyncline, believed that they were depressed by the weight of sediments deposited in
them. Another American geologist of great distinction, J. D. Dana, argued

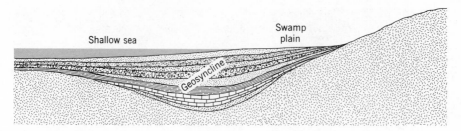

FIG. 9-35. Geological cross-section of a geosyncline filling with sediments derived principally from the erosion of a highland source to the right. The depth of water in the geosyncline is small. The great thickness of sediments is made possible because of continuing subsidence of the floor of the geosyncline as sediments accumulate.

instead that geosynclinal subsidence occurred from a different cause and that sediment filled the geosyncline as it subsided. In the years following both Hall and Dana, other scientists argued further against Hall's concept on the basis that it was contrary to the flotation mechanism inherent in isostasy. The argument from isostasy is contained in the idea that you cannot sink a rowboat by filling it with sawdust. In other words, a floating body floats because it is less dense than the medium that supports it. It will sink only until it displaces its weight in the fluid, as we noted earlier.

Suppose (Fig. 9-36a) we place a 40,000-foot-thick sedimentary pile of density 2.4 grams per cubic centimeter on top of the denser upper mantle whose density is 3.3. The intervening crust need not be considered here, as it is the more plastic mantle that yields to the load. The ratio of the densities of sediment to mantle is 2.4/3.3 or about 0.7, or a given volume of sediments would weigh only 0.7 the same volume of upper mantle

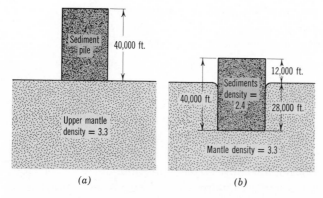

FIG. 9-36. Illustration of the argument that simple loading of the crust by sediments would not be adequate to cause a geosynclinal depression. See text for explanation.

rock—or the weight of a given volume of sediments would be balanced by 0.7 that volume of mantle rock. Hence a sedimentary column 40,000 feet thick would be balanced by a mantle column of the same area only 28,000 feet thick. Thus the sedimentary column would sink only 28,000 feet into the mantle, leaving 12,000 feet projecting above the surface. According to isostasy, this means that a sediment pile 40,000 feet thick simply couldn't depress the mantle by a like amount and could not accumulate in the shallow water that observations indicate must have existed. Only an initial depression with water more than 12,000 feet deep could have allowed such an accumulation—but this is contrary to observation. Thus, if isostatic balance prevails in the crust, simple load depression by sediments cannot explain geosyncline origin.

Despite this objection, the fact that so much of the geosyncline sediments were deposited in shallow water has made it tempting to explain the subsidence as a consequence of sedimentation. Mechanisms other than loading have been turned to. One suggestion (A. Scheidegger and J. O'Keefe) compensates the sediment load on the crust by the lateral displacement of a portion of the upper mantle. The necessary room for subsidence is thus provided. Whether such a movement or any related process can actually occur is yet to be demonstrated. According to another suggestion, the pressure from accumulating sediment might cause the rock material in the upper mantle to change to a denser phase, causing a shrinkage in volume that would provide room for subsidence.

It may yet turn out that a process independent of sedimentation causes geosyncline downwarping. For example, in Chapter 8 on sea-floor spreading, we examined evidence that the ocean crust moves horizontally toward the continents from an ocean-ridge system, and then bends downwards adjacent to either an island arc or a continental margin. If the ocean trench marginal to island arcs can be explained from depression as a crustal plate is deflected downward, it has been argued that geosynclines near continental borders may have a similar history. Sedimentation from the large continental source may roughly keep pace with the downwarping of the geosyncline, thus accounting for the shallow-water nature of the sediments.

Deformation of Geosynclines

Remember that a major goal in earth science is the unravelling of the natural history of the earth. This goal involves observation of existing rocks and rock structures and then the task of explaining their origin. We looked at the observable elements of geosynclines—the first stage in mountain building—and then considered the theoretical elements of their origin. Here we will examine further some essential elements of fold mountains following the geosyncline stage, and then look into the origin of the deformation of the geosyncline sediments.

In the course of mountain building, the thick pile of sediments deposited in the subsiding geosyncline becomes strongly folded and uplifted. The displacement of huge wedges of rock usually accompanies the folding. Fracturing and displacement of a crustal block is known as faulting, and the break itself is called a fault. Although a variety of motions of one block relative to another is possible in crustal deformation, *thrust faults* are the most important in geosynclinal deformation. These faults involve a low-angle *fault plane* (fracture plane), with the overlying rock wedge moving upwards relative to the block beneath, as illustrated in Fig. 9-37.

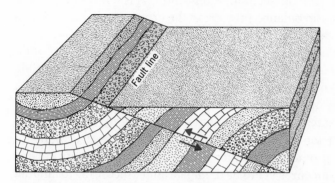

FIG. 9-37. Diagrammatic illustration of a thrust fault. The movement of the wedge above the fault relative to that below may be from feet to tens of miles.

Rock that was originally continuous may be displaced for many tens of miles along the fault surface. A thickness that may be a few tens to tens of thousands of feet of rock may be involved in the displacement. And the fault line along the earth's surface may extend for a few hundred feet to several hundreds of miles. The trace of the major thrust fault along the exposed face of a mountain is quite distinct in Fig. 9-38, where the exposure is not masked by the soil and vegetation that would be present in more humid regions.

Giant thrust sheets are intimately involved with the folding deformation. As we noted, scores of miles of movement may have occurred, as with the Lewis thrust visible in Glacier National Park, in the Northern Rockies of Montana (see Fig. 9-39). Folds in the Alps are contorted into almost unbelievably complicated forms (Fig. 9-40*a*). Slicing through such structures are great thrust faults like those in Fig. 9-40*b*, where displacements up to 75 miles can be shown to have occurred.

The features of mountains described here lead to a fundamental question that we should consider before trying to wrestle with the problem of the cause of the deformation.

In Fig. 9-41 we see the before-and-after stages of rock that has been folded. In both cases, it is quite clear that the folded and faulted rock

FIG. 9-38. View of a great thrust fault in Nevada. The upper tree-covered wedge represents relatively fertile Permian limestone lying above much younger and quite barren Jurassic sandstone. The dip of the thrust fault plane is very gentle. (Photo by Chester Longwell.)

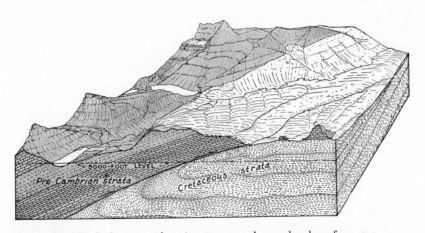

FIG. 9-39. Block diagram showing topography and subsurface geologic structure associated with the huge Lewis thrust fault in Glacier National Park, Montana. Chief Mountain (lower right corner labelled "C") represents a remnant of the overlying thrust "sheet" that has been isolated by erosion from the main body of faulted rock. (After Longwell, Knopf and Flint, John Wiley & Sons.)

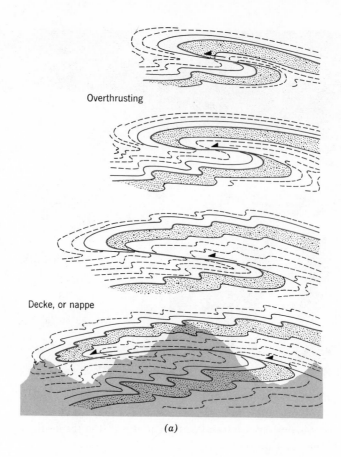

Overthrusting

Decke, or nappe

(a)

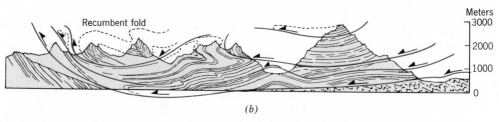

Recumbent fold

Meters
3000

2000

1000

0

(b)

FIG. 9-40*a*. Schematic illustration of complicated fold structures in the Alps Mountains. (After A. Holmes and A. Heim.)

FIG. 9-40*b*. Complicated thrust faulting and folding in the Alps. Modified from A. N. Strahler after A. Heim in the *Earth Sciences,* Harper & Rowe.)

has undergone shortening. But what about the basement, or crust beneath the sediments? Are only the sediments involved or has the crystalline crust also been shortened? For example, a tablecloth may wrinkle and distort above the table, with the latter unaffected.

Because the crust of the earth as a whole cannot shorten without the entire volume of the earth also shrinking, we are led to believe that the folding and faulting of geosynclinal sediments do not involve the entire crust. Note the folding and deformation of sedimentary rock above the relatively undisturbed basement rocks in the Swiss portion of the Jura

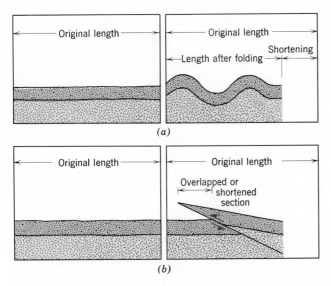

FIG. 9-41. Illustration of the shortening of the crust involved in folding and thrust faulting.

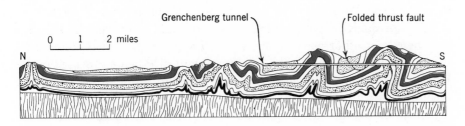

FIG. 9-42. Folding in the Swiss Jura Mountain section of the Alps showing how deformation may affect the weak sedimentary rocks above the stronger crystalline crust.

Mountains illustrated in Fig. 9-42. And so we are led to the question: what is the nature of the force(s) and process(es) that can deform tens of thousands of feet of sedimentary rock in a process that appears to be superficial to the crust beneath?

Cause of Geosyncline Deformation

A fundamental cause of deformation of the type occurring in geosynclines is the familiar *compression*, illustrated in Fig. 9-43, which results from two forces acting toward each other in a straight line, as in (a). The diagrams, b and c, show how horizontal compression can produce folds and thrust faults with the required shortening effect already referred to. For many years the simplicity of this explanation gave strong support to horizontal compression as the direct cause of geosynclinal deformation.

The question never answered, however, is how the strong horizontal compression is developed in the crust. Although the problem is by no means solved, we can examine some of the mechanisms from which compression may result.

Contraction. We must note at the outset that no completely satisfactory answers to the questions raised regarding orogeny have yet been provided. Historically, the earliest cogent explanation was based on a hypothesis of a shrinking earth. The analogy was made with an apple whose skin shrinks and shrivels as the core and calyx dry out. But innumerable geologic observations were completely contrary to the analogy of the earth's crust to the skin of an apple.

Continental drift. Another mechanism of geosynclinal birth and deformation was included in the initial Wegnerian theory of continental drift (see Chapter 8). According to this concept, deformation on the "windward" sides of drifting continents as they plowed through the resisting subcrustal medium would deform the crust and any sediments collected in geosynclinal troughs.

Plate tectonics. The theory of lithosphere plates, in nearly continuous motion (Chap. 8) has provided the most unifying concept available to explain the deformation of geosynclines and of the earth's crust. This is accomplished by the pushing of one oceanic plate against either a continental or other ocean plate, or by the collision of two continental plates. For example, when the supercontinent of *Pangaea* formed in the Paleozoic Era, the many continents came together from earlier separated locations. The great chain of fold mountains from the Alps to the Himalayas can be explained by the pushing of Africa and India into Eurasia. The Appalachian Mountain system of eastern North America can be explained by the collision of North Africa with North America. The mountain systems of western North American are more readily explained by the pushing effects of the Pacific plate into the western part of the North American plate. Although still in its infancy, it is likely that the theory of plate tectonics will explain most features of terrestrial deformation.

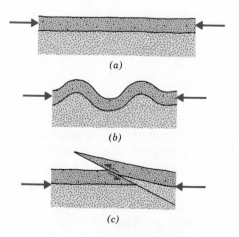

FIG. 9-43. Types of deformation in (*b*) and (*c*) resulting from compression of an original structure as in (*a*).

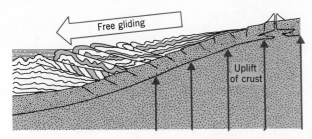

FIG. 9-44. Folding of rocks from the effect of gravitational gliding. (After A. Hills.)

Gravity sliding–or gravity tectonics. This is one of the most compelling mechanisms advanced to explain the details of deformational features that accompany mountain building. Suppose uplift occurs along all or part of an elongated geosyncline after a fairly thick accumulation of sediments has taken place. After a critical amount of uplift, depending on sediment type, thickness, and amount of water present under pressure with the rock, the strata will slide downslope, developing all types of wrinkles, folds, and thrust faults as illustrated in Fig. 9-44. If uplift is episodic, with erosion occurring between stages of sliding, the crystalline or more ancient rock basement may be exposed in the region of maximum elevation.

Note that vertical rather than horizontal forces are the motivation for geosynclinal deformation according to this model. The folding and faulting of the thick sedimentary pile are not the result of strong horizontal compressional forces but are rather the consequence of essentially vertical forces—first the uplift and tilting of the sedimentary mass, followed by slumping in response to the vertical pull of gravity. In this explanation, horizontal compression is a secondary and simple consequence of friction and resistance as sediments slump downslope. Note also (Fig. 9-44) that the sedimentary strata can actually shear along the basement rock over which they are sliding. Although the initial sedimentary expanse becomes shortened, the effect is superficial, since no shortening of the true crust is required. The structures in Fig. 9-40 are explainable rather readily by gravity tectonics.

But what causes the uplift responsible for the sliding? Once again we resort to plate tectonics. In the push of one plate against another crumpling and uplift occurs as one plate is thrust beneath the other. Gravity sliding then occurs in the uplifted continental margin.

TENSIONAL FORCES AND VERTICAL MOVEMENTS AND THEIR EFFECTS

Despite the importance of compression and regardless of its origin, we cannot omit consideration of one other force of great importance in

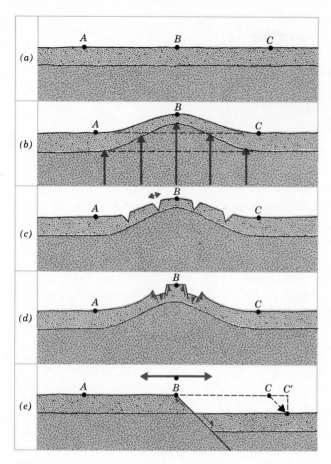

FIG. 9-45. Examples of deformation from tension. In (*a*) a layer is undeformed; in (*b*) differential vertical forces cause stretching and thinning of the layer; in (*c*) tensional forces shown by the arrow cause a layer to be pulled apart to form *tension joints* or fractures; in (*d*) rock displacement or faulting occurs from tension; (*e*) is a view of a single *normal* or *gravity* fault—the type that results from tension.

causing crustal deformation. Tension, the opposite of compression, involves two forces in line pulling away from each other. If compression produces shortening, tension causes stretching or elongation in the crust or some part of it.

A major cause of tensional deformation is the vertical movement referred to earlier in this chapter, whether from isostatic adjustments, phase changes, or other effects. We have already reviewed many of the strong evidences that uplift has occurred frequently and in many regions.

In Fig. 9-45 a crustal layer is shown undeformed in *a*. In *b*, the layer is stretched by thinning after vertical uplift; in *c* it is stretched by being pulled apart to form a *tension joint* or fracture. Points *A* and *B*, originally in contact, are now separated horizontally, indicating local crustal elonga-

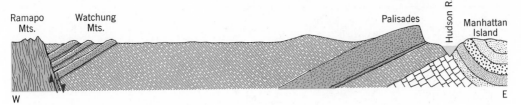

FIG. 9-46. Schematic section from the Ramapo mountains to Manhattan Island showing the down faulted and tilted sedimentary basin. The Palisades sill was intruded in the Triassic Period; the Watchung mountains are formed of lava flows. The sketch is not drawn to scale in order to emphasize geologic relations. The west-east distance is actually 30 miles while the breadth of the top of the Palisades is about a mile. The fault on the east face of the Ramapo mountains must involve a total displacement of at least 18,000 feet.

tion. If the tension (horizontal stretching) is more severe, actual faulting may occur as in (e), in which case the crustal elongation can be much greater. The detail of a single fault is illustrated in (d) where it is quite apparent that the horizontal distance ABC has increased by the amount CC^1. A fault of this type, in which the block overlying the fault plane moves down rather than up as with thrust faults, is called a *normal fault*. Because the motion can be interpreted as a down-dropping of this overlying block as tension pulls the crust apart, these faults are also called *gravity faults*.

Innumerable features are present that tell us that tension is an important cause of crustal deformation. In eastern United States the sedimentary basin illustrated in Fig. 9-46 is bounded by the Palisades sill on the east and the great Ramapo fault on the west and north. The relative motion between the uplifted crystalline Ramapo Mountains to the west and the down-dropped block to the east has been estimated to be between 18,000 and 30,000 feet. Triassic sediments accumulated in the subsiding trough. The Basin and Range country in western United States (Fig. 9-47), covering one tenth the area of the country (or about 300,000 square miles), is characterized by parallel mountain ranges, 50 to 75 miles in length, of normal-fault origin. The crests of the ranges are usually quite sharp and straight, as are the contacts between the mountain slope and the adjacent floor (Fig. 9-48). Such straight sharp features are typical of rather young fault-block mountains.

The Sierra Nevada (Fig. 9-49), which forms the western boundary of the Basin and Range Province and includes Mt. Whitney, the highest peak in the original 48 states, is itself a huge mountain system formed by a closely spaced series of normal faults having an extent in a northerly direction of 400 miles and a width about 80 miles.

FIG. 9-47. Relief sketch map of the great basin of southwestern United States. The east face of the Sierra Nevada Mountains is a major gravity fault zone. Smaller ranges within the basin are gravity faults, many of relatively recent origin. (Drawn by Guy-Harold Smith in Fenneman's *Physiography of Western United States*, McGraw-Hill Book Co.)

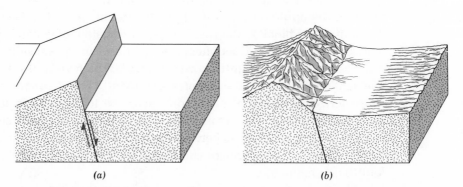

FIG. 9-48. (*a*) Sketch of the relative motion and topography resulting from a normal or gravity fault. The resulting mountain is often called a *block mountain*. (*b*) Stream erosion quickly modifies the appearance of a fault mountain even as uplift occurs resulting in this more typical appearance. The mountain is characterized by a straight base which indicates the fault trend.

FIG. 9-49. The east face of the Sierra Nevadas showing the famous cloud that forms from the disturbance in the wind flow that crosses the mountain from west to east. (Photo by Symons Flying Service.)

We have already referred to the great narrow valley at the crest of the Mid-Atlantic Ridge system (Fig. 274), a valley now explained as originating from the tensional pull of the rising and spreading crust rock. For much of its extent, the Rhine River flows in a great *rift* valley, also explained as originating from the down-dropping of a long narrow crustal block to form a structural feature called a *graben* (Fig. 9-50). Most prominent of continental grabens are the rift valleys of East Africa (Fig. 9-51) that contain the lakes Nyassa, Tanganyika, and Rudolf, and the Red Sea.

The examples given here, although only a few of many others that could be cited, are certainly convincing that tensional forces must be active in the earth's crust and must be considered one of the major mechanisms involved in deformation.

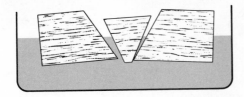

FIG. 9-50. Model of floating wooden blocks illustrating the formation of a rift valley or graben. As the blocks are pulled apart under tension, the inner wedge subsides. (After A. Holmes.)

MOUNTAINS FORMED OF CRYSTALLINE ROCK

Some of the loftiest mountains in the world are composed of crystalline igneous (mostly granitic) and metamorphic rocks formed from prior sedimentary rocks. Such high mountains owe their prominence to both recency of uplift and resistance of their crystalline rocks. In western North America, where uplift has been relatively recent geologically, the greatest elevations on the continent occur. In eastern North America, where the crystalline rocks form mountains of much greater antiquity, elevations and relief are correspondingly lower.

Mountains of crystalline igneous rock nature are primarily great granite batholiths that represent the cores of mountain ranges whose sedimentary rock covers have been eroded off. In Chapter 5 we considered some of the problems related to the origin of granite, namely magmatic intrusion versus granitization in place. From the location of granites now at the surface but originally deep in the heart of a geosyncline, we can relate the origin of these granite cores to the process of mountain building. High temperatures, possibly with the addition of new mineral material, generate granites either by the granitization process or by some giant crucible process in which original sedimentary rocks are melted and recrystallized to form the observed granite masses after prolonged erosion.

After one or more cycles of uplift and erosion following the initial mountain-building event, a mountain range formed of the original deep granite core is developed. Often, in close geologic occurrence with the igneous rocks are assemblages of highly metamorphosed gneisses and schists. These were originally sedimentary rocks also deeply buried and then metamorphosed in the process of mountain building. The low Piedmont range, which is some 100 miles broad and 750 miles long, extending from Maryland through Virginia and the Carolinas into Georgia, represents just such a belt of igneous and metamorphic rocks. Uplifted crystalline rocks in the Rocky Mountains, referred to on p. 201, form another but loftier example. Many fault mountains, such as the high Sierras and ranges in the Basin and Range Province, are examples of igneous and metamorphic rock complexes whose prominence is due to both uplift through faulting plus great rock resistance.

In addition to the rock masses that represent the cores of known mountain chains, broad terrains of crystalline rocks known as *shields* stand out as the heartlands of the continents. The rocks of greatest age in the earth's crust are found in these shields, whose distribution is shown

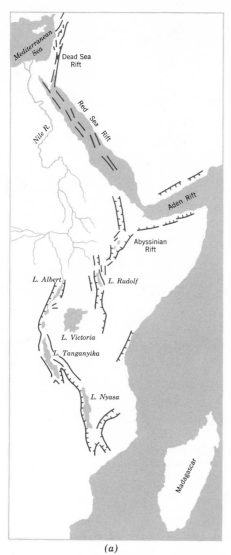

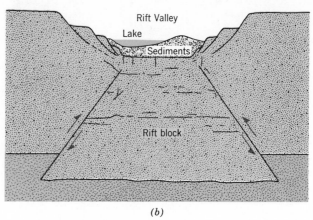

(b)

(a)

FIG. 9-51a. Distribution of the east African rift valleys the deeper ones being occupied by elongated north-south lakes.

FIG. 9-51b. A section across Lake Albert in (a), compare to the model in Fig. 9-50. (After Wayland.)

in Fig. 9-52. The exact nature of the mountains related to the origin of the granites, gneisses, and schists that compose the shields is shrouded in the mists of antiquity, but careful geologic field work is slowly unraveling the history of these regions.

Many geologists believe that the shields are original nuclei of the continents, which then grew slowly by accretion of low-density siliceous rocks that separated from the denser basaltic upper mantle. On the other hand, the continental margins are at present and have for some time in the past been the unstable and mobile parts of the continent. Repeated uplift and erosion has certainly recycled the rocks peripheral to the shields and destroyed any evidence of possibly earlier great age. Detailed surface and deep-drilling explorations should help clarify this problem, which is of such great geologic significance.

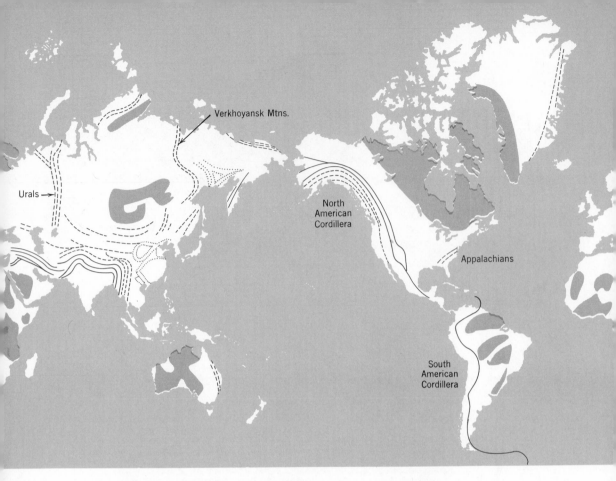

FIG. 9-52. The distribution of continental shields and main mountain chains. (After C. O. Dunbar and Longwell, Knopf and Sanders, John Wiley & Sons.)

STUDY QUESTIONS

9.1. From a geologic point of view, what is the difference between a mountain and a rugged plateau?

9.2. Distinguish between "cinder cones" and "lava cones."

9.3. Give examples of the different types of volcanic cones.

9.4. Explain how the source depth of Hawaiian volcanoes was determined.

9.5. Describe the types of material ejected by volcanoes.

9.6. Describe the global patterns of volcano distribution.

9.7. How is much of the marine volcanic activity related to tectonic plates or sea floor spreading?

9.8. Discuss the problem of the heat source involved in the melting of rock to form magma.

9.9. Explain briefly how we know that vertical movements of the earth's crust have occurred in the recent as well as the geologic past.

9.10. Explain the importance to geology of the surveying observations of Sir George Everest.

9.11. Distinguish between the Pratt and Airy models of isostasy.

9.12. What is meant by the term "phase change" as applied to the earth's composition?

9.13. How is isostasy explained in terms of phase change?

9.14. What do seismological observations tell us in support of either the Pratt or Airy models?

9.15. By means of diagrams, show the difference between horizontal and plunging folds.

9.16. Show by one or more diagrams how parallel ridges and valleys commonly develop from the erosion of fold rock structure.

9.17. Explain the term "geosyncline."

9.18. List several great geosynclines of the past. How do you know these regions were submerged to form geosynclinal downwarps of the crust?

9.19. Discuss the problem of origin and maintenance of geosynclines and their relationship to sea floor spreading.

9.20. How are mountain chains related to the movement of continental plates?

9.21. Summarize briefly the principal ways in which geosynclinal sedimentary rocks may have been deformed.

9.22. Distinguish between tension and compression. What is the general effect of each on crustal rocks involved in deformation?

9.23. What are the types of faults that form in response to compression and tension, respectively?

9.24. Give some examples of major gravity faults.

9.25. What is meant by continental shields? Where is the main shield of North America?

chapter 10

The San Andreas Fault is commonly marked by a long gash-like feature in the landscape such as is visible on the arid Carrizo Plains (near Paso Robles, California). The photograph is an oblique view to the north. The prominent fracture, having a trend from north-northwest to south-southwest, occurred in 1857. (U. S. Geological Survey.)

Earthquakes and the Earth's Interior

To those who live in seismically active regions, earthquakes are of most practical importance. To scientists, earthquakes provide the most powerful tool available for the study of the earth's interior. Most of our knowledge of what the earth may be like below the shallow zone of direct observation comes from the interpretation of seismic or earthquake waves.

When large segments of the earth's crust or upper mantle break and slip, energy in the form of *seismic waves* is radiated in all directions from the fracture. The earth tremors produced by the passage of these waves are called an earthquake. To those who experience even a moderately severe earthquake, the solidity of *terra firma* seems to disappear as the ground trembles and shakes. When the primary break is at or near the earth's surface, those in the vicinity experience the most terrifying of natural phenomena. Strong motions of the ground destroy one's sense of

FIG. 10-1. "Topsy-Turvy" landscape showing the badly dislocated, snow-covered terrain and structures in Alaska following the 1964 earthquake. (Life photo by Bill Ray.)

balance, topple buildings, disrupt roadways, and generally derange the engineering works of man. Locally, the face of the earth may be altered considerably. Landslides scar the hillsides, dam watercourses, and may change the pattern of streamflow. When the earthquake directly disturbs the ocean bottom, huge waves (*tsunami*) may be generated that roll across the oceans at jet-plane speeds. Upon reaching shorelines, even many thousands of miles away, these waves build up to disastrous heights, bringing death and destruction to unprepared residents.

SOME GREAT EARTHQUAKES

Five major earthquakes from 1908 to 1935 were responsible for the deaths of more than a half million people, and tens of thousands of others were injured. In the most disastrous shocks, entire villages and cities were destroyed, in some cases from the direct shaking and in others from landslides induced by the shaking. In a single devastating earthquake that struck Calcutta, India on October 11, 1737, 300,000 lives were lost—more than the present population of the state of Alaska.

FIG. 10-2. Damage from the San Fernando earthquake of February 1971 (near Los Angeles). A highway overpass (at interchange of Routes 5 and 210) has completely collapsed due to crustal shaking and the actual change in distance between the north and south abutments. Just beyond the fallen highway, the deck over the railroad tracks has also collapsed. The San Gebriel Mountains lie in the background. (U. S. Geological Survey photo by R. E. Wallace.)

In 1920, in the province of Kansu, China, a city built in and on enormous deposits of loess (fine angular wind-blown silt) was completely destroyed by great slides as the coherence of the bluffs of loess was destroyed; 100,000 lives were lost. A similar toll occurred in the giant shocks of 1908 in Messina, Italy, and 1923 in Kwanto, Japan. A severe earthquake in Jamaica, West Indies in 1962 destroyed much of the city of Port Royal, where the toll was higher than 20,000 lives as a huge tract of the city was hurled into the sea in a landslide. In 1962, the beautiful Mediterranean seaport of Agadir was almost completely shaken to the ground in about a quarter of a minute. Out of a population of 33,000, 12,000 were killed and 12,000 injured.

Among the records of natural disasters, the Lisbon earthquake of 1755 stands high on the list. Suddenly, on All Saints Day (November 1), a frightening shaking and twisting of the ground occurred to the accompaniment of an ominous thundering noise. The trembling of the ground continued for about six minutes, after which the city was a collection of ruins and rubble. Along the waterfront, the sea at first withdrew,

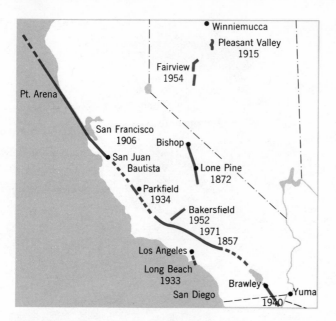

FIG. 10-3a. (above) Map showing the San Andreas fault in Western California. The fault is traceable from the Pacific Ocean northwest of San Francisco to the Mexican border. The dates of some strong earth shocks are given. Broken lines indicate where no historical movement has occurred although the fault is recognizable.

FIG. 10-3b. (opposite, above) The San Andreas fault is really a zone of multiple, closely spaced faults as evident by the five linear fractures that trend left to right (SSE-NNW) across the view. The most recent displacement in 1857 caused a displacement towards the north of the streams to the west of the fault. Note especially the large displacement of the main stream where it crosses the 1857 fracture. The view is westward. (U. S. Geological Survey.)

FIG. 10-3c. (opposite, below) The Carrizo Plains fracture of 1857 is also well marked in this view towards the east. Prominent stream displacement is again visible. (U. S. Geological Survey.)

exposing much of the harbor bottom, and then roared back with a wall of water many tens of feet high, drowning everyone and washing over everything in its path. Two more major shocks followed the initial one, delivering the final blows to the already devastated city. The effects of building collapse, landslides, fire, and seismic sea waves decimated the original population of 235,000 people by 60,000, to say nothing of the thousands injured and left homeless.

Some typical earthquake effects are illustrated in Figs. 10-1 and 10-2.

No history of earthquakes would be complete without references to the San Francisco shock of 1906 and the Good Friday earthquake in Alaska in 1964. The former is something of a historical classic in the United States. Although the greatest damage occurred in San Francisco, the earthquake

FIG. 10-4a. Fault displacement related to the Alaskan earthquake of March 28, 1964. The reef to the left (NW) of the fault line has been brought up from 4 to 5 meters (13 to 16½ feet). The fault can be traced, as shown, across the hill in the background. (Photo by G. Plafker, U. S. Geological Survey.)

involved horizontal slipping along much of the San Andreas fault, a continuous fracture running from Point Arena, north of San Francisco, southeastward to the Mexican border (Fig. 10-3a and b). The 10 to 20 feet of motion at the time were but one of a long sequence of slips during some one hundred million years, in the course of which the Pacific side of the fault has been estimated to have been displaced about 350 miles to the north relative to the continental side as the Pacific tectonic plate slides northward relative to the American plate.

The San Francisco shock was all the more terrifying because it began about 5 a.m., when most of the populace was still asleep. In one minute, San Francisco and many of the surrounding cities and villages were a shambles. Much of what was left in the main city was soon destroyed by the three-day fire ignited by the effects of the shock. Fewer than 500 persons perished, but the damage was estimated at nearly a half billion dollars.

The Alaskan Good Friday earthquake (March 28, 1964), which originated in Prince Welham Sound at the head of the Gulf of Alaska, was one of the strongest shocks of modern times (Fig. 10-4a and b). The zone of deformation extended for 500 miles southwestward from the coast. The area of deformation of nearly 77,000 square miles is the largest

FIG. 10-4b. Fault scarp (or cliff) at Hanning Bay, Alaska. The scarp of 13.5 feet extends in a northeast-southwest trend across ground that was level prior to the Good Friday earthquake. (Photo by George Plafker, U. S. Geological Survey.)

FIG. 10-4c. The southwest tip of Mantague Island, Alaska seen three days after the great earthquake. The white band at the base of the sea cliff, now 26 feet above water level, is dried calcareous shell material formed in the intertidal zone prior to the earthquake. (Photo by George Plafker, U. S. Geological Survey.)

FIG. 10-5. Harbor chaos in Kodiak, Alaska from the tsunami resulting from the major "Good Friday" earthquake of March 1964. Many fishing boats were washed onto land from the harbor in the left background. (U. S. Navy Photograph.)

associated with any known earthquake of historic times. On Montague Island, near the axis of deformation, a vertical displacement of 33 feet occurred, while offshore from the island a displacement of about 49 feet was measured. Most of the zone of maximum uplift was probably beneath the ocean, as indicated by the huge tsunami generated by this displacement (Fig. 10-5). These waves radiated over much of the Pacific Ocean, with disastrous effects along the coast of California, and reached a height of 26 feet along nearby shores.

So strong was the ground movement that very-low-frequency (inaudible) sound was recorded at Berkeley and San Diego, California, and even in New York City the seismic-wave motion was still strong enough to generate sound recorded on sensitive laboratory instruments. Along the Gulf Coast of the United States, the rocking motion of the ground threw canals and swimming pools into oscillations. "But it is an ill wind that blows no good." Were it not for these great earthquakes, many of the most fundamental mysteries of the earth's interior might never be solved!

TSUNAMI—SEISMIC SEA WAVES

Much of the devastation and terror that follow marine earthquakes arise from the huge ocean waves or tsunami that often accompany them. Some tsunami are generated directly by the vertical movement of a part of the sea floor. Others result from large submarine landslides triggered by the intense shaking of a major earthquake. Whatever the cause, the result is a series of large sea waves that travels rapidly across the ocean. Because these waves can travel through thousands of miles of ocean without suffering much decrease in energy, they often strike completely unprepared coastal regions far from the actual earthquake.

In view of their catastrophic effects, it is necessary to predict the times of tsunami arrival at distant shores. To do this, the time and place of the earthquake must be determined quickly and then the speed of wave travel must be calculated. Tsunami travel with the speed of all long-wavelength ocean waves according to the formula:

$$V = \sqrt{gh}$$

where V is the velocity, g is the constant value for the acceleration of gravity at the earth's surface (32 ft per sec per sec), and h is the depth of water. Most tsunami occur in the Pacific Ocean, whose average depth is about three miles (15,800 ft). The speed of a tsunami crossing the Pacific Ocean is

$$V = \sqrt{(32 \text{ ft/sec}^2) \ (15,800 \text{ ft})}$$
$$= \sqrt{505,600 \text{ ft}^2/\text{sec}^2}$$
$$= 711 \text{ ft/sec or 485 mph}$$

We can see from this simple calculation that tsunami fairly race across the ocean at jet-plane speeds. In mid-ocean, where the water is still deeper than our average value, the speed can reach 520 mph. From the tide gage record of the tsunami following the Alaskan Good Friday shock (Fig. 10-6) at Maui Island, Hawaii, we see that the period or time between successive waves was close to 30 minutes. Other tsunami have periods from 10 to 60 minutes. Because the wavelength (crest-to-crest distance) is found by the relationship:

$$L = VT$$

where V is the speed and T the period of the tsunami, the wavelength of the main waves in Fig. 10-6 is

$$L = (480 \text{ miles/hr}) \ (0.5 \text{ hr})$$
$$= 240 \text{ miles}$$

The wavelength of tsunami in general, then, is of the order of hundreds of miles. In the open ocean the height of the tsunami reaches only a few inches to a few feet. When spread over hundreds of miles, such

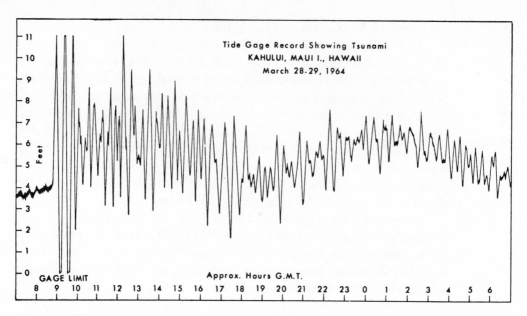

FIG. 10-6. Tide gage record at Maui Island, Hawaii following the Alaskan earthquake of March 28, 1964. The first three waves reached 11 feet in height. (Courtesy U. S. Coast and Geodetic Survey.)

waves are not recognized at sea, but when these long, high-velocity waves reach shallow coastal waters, both their wave length and speed decrease, causing the height to increase enormously as the wave energy gets "crowded" into a smaller distance. Thus at Maui, thousands of miles away from the epicenter of the Alaskan earthquake, waves 11 feet high were developed at the shore (Fig. 10-6).

At times, the trough rather than the crest arrives first, causing an actual withdrawal of the sea. This is followed by the first crest, which often rushes in as a great breaking wall of water toppling all in its path. The effects of the wave from the Lisbon earthquake have already been described. A major earthquake occurred near the coast of Chile on May 21, 1960. In addition to tremendous local damage, great destruction was wrought all along the South American coast by the seismic sea wave that occurred. The tsunami then traveled 6600 miles across the ocean to the Hawaiian Islands, where waves that reached 33 feet in height put much of Hilo, Hawaii out of commission. The waves continued on to the coast of Japan, 10,600 miles away, where their onslaught affected 150,000 Japanese, killing 180 of them and destroying most of the homes of the others. From a knowledge of the wave-arrival times and the time of earthquake occurrence, the tsunami had an average speed of 442 mph in traveling to Hawaii and 480 mph to Japan, matching well with the calculation given above.

Over the years, Japan has probably suffered the most from seismic sea waves, which have repeatedly overwhelmed the Japanese Islands, destroying whole villages in a single wave onslaught. The word *tsunami* is in fact a Japanese term.

In order to minimize the damage from tsunami, the United States Coast and Geodetic Survey has developed an effective Tsunami Warning Service with headquarters near Honolulu, Hawaii. Reports of all major Pacific Ocean earthquakes are quickly channeled to this unit, which then assesses the severity of the shock and issues necessary tsunami warnings to coastal and island regions. The value of the service has already been proven.

RECORDING EARTH MOTIONS

In order to make scientific use of earthquake waves, we must first record them. This is done with the use of a sensitive instrument called a *seismometer*. When connected to a recorder, the combined unit is called a *seismograph*. In principle, the seismometer is very simple; it makes use of Isaac Newton's principle of inertia, according to which an object at rest tends to remain at rest unless acted on by a new force. We can very easily illustrate the principle of a seismometer as shown in Fig. 10-7. If you suspend a heavy object from your outstretched arm and then swing your arm fairly rapidly from left to right, the weight will tend to remain motionless because of its inertia and because very little new force is transmitted to the weight by the string.

Now transfer the principle to the still-schematic model in Fig. 10-8, which shows the inertial unit attached to a horizontal boom in turn hinged to a vertical mast that is secured to the ground in some way. When earthquake waves pass a station, ground motion can be resolved into three components, north-south, east-west, and up-down (vertical motion). As

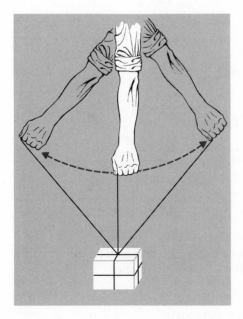

FIG. 10-7. If you swing your arm from side to side with a heavy weight suspended from your hand, the weight will tend to remain motionless. This simple demonstration illustrates the principle of the seismometer in Fig. 10-8.

oriented in Fig. 10-8, if horizontal north-south motion occurs, the mast will move with the ground, but the weight will tend to remain stationary because the boom has a hinged rather than a solid connection to the mast. If the instrument is rotated through 90°, it will then respond to east-west motion. If the hinge between the boom and the mast is horizontal so that only vertical movement can occur, the response of the instrument will be limited to the up-down component of wave motion. In practice, a regular seismograph station contains a minimum of three instruments arranged to record the one vertical and the two horizontal components of motion. Modern seismometers are illustrated in Fig. 10-9.

In actual operation, modern seismometers contain a variety of additions to the features illustrated schematically in Fig. 10-8, all of which greatly increase the efficiency and usefulness of the instrument. One important addition is the damping element. The suspended weight in Fig. 10-8 is really a form of pendulum that will oscillate with its own *natural period* when disturbed, as will a weight suspended from a string (the simple pendulum). In order for the instrument to record faithfully the true ground motion rather than its own oscillations, the motion of the weight must be damped. This is accomplished in most modern instruments by a sheet of copper attached to the boom, which is then surrounded by a strong magnet. Movement of the boom in the magnetic field sets up currents in the copper opposite to the motion of the boom, which is thus

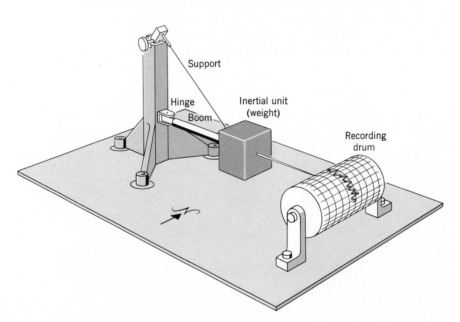

FIG. 10-8. Principle of the seismometer. When the earth shakes, the recording drum and support of the inertial unit move relative to the weight which tends to remain stationary. The record on the drum shows the relative motion.

FIG. 10-9*a*. Vertical seismometer. The cylindrical mass supported by the spring at the end of the horizontal boom is so constrained as to move only in the vertical direction. The instrument thus responds only to the up-down or vertical component of ground motion as seismic waves pass. (Lamont-Doherty Geological Observatory.)

FIG. 10-9*b*. Horizontal seismometer. The cylindrical mass and boom are free to move only in the horizontal direction. The horizontal cylindrical case attached to the frame encloses the magnet and the coil.

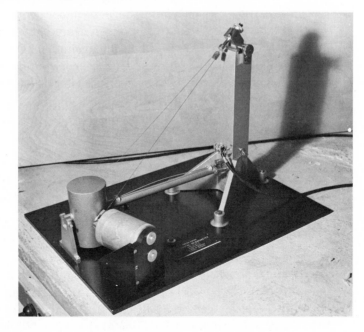

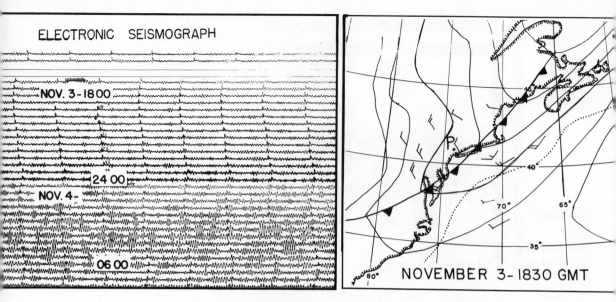

FIG. 10-10. Seismogram (left) showing the development of microseisms following 2400 hours on November 3, 1952. This corresponds to the time a meteorological cold front moved from west to east across the east coast of New England (right). The record was made at Palisades, New York (*P* on the map).

quickly brought to rest. On older instruments, a small extension attached to the boom rested in a dashpot of oil, providing the necessary resistance to damp the natural vibrations of the boom.

Recording of the seismic *signal* has been accomplished in many ways. On early instruments, a sharp scriber attached to the inertial element rested directly on a sheet of soot-covered paper wrapped about a rotating drum, as in Fig. 10-8. When the ground shakes, the drum moves with it so that the scriber traces a record showing the relative motion between drum and weight. A few instruments of this type are still in operation. Modern seismographs obtain greater magnification of ground motion through the use of either optical or electronic amplification systems.

SEISMICITY OF THE EARTH

Distribution and Depth of Focus of Earthquakes

The earth is a most dynamic planet. Its crust is in a nearly continuous state of agitation. Most of this trembling is in the form of *microseisms,* which are small seismic waves produced primarily by great ocean storms whose wave effects disturb the ocean bottom. An example of strong microseisms generated by a storm moving out over the Atlantic Ocean from Eastern North America is shown in Fig. 10-10.

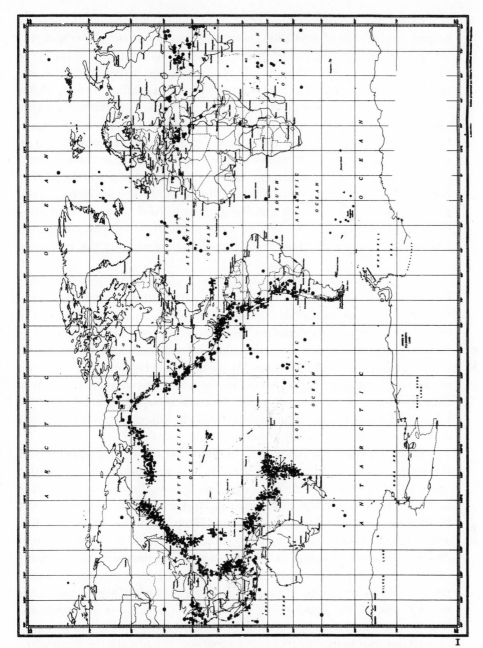

FIG. 10-11. Distribution of earthquake epicenters for the year 1955. Most of the epicenters cluster around the Pacific Ocean and in the Alpide zone (see text). A small but important group of earthquakes are located along the mid-ocean ridges (Recall Chapters 3 and 8).

Every day, many earthquakes can be detected on standard seismographs, and scores of small *microearthquakes* are recorded on specially sensitive seismographs. All of these indicate the great unrest of the earth. A description of the earth's seismicity involves the distribution of earthquake *epicenters*. These are points on the earth's surface directly above earthquake foci. Of great geologic importance is the fact that when earthquake *epicenters* are plotted on a map, as in Fig. 10-11, most are seen to fall into two major zones. The circum-Pacific zone contains about 90 percent of the world's earthquakes. The second, or Alpide zone, extends from the Azores Islands through the Mediterranean Sea and Near East, and after passing the northern border of India, continues through Sumatra and Indonesia to join the circum-Pacific belt in New Guinea. In addition to these two major earthquake zones, a statistically small percentage, but geologically important class of earthquakes originates in the midocean ridges.

The geologic importance of the zonal distribution just described becomes especially significant when it is realized that the distribution of the world's active volcanoes is nearly identical with it. Certainly one of the most geologically interesting and active portions of the earth lie in these coincident regions of earthquake and volcanic activity.

The significance of the seismicity picture must be considered in terms of the focal depth of earthquakes. In the now-classical classification of Gutenberg and Richter, earthquake foci are grouped as shallow—0–70 kilometers (0–44 miles); intermediate—70–300 kilometers (44–188 miles); and deep—greater than 300 kilometers (188 miles).

When looked at in detail, it is evident that the earthquakes of both the circum-Pacific and the Alpide Zones follow prominent arcuate patterns. In the case of the Pacific Zone they follow directly the pattern of tectonic plates described in Chapter 8. And when earthquakes are distinguished as shallow, intermediate, and deep, as in Fig. 10-12, a further striking fact is revealed. Recall first that island arcs are broad, arcuate chains of islands marginal to continents and distributed so that they are convex toward the ocean and concave toward the continents. Shallow-focus earthquakes, often so shallow as to lie at the surface, occur on the convex side of the arcs, either under the islands, or in the profound trenches or foredeeps just seaward of the islands. Epicenters of earthquakes possessing deeper foci occur progressively toward the continental or concave portion of the arcs. If the actual foci of the earthquakes are plotted in cross sections perpendicular to the arcs, the effect is shown even more strikingly, as in the diagrams in Fig. 10-13, which shows the results of recent very precise locations of earthquake foci. This pattern can now be explained by the descent of an ocean plate in the region of ocean deeps and island arcs.

The points representing foci of particular earthquakes lie in a curve that intersects the surface in the foredeep region on the outer margin of each of the arcs. Most of the catastrophic earthquakes along continental

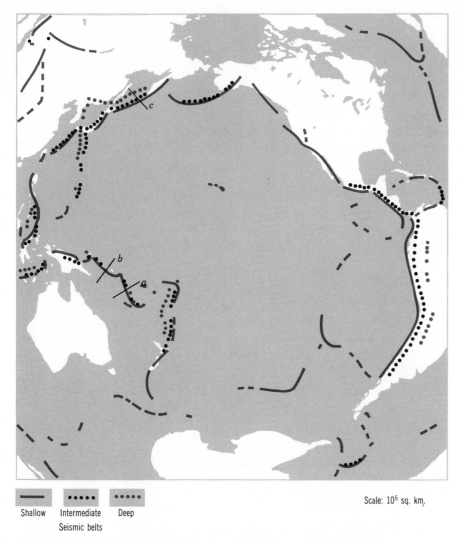

| Shallow | Intermediate | Deep | Scale: 10^6 sq. km. |

Seismic belts

FIG. 10-12. Distribution of shallow-, intermediate-, and deep-focus earthquakes around the Pacific Ocean. (After Gutenberg and Richter, Princeton University Press.)

margins are in the shallow-focus zone of this pattern. And most of the tsunami, described previously, originate in crustal fracturing in the ocean deeps marginal to the island arcs. We can certainly conclude that the most active and unstable parts of the earth crust must be the earthquake-and volcano-disturbed regions of the circum-Pacific and Alpide belts. The other active zone is the midocean ridge system (see Chapter 3). For the most part, the epicenters of the ridge shocks coincide with the remarkable depression, or rift, that marks the axis of the great ridge system. The striking relation of marine earthquake epicenters to the Mid-Atlantic ridge-rift system is shown in Fig. 10-14.

For many years, the relationship of earthquakes to marine structures

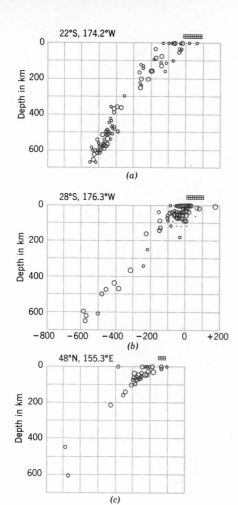

FIG. 10-13. The distribution of earthquakes foci with depth as plotted in vertical sections across the (*a*) Tonga Arc, (*b*) Kermadec Arc, and (*c*) Kuril-Kamchatka Trench. The positions of these sections are shown by lines *a, b* and *c,* respectively in Fig. 10-12. (After L. Sykes, Journal of Geophysical Research.)

was puzzling, but continued geophysical exploration of the sea bottom in the 1950s and 1960s has shown that all of these active regions are really the margins of tectonic plates.

Earthquake Magnitude and Energy

Newsworthy earthquakes are usually described according to the magnitude scale devised by G. F. Richter and known as the *Richter Scale.* The *magnitude* is a number that indicates the difference in severity between a particular earthquake and a reference or zero-level earthquake. The comparison is made as follows: The zero-level earthquake is one whose largest amplitude wave is $\frac{1}{1000}$ millimeters when recorded on a standard seismograph at a distance of 100 kilometers from the epicenter. The zero earthquake amplitudes have been determined by observation out to about 600 kilometers (375 miles) from the epicenter. For such relatively close earthquakes the amplitude of the highest wave is measured; from this value is subtracted the known value of the zero earthquake. To avoid the use of the large numbers involved in the large amplitude range of earthquake waves, the $\log_{10}$ of the amplitude is used rather than the actual amplitude in millimeters. Thus the magnitude, M, is expressed as

$$M = \log A - \log A_0$$

where A is the amplitude of the largest wave of the earthquake and A_0 is the amplitude of the zero earthquake.

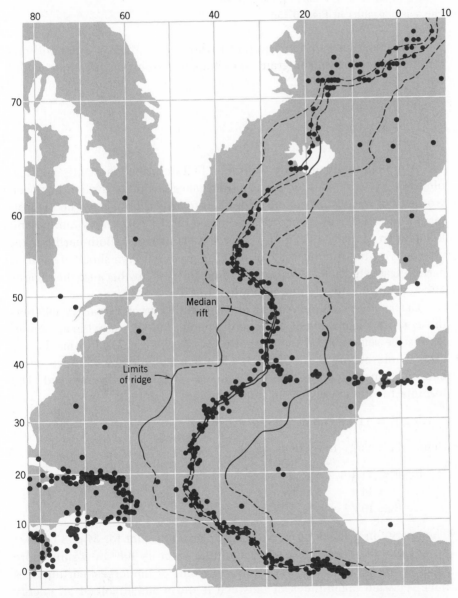

FIG. 10-14. The relationship of earthquake epicenters to the axial rift of the mid-Atlantic Ridge. (Courtesy B. C. Heezan.)

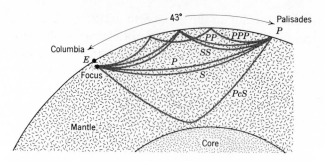

FIG. 10-15*a*. Paths of some of the earthquake phases marked on the seismographs in Fig. 10-15*b*.

The magnitude of an earthquake whose largest wave is 100 millimeters on a standard seismogram 100 kilometers from the epicenter would be

$$M = \log 100 - \log \frac{1}{1000}$$
$$= 2 - (-3)$$
$$= 5$$

Although the magnitude technique first described is strictly applicable to fairly close shallow-focus earthquakes, a similar system has been worked out for distant and also for deep-focus earthquakes. The highest-magnitude shocks ever recorded occurred off the coast of Columbia and Ecuador in January 1906 and in March 1933 in Japan. Both earthquakes are rated on the Richter scale at 8.9. The great Alaskan shock, described earlier, was rated at 8.5. Fortunately, their toll was low because the regions affected did not involve large populations.

Destructive earthquakes have magnitudes greater than 7, whereas those with a magnitude of 2 are barely perceptible close to the epicenter. It must be remembered that magnitudes are given as logarithms, hence, each whole magnitude number is ten times greater than the preceding one. In terms of energy, the difference is even more pronounced. The formula for the energy (E) of an earthquake in ergs is

$$\log E = 11.4 + 1.5M$$

Hence the energy in a shock of $M = 2$ is

$$\log E = 11.4 + 1.5 \times 2$$
$$= 14.4$$
$$E = 10^{14.4} \text{ ergs}$$

The energy in a shock of $M = 1$ is $10^{12.9}$ ergs and that for $M = 3$ is $10^{15.9}$ ergs. Thus, the energy of each magnitude value is $10^{1.5}$ or a little more than 30 times that of the preceding one. Some of the largest earthquakes have reported magnitudes of between 8 and 9 and involved an energy release of the order of 10^{24} ergs. This energy release is enough to run many of the major cities of the world for nearly a year!

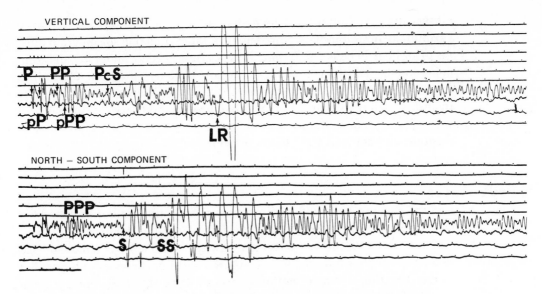

FIG. 10-15*b*. Seismograms from the vertical (top) and north-south (bottom) components of the Palisades seismograph from the earthquake in Columbia, South America June 29, 1967. Types of *P* and *S* phases are labeled. The beginning of the surface wave train is marked *LR*. (Courtesy R. B. Simon, Colorado School of Mines.)

Fortunately, great earthquakes (*M* above 8) with foci close to the surface are not too frequent; they occur once or twice a year somewhere in the active earthquake belts. But to residents of these regions, the threat of a disastrous shock is ever present. As the earthquake magnitude decreases, the frequency of occurrence rises enormously. For the range 6 to 7, about 200 shocks are recorded each year. If we consider all earthquakes down to magnitude zero, the annual number would be more than a million.

SEISMIC WAVES

The record of an earthquake as seen on a seismogram may look like a confused series of wiggly lines to the inexperienced observer. But from these wiggles much of the nature of the earth's interior has been deciphered. Before any interpretation can be undertaken, the seismogram must first be "read" in order to identify the different classes or *phases* of earthquake waves present. Then, from a knowledge primarily of the speeds of these waves, interpretations of earth structure and composition can be made.

Fig. 10-15(*b*) shows a combination of the vertical and horizontal seismogram components recorded at the Lamont-Doherty Geological

Observatory of Columbia University in Palisades, New York, from an earthquake in Colombia, South America. More than one seismogram must be examined because different phases are recorded better by different components. The epicenter lay at an arc distance of 43° or 2983 miles from Palisades. Many separate arrivals or wave phases are present in the record, but they all fall into two broad groups, *body waves* and *surface waves*. After a consideration of these waves, we will return to this seismogram and sort out the phases present.

Body Waves

Body waves are so called because they radiate throughout much of the earth from the source or focus. The fastest are the primary (*P*) waves, which are true sound waves traveling away from the focus. They are also known as *longitudinal* waves because the earth particles vibrate parallel to the direction of wave motion in transmitting the wave energy, as in Fig. 10-16. Primary waves are also known as *compressional* waves because the transmission of energy during wave travel is accomplished by a sequence of compressional (squeezing together) and rarefactional (separation) pulses as shown in Fig. 10-16. *P* waves can travel through any type of medium whether solid, liquid, or gas. Like normal sound waves, they travel fastest in solids and slowest in gases.

Secondary (*S*) waves, the other type of body waves, travel slower than *P* waves and thus arrive somewhat later. They are also known as *transverse* waves because the earth particles carrying the energy move at right angles to the line of wave progression as in Fig. 10-17. Here, particles evenly spaced along a line are displaced into the undulatory motion shown, as transverse waves pass. Although the line of motion is always

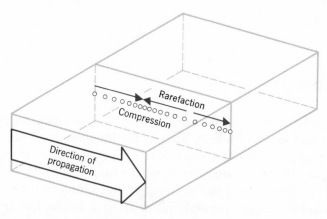

FIG. 10-16. Longitudinal ground particle motion showing regions of compression and rarefaction during the propagation of a *P* wave (large arrow). Particles all vibrate in the line of propagation. Fine arrows show relative motion of the ground particles.

perpendicular to the line of wave travel, the particle motion may lie in either the vertical or horizontal planes drawn, or in any one of the infinity of intermediate planes. Note that the motion in one half of a wave is all in the same direction; in the adjacent half it is in the opposite direction. Such movement is known as shear, and the waves are thus also called *shear* waves. Only rigid (solid) materials can support shear motion—try cutting water with a pair of scissors (shears). We will soon see the great importance of this limitation on *S* wave propagation.

The behavior of *P* and *S* body waves after emission from the focus is illustrated in Fig. 10-15*a* on the basis of the Palisades seismogram referred to earlier (Fig. 10-15*b*). The Colombia earthquake occurred on June 29,1967 at 10 hr 24 min 25 sec. The first motion on the Palisades record that occurred about 7.5 minutes later at 10 hr 32 min 15 sec is identified as the *P* wave whose path is shown in Fig. 10-15*a*. The curved path of this and all other body waves is a consequence of the increasing wave speed found with deeper penetration in the mantle. Normal refraction, exactly similar to that of light waves in passing from lower- to higher-speed optical media, causes a bending of all body waves, giving them a path always concave upward.

How do we know that this is the *P* wave rather than some slower phase, especially since all earthquakes do not produce good *P* wave motion? The identification can be made because the speeds of *P* waves have been well established for all depths in the earth. This leads to determinations of the travel times of *P* and other waves for all possible

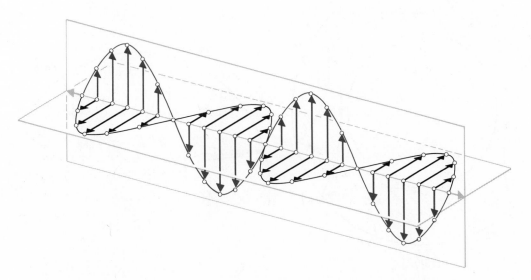

FIG. 10-17. Transverse particle motion during propagation of an *S* wave along the line shown by the long arrow. The particles can oscillate in any plane perpendicular to the wave path. Here motion in the vertical and horizontal planes is shown, but for a particular earthquake only one plane of vibration occurs of all those that are possible.

distances from earthquake epicenters. A travel time of 7.5 minutes and a few seconds is appropriate for the P waves for the distance between Palisades (P) and the Colombia epicenter (E).

Other P phases are identified as PP and PPP, which are P waves that have undergone one and two reflections from the surface, respectively. On the seismogram, "S" identifies the arrival of the principal transverse wave whose path is indicated in Fig. 10-15(a). The SS phase, which is an S wave with one internal reflection, is exactly analogous to the PP phase.

Whenever a P wave is reflected or refracted at some boundary surface, both P and S waves are generated at the interface, which may be at the surface, the base of the crust, or between the mantle and core. The P and S phase is thus a wave that has traveled down to the core as a compressional (P) wave and upon reflection has been converted to a shear (S) wave.

Figure 10-18 is a general summary of these and other body wave phases that radiate from an earthquake focus. In general, in addition to P and S designations, c refers to any wave reflected from the core, K refers to a P wave in the earth's outer core, i to a P wave reflected from the inner core, and I to a P wave passing through the inner core. Many of these phases appear on the interesting Palisades seismogram (Fig. 10-19) of the deep-focus shock of August 20, 1965 in the Banda Sea. Because

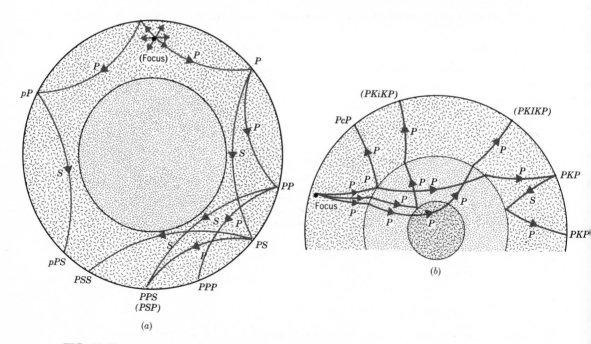

FIG. 10-18. A summary of the many wave paths that may radiate from an earthquake focus. The single shock generates all of the phases shown, but two views are given (a and b) to avoid overlapping of wave paths. (After K. E. Bullen.)

of the distance and depth of focus of 326 kilometers (204 miles), many phases that have been multiply reflected from the surface, in addition to others that penetrated the core, are present.

Dozens of phases representing all combinations of reflected and refracted waves have been identified, all of which add to our knowledge of the earth's interior. We will see soon how the interpretation of body waves has established the boundaries of the crust, mantle, core, and inner core. But before that we must consider the second major wave group—the surface waves.

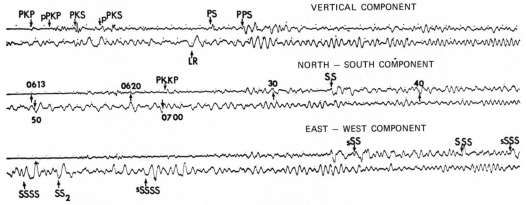

FIG. 10-19. Palisades seismograms for the vertical (top), north-south (center), and east-west (bottom) components seismometers for the deep-focus shock of August 20, 1965, in the Banda Sea. A large number of phases are present because of the deep focus and the long distance from the epicenter. Many of the phases whose paths are shown in Fig. 6-18 are recorded here. (Courtesy R. B. Simon, Colorado School of Mines.)

Surface Waves

On the seismogram in Fig. 10-15b, the large waves (LR) on the right side of the record are surface waves. These are the slowest and usually the largest amplitude waves on a seismogram. Before their true nature became known, they were designated L waves in view of their size compared with P and S waves. Surface waves are called so because they have their largest amplitude at the earth's surface and actually travel around the earth rather than through it, as with body waves. We now know that the large-amplitude oscillations following P and S waves are composed of two very different types of surface waves, both of which are particularly important in working out the layered structure of the earth. Lord Rayleigh, the eminent British physicist of the 19th century, was the first to predict the characteristics of one group of surface waves that now bears his name.

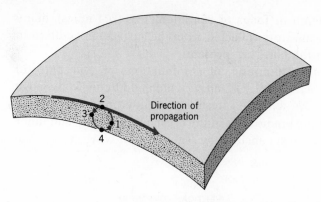

FIG. 10-20. The retrograde particle motion that occurs with Rayleigh wave propagation. The particle moves in a vertically-oriented ellipse. At the top, the particle moves opposite to the direction of wave propagation.

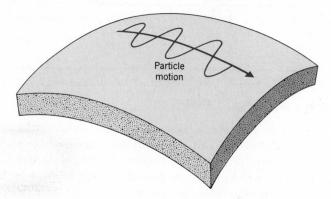

FIG. 10-21. Ground particle motion associated with Love waves. The motion is transverse to the direction of propagation (arrow) and always in the horizontal plane.

As Rayleigh waves travel, the earth particles move in an elliptical path confined to the vertical plane, which includes the direction of wave propagation as in Fig. 10-20. Notice that the ground particle, when at the top of its orbit, moves opposite to the direction of wave travel. The particle motion is thus retrograde. The velocity of Rayleigh waves is related to the speed of the transverse or shear waves, being 0.9 times this speed in whatever part of the earth they are traveling.

Love waves, named after another distinguished British mathematician who predicted their behavior, are the other class of waves whose dominant motion occurs at the earth's surface. As Love waves travel around the earth's surface, the ground particle in transmitting the wave energy undergoes shear motion confined to the horizontal plane (Fig. 10-21). This motion is the same as that with S waves except for the limitation to the surface plane of both the path of the wave and the movement of the ground particle.

Because Rayleigh and Love waves travel along the earth's surface and have their largest amplitude there, the more shallow the earthquake focus, the stronger are the surface waves. Very deep-focus waves generate weak or no surface waves. As the distance of the observer from the earthquake increases, these waves become increasingly strong compared with body waves. Body waves lose amplitude faster because their energy is spread through the three-dimensional earth, whereas surface-wave energy spreads over the two-dimensional surface. Note, for example, the prominent surface waves in the seismograms of Fig. 10-22.

Because earthquakes are so important in understanding the earth's interior, we will go through some simple examples of how they have been used for this purpose, and consider (1) how earthquake epicenters are located, (2) how body wave study has revealed the crust-mantle boundary and the presence of the core and inner core, and (3) how surface waves are used to determine the thickness of the crust.

LOCATING EARTHQUAKES

Fortunately, most earthquakes are not destructive because they either occur below the surface, are of relatively low magnitude, or are remote from habitations. But a very accurate knowledge of an earthquake epicenter is necessary before useful interpretations can be made. The basis for determining the epicenter is the known difference between P- and S-wave speeds. Let us go through a simplified exercise in epicenter determination for the damaging earthquake of July 21, 1952. The minimum number of seismograms necessary for a determination is three. In this case we will use those from U. S. Coast and Geodetic Survey stations in Columbia, South Carolina, Sitka, Alaska, and Honolulu, Hawaii. In Fig. 10-22, records from these stations contain time breaks at one-minute intervals. Time to the nearest minute is marked on selected time breaks, as well as the first motions identifiable as P and S. The large-amplitude waves following S are surface waves.

In Table 10-1, the arrival times of P and S waves are given as measured on the records. The third column gives the time difference or S minus P. The shorter the distance from the epicenter, the smaller is this value. The greater the distance, the larger is the value. Since the speeds of P and S have been established, travel-time curves like those in Fig. 10-23 can be constructed. In this diagram, the travel times of P and S waves are plotted for distances up to 3500 miles. The difference in travel time $(S - P)$ for any distance is thus the vertical separation between the two curves. For example, at 2000 miles from the epicenter, the S wave will arrive 4.63 minutes after the P wave. Conversely, if we measure an $S - P$ time of 4.63 minutes on a seismogram, the station is 2000 miles from the epicenter. The origin time of the earthquake can also be determined from

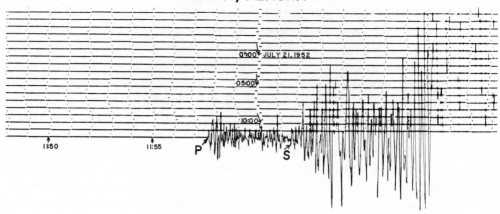

SITKA, ALASKA

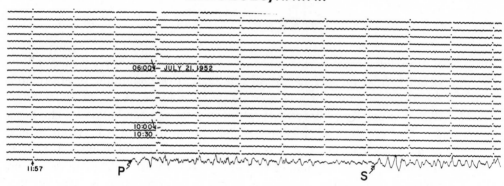

HONOLULU, HAWAII

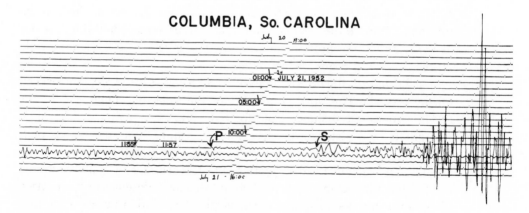

COLUMBIA, So. CAROLINA

FIG. 10-22. Portions of seismograms from Sitka, Alaska, Honolulu, Hawaii, and Columbia, South Carolina for the shock of July 21, 1952. *P* and *S* wave arrivals are identified for the purpose of locating the epicenter.

TABLE 10-1

	P Arrival Time	S Arrival Time	S − P	Distance to the Epicenter (miles)
Columbia	11:58.5	11:63.5	5.0 m	2200
Sitka	11:57.6	11:61.8	3.8 m	1450
Honolulu	11:59.5	11:62.2	5.7 m	2580

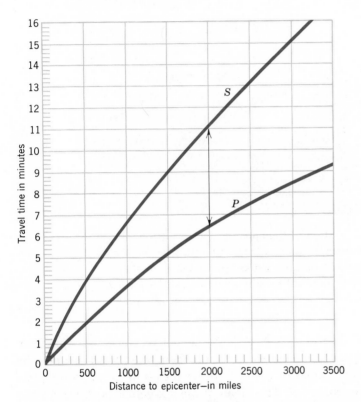

FIG. 10-23. Travel time curves for P and S waves used to determine the distance to an earthquake epicenter when P and S arrival
times are known. For example, for a vertical or time separation
between the curves of 4.63 minutes (11-6.3 min), the epicentral distance is 2000 miles.

these curves. The 2000 mile or 4.63 S − P value cuts the S curve at 11
minutes before the time of the S wave arrival and 6.37 minutes before
the P wave arrival.

Using the S − P values in Table 10-1, we can determine from the
travel-time curves the distances from each station to the epicenter; these
are given in the last column of Table 10-1. We now turn to the map in
Fig. 10-24 where the three stations are located. Using the scale of miles,
we draw arcs at the appropriate distance from each station (from Table
10-1). The arcs intersect with a small "triangle of error" close to the city

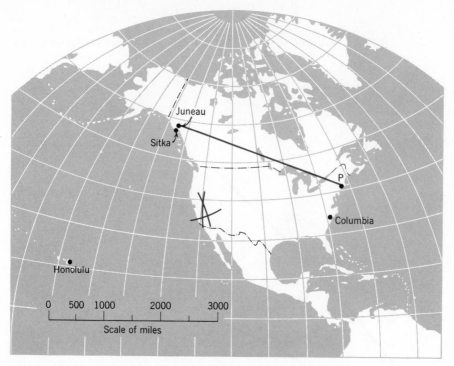

FIG. 10-24. Map showing the location of the epicenter of the earthquake shown in Fig. 10-22. The epicentral distances from Table 10-1 are drawn as arcs of circles about the respective stations (Sitka, Honolulu, Columbia). These arcs rarely intersect in a point but usually yield a small "triangle of error" which becomes reduced as the data from more stations are used.

of Bakersfield, California (35°N, 119°W), which was near the epicenter. The time of origin of the shock can also be computed from our data by the use of the travel times of *P* and *S* for each station for the three pairs of *P* and *S* waves recorded. The mean of the six values is 11 hrs 52 min 27 sec on July 21, 1952.

In actual practice, data from tens of seismograms are used rather than the minimum of three, and locations are obtained either on a very large globe to avoid scale inaccuracies inherent in a map like the one in Fig. 10-24, or by a high-speed digital computer that calculates a precise epicenter location. The exact origin time of the earthquake is another value of great importance in interpreting the data.

BODY WAVES AND THE EARTH'S INTERIOR

Crust-Mantle Boundary

The discovery of the crust as a distinct layer was made in 1909 by A. Mohorovičić, a Serbian seismologist. He did this by studying seismograms from a Balkan earthquake. The seismograms were arranged according to increasing distance from the epicenter, and the arrival times of the P wave were noted and were plotted against the distance from the epicenter, as in the lower part of Fig. 10-25. Mohorovičić made two new and rather strange observations. The first was the break in the travel-time curve between the positions of, say, stations 4 and 5. Secondly, he observed that at station 5, the first arrival was a weak P wave followed by a stronger P wave.

These observations were interpreted as in the top diagram. From the epicenter to point 4, the P waves were assumed to follow the normal paths shown. But to explain the change in the slope of the travel-time curve and the two P arrivals at station 5, a velocity discontinuity was assumed, with lower velocities in the overlying rock. The ray (line perpendicular to wave) refracted into the lower layer would have an average velocity so much higher than that of the direct ray that it would reach station 5 slightly earlier but would be weaker due to energy loss on twice crossing the discontinuity. Beyond station 5, only the weak refracted P waves can arrive. The break in slope between stations 4 and 5 was explained by the higher average wave velocities for P waves refracted below the velocity discontinuity.

The cause of the velocity difference was taken to be the difference in density of the upper layer, termed the *crust*, from that below, now known as the *mantle*. The lower boundary of the crust has since been known as the *Mohorovičić discontinuity*, sometimes shortened to "Moho" or "M" discontinuity. Further investigation by others has shown the crust to be a worldwide layer with varying thicknesses—about 30 to 40 kilometers (19–25 miles) beneath the continents and about 5 kilometers (3 miles) beneath the oceans.

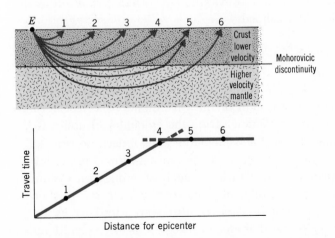

FIG. 10-25. How P waves revealed the velocity discontinuity known as the Mohorovičić Discontinuity or separation between crust and mantle (see text).

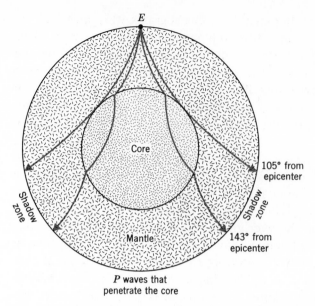

E

Core

105° from
epicenter

Shadow
zone

Shadow
zone

Mantle

143° from
epicenter

P waves that
penetrate the core

FIG. 10-26. The effect of the earth's core on the propagation of
P waves. No direct *P* waves are recorded beyond a distance of
105° from an epicenter. Beyond 143° *P* waves that penetrate the
core are recorded but these show travel times different from the
expected values based on normal *P* waves. The absence of *P* waves
in the shadow zone between 105° and 143° plus the "late" arrival
of *P* waves beyond 143° indicate the presence of a core of different velocity, properties from the mantle.

Precise determinations over small areas have been made by artificial
earthquakes produced by explosions with the same interpretational
scheme as that employed by Mohovičić. Average thicknesses beneath
large regions have been computed from a study of surface waves following
a procedure to be described shortly.

The Mantle-Core Boundary

If an epicenter is at E in Fig. 10-26, the *P* phase is not observed all around
the earth regardless of the strength of the earthquake. The last normal
P wave is recorded at an arc distance from the epicenter of about 105°,
and between 105° and 143° a *P*-wave shadow zone exists. Beyond 143°
the *P* phase reappears but is much weakened in amplitude and delayed
in time over the expected arrival time for such an epicentral distance.
The pattern is symmetrical on opposite sides of the earth for all shocks
no matter where the epicenter is. This was discovered initially by R. D.
Oldham in 1906; he concluded that a central region of low velocity must
be present in order to refract the rays as shown in Fig. 10-26.

The seismologist B. Gutenberg refined these observations and deter-

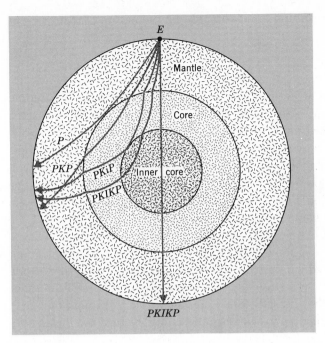

FIG. 10-27. The recognition of the earth's inner core from the identification of the PKiKP and PKIKP phases.

mined the size of the low-velocity core. By trial and error he was able
to determine the P-wave velocity in the core necessary to give the P arrival
times observed for penetrating the core.

In the continuing study of earthquakes whose waves penetrated the
central core, a very striking and significant fact emerged. No S-wave
motion was recorded in the shadow zone from earthquake foci on the
opposite side of the earth. Recall that S (or shear) waves can only propagate through rigid materials or solids. The conclusion from the S-wave
deficit was that the core is not solid, at least as in regard to having enough
rigidity to support S waves. But if it is a liquid, it is not a familiar kind.
The tremendous pressures in the deep interior would keep a rock quite
rigid at surface temperatures, but at the temperature of about 10,000°F
believed to prevail in the core, even this high-density material is in a
molten form.

The lack of rigidity also explains the reason for the slowing of P
waves, resulting in the shadow zone. It is interesting to consider the
scientific chain of reasoning involved in the examples already given and
still to be considered. The facts in the reasoning are (1) the presence or
absence of particular earthquake phases, (2) the knowledge that an earthquake occurred at a particular time and place, and (3) the travel time of
the waves. From this information is figured the wave velocity—a computed fact. From here on, deductions are made on the basis of established

seismologic theory resulting in information on the nature of matter in the interior.

The Inner Core

For many years this structure of crust, mantle, and core was accepted as the gross zonation of the earth as available from a study of seismic waves. But as higher-quality seismographs were developed it was discovered that faint P waves could indeed penetrate into the shadow zone. After a very painstaking study of many seismograms from different stations, Miss Inge Lehman, a Danish seismologist, accounted for the strange P phase by developing a new earth model containing an inner core with seismic velocities higher than that of the outer core. The effect of the higher velocity on the inner core would be to reflect and refract P waves into the shadow zone as shown respectively by the two paths PKiKP and PKIKP in Fig. 10-27, which uses the nomenclature described under "Body Waves." As a further indication of the higher velocity nature of the inner core, it is observed that PKIKP phases coming directly through the earth arrive earlier than expected for waves penetrating the uniform liquid core of the older model.

In view of the extremely high pressures in the central core region and the observed higher P velocities compared with those of the "outer core," it has been suggested that the inner core is solid. One way to establish this hypothesis is to find phases whose travel times are appropriate to shear waves while in the inner core. This is a subject of current research.

SURFACE WAVES AND THE EARTH'S INTERIOR

Application of surface wave data to the study of the earth's interior involves a rather subtle use of seismic wave speeds. This powerful method of applying Love wave and Rayleigh wave theory to the determination of the earth's layered structure was developed by Maurice Ewing of the Lamont-Doherty Geological Observatory soon after the close of World War II. So important is the method that we will consider it in some detail and then review one example of its application.

Variation of Surface-Wave Speed

Before we study this procedure, we must review some important characteristics of waves in general and seismic surface waves in particular. In the series of waves in Fig. 10-28, the wavelength (L) is defined as the distance between any two successive wave phases, such as the two crests

separated by the distance L. The period of a wave is the time it takes a full wavelength to pass a stationary point. If the wave of length L takes 10 seconds to pass, its period is 20 seconds, and so on.

Also, we noted earlier, that surface waves have their maximum amplitude at the earth's surface. The wave amplitude then decreases very rapidly below the earth's surface. *The depth to which a surface-wave oscillation extends depends on the wavelength; the longer the wavelength, the greater is the depth of penetration of the oscillatory motion.* This behavior is exactly the same as with common water waves and is of fundamental importance in the interpretation of surface waves.

Recall that the speeds of surface waves are related closely to the speed of transverse waves in the same rock medium. In general, the speed of transverse waves increases with depth. Since both Love and Rayleigh waves travel in a layer of variable thickness depending on their wavelengths, it follows that the speeds of surface waves must depend on their wavelength. Waves of small length are restricted to thin surface layers where the S-wave speed is small; thus they have a relatively low velocity. Long surface waves travel in thicker layers and have a higher speed related approximately to the average speed of the S wave in that layer. Actually, a continuous range of surface wave speeds exist for their range of wavelengths.

Surface wave speed also varies with rock density such that for layers of the same thickness, wave speed is greater in layers of higher density. In this treatment we will assume that we know the density to a good approximation and simply examine the use of surface waves in determining thicknesses.

Wave Dispersion

When an earthquake occurs, surface waves of many wavelengths are generated simultaneously. As they travel, the waves separate into trains of waves with longer waves normally moving ahead faster and the shorter, slower waves lagging behind. *Dispersion* refers to this variation of wave speed with wavelength. How does dispersion show up on a seismogram? As surface waves travel away from the epicenter, waves of greater length, having higher speeds, outdistance shorter waves of lower velocity. Close to the epicenter, waves of nearly the same length have so nearly the same speeds that they are not separated into distinct components. But with increasing distance these waves do become resolved. Thus the longer the path, the greater is the stretching of the train of surface waves into

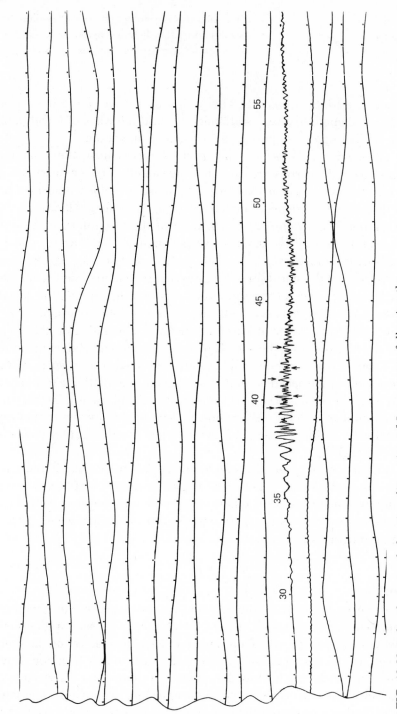

FIG. 10-29. Palisades record showing dispersion of Love waves following the shock in Alaska on July 10, 1958. Time is marked on the record at 5-minute intervals during the arrival of the earthquake waves.

individual components, with the longest waves arriving first. Such a series of waves is called a *dispersive* wave train. Each subsequent wave is of slightly shorter wavelength. Note the Love wave train in Fig. 10-29 recorded at Palisades on July 10, 1958 from a shock in the mid-Atlantic Ridge. On this record, time marks are again made at one-minute intervals. On a seismogram we cannot see the wavelength, but we do see the wave period, which is directly related to wavelength as in the formula $P = L/V$, which states that the period is equal to the wavelength divided by the velocity. Thus the periods of the first two waves are about 70 and 55 seconds, respectively, and so on.

Velocity-Dispersion Curves

If we have located an earthquake epicenter and determined the time of origin with good accuracy, we can then obtain a good measurement of the distance to a given recording station. We can also determine the time of arrival of each surface wave peak on the seismogram, and by subtracting the earthquake origin time from these values, we obtain the travel time of each wave. By dividing the distance by this travel time we can compute the velocity of each wave. We can also measure the period of each wave directly from the record. If we plot the period vs velocity of each wave, we get the velocity-dispersion curve, which looks something like the example in Fig. 10-30. Since waves of shorter period have lesser velocity, the slope of the curve must be downward toward the origin as shown.

The procedure just described provides us with an experimentally-determined dispersion curve. But, from surface wave theory it is also possible to calculate what the dispersion curves should look like for particular thicknesses of the earth's crust. A collection of such curves for Love waves calculated for crustal thicknesses of 30, 35, and 40 kilometers is shown in Fig. 10-31. If on the same diagram we plot the dispersion curve determined from a particular earthquake, the model whose curve most nearly matches the experimental one can be considered to be the most realistic interpretation of the structure being investigated.

Let us briefly review one application of the procedure. Fig. 10-29, already referred to, shows the Love waves recorded at Palisades, New York from an earthquake near Juneau, Alaska. The heavy line in Fig. 10-24 shows the wave path from the epicenter (E) to Palisades (P) over the distance of 4640 kilometers. Wave period for waves 1 to 9 in the dispersive wave train can be measured with fairly good accuracy directly on the record. Since the earthquake occurred at 06 hr 15 min 54 sec GMT we can determine the travel time of each wave after noting its arrival time on the seismogram. The distance (4640 km) divided by the travel time yields the velocity of each wave. In Table 10-2 we have a summary of the period and velocity of each numbered wave as determined directly

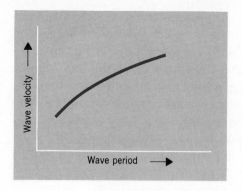

FIG. 10-30. Example of a dispersion curve of velocity vs period which shows that velocity decreases as wave period decreases.

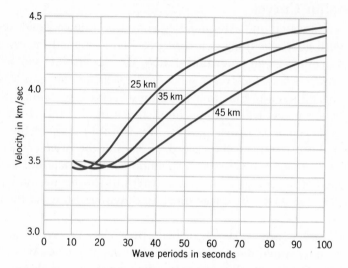

FIG. 10-31. Dispersion curves for Love waves traveling through a continent. The curves have been calculated for a fixed crusted density and the different thicknesses shown. The points plotted have been determined from the record of Fig. 10-30.

TABLE 10-2 Wave Period and Velocity for the Numbered Love Waves in Fig. 6-19

Wave Number	Period (sec)	Velocity (km/sec)
1	70	4.26
2	50	4.01
3	40	3.85
4	35	3.72
5	30	3.60
6	25	3.55
7	20	3.48
8	15	3.44
9	10	3.42

from the seismogram. The results are plotted as points in Fig. 10-31. By interpolation, we can estimate that the average thickness of the North American crust along the wave path is close to 21 miles (33 kilometers).

Summary of the Earth's Interior

The interpretation of seismograms has led directly to very accurate estimates of the dimensions of many of the earth's structural dimensions. Additional considerations have led to good determinations of the density of the layers. Then reasoning from other lines of scientific evidence, when added to that derived from seismology, led to further conclusions about the chemical composition of the interior below the level of observation.

Some of the additional information comes, interestingly enough, from outer space. Many of the meteorites that strike the earth's surface are composed almost completely of iron, or of a mixture of iron and iron-magnesium minerals, much like those of basic and ultrabasic igneous rocks. There is reason to believe that these fragments from space are remnants of a disrupted planet and that the different classes of meteorites represent a profile of the planet's interior, with the heaviest pure iron types having come from the core region.

Observations of volcanic rock thought to be extruded from the upper mantle plus meteorite observations and rock densities from seismic interpretations indicate that the upper mantle is composed of ultrabasic rocks like peridotite and/or eclogite. These are rocks somewhat richer in iron-magnesium concentrations than the normal basic igneous rocks such as basalt and gabbro, referred to in Chapter 5. Refer also to the summary diagram of the earth's interior in Fig. 4-1.

Results of some of the most recent research on the nature of the earth's interior is summarized in Table 10-3. Here we see some of the main physical and chemical properties of the gross structural elements within the earth. Knowledge of the crust and mantle has been refined into smaller divisions than those shown here, but further details are beyond our immediate scope.

ORIGIN AND PREDICTION OF EARTHQUAKES

The benefits of earthquake prediction to humanity would be vast. To science, prediction would permit setting up experiments at the proper places and times to learn more about the earth. But before prediction must come some understanding of their cause.

In general, earthquakes are caused by the movement of large masses of rock on opposite sides of a fracture plane. The movement may involve inches or less to displacements of tens of feet at a time.

TABLE 10-3 Summary of Properties of the Earth's Interior

Layer	Thickness		Density (gms/cm³)	Composition
	km	miles		
Crust				
Continental	32	20	2.7	Aluminum silicate rocks (acid rock)
Oceanic	5	3	3.0	Iron-magnesium silicate rock (basic rock)
Mantle	2900	1800	3.3 to 5.6	Ultrabasic rocks at top grading downward to Fe-Mg oxides
Outer core	2100	1300	9.5 to 12	Liquid Fe-Si and Fe-Ni alloy
Inner core	1400	900	13+	? Solid, Fe-Ni alloy

Elastic-Rebound Mechanism

It is now accepted by seismologists that most earthquakes result from the *elastic rebound* of rock that has suffered strain or distortion prolonged sufficiently until fracturing and movement (faulting) occurs. This mechanism of strain release is illustrated in Fig. 10-32, which shows schematically a rock unit in three stages of deformation. *A* shows the layer prior to deformation, *B* during deformation or strain, and *C* after fracture and elastic "snapping back" to the unstrained position. The break along which movement occurs is called a fault. Often, the elastic recovery involves an initial large return movement causing the main earthquake, followed by continued smaller movements if initial recovery does not relieve all of the strain. The latter movements cause many of the "aftershocks" that often accompany major earthquakes. The strain and resulting movements may be oriented vertically, horizontally, or in any intermediate position.

The building up of the strain may be the direct cause of the fault and earthquake, but it is certainly not the ultimate cause. The fundamental cause lies in stresses set up as tectonic plates slide past or beneath each other. Much of the mystery has been removed by the unifying concept of plate tectonics. Earthquakes at ocean ridges and trenches are now explained by effects of rising of warm rock and the descent of plates, respectively. Movements along major zones such as the San Andreas fault are related to the meeting of the Pacific and American plates.

The accurate location of earthquake foci has established the source of deep focus earthquakes to depths reaching beyond 440 miles, or well into the mantle. Interpretation of the nature of *P* and *S* waves from these deep sources indicates that they, too, originate from displacements along a surface; they too can be explained by the elastic-rebound mechanism. It came as something of a surprise to realize that large units of rock under the tremendous pressures and high temperatures of the mantle could fail

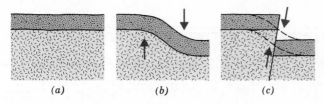

FIG. 10-32. The principles of elastic rebound as a cause of earthquakes. Three stages of deformation are shown. In the last stage rebound after rupture initiates the earthquake.

by fracture and displacement. This conclusion in turn has been very helpful in gaining a better understanding of the mechanical properties of the earth's interior.

Earthquake Prediction

At the present time many experimental programs are gathering data of the kind that may produce adequate warning of impending earthquakes. Some of the programs attempt to measure the very minute strains developing in rocks near the surface. The purpose of this is to evaluate the possibility of recognizing some threshold strain level following which a significant shock would be triggered. Other programs are recording the many microearthquakes that occur in active seismic regions. There may be up to 200 of these minute shocks in one day. They are being studied with the hope that characteristics of microearthquakes preceding strong shocks may indicate when major displacements are about to occur. The solution to this most vital problem is probably still a long way in the future.

LUNAR SEISMOLOGY

Because seismology has been so important a tool in studying the interior of the earth, it seemed natural to apply the technique to the study of the moon's interior. One of the most important scientific missions of the early landings was the installation of the first seismographs to operate on the moon. More than 160 separate seismic events were recorded by the instrument placed by the crew of Apollo 12 (November 1969). At least 26 of these events were small moonquakes whose magnitudes were between 1 and 2 on the Richter scale. Since all of the moonquakes occurred close to the time of lunar perigee (see p. 23), the large tidal stresses set up in the moon by the earth are believed to have caused the quakes.

Most of the natural events other than the moonquakes are explained as impacts of meteoritic objects. Two artificially caused events have actually been the most interesting. After the lunar module returned the crew of Apollo 12 to the main spacecraft on November 20, 1969, it was adjusted to crash back on the lunar surface. A similar experiment was

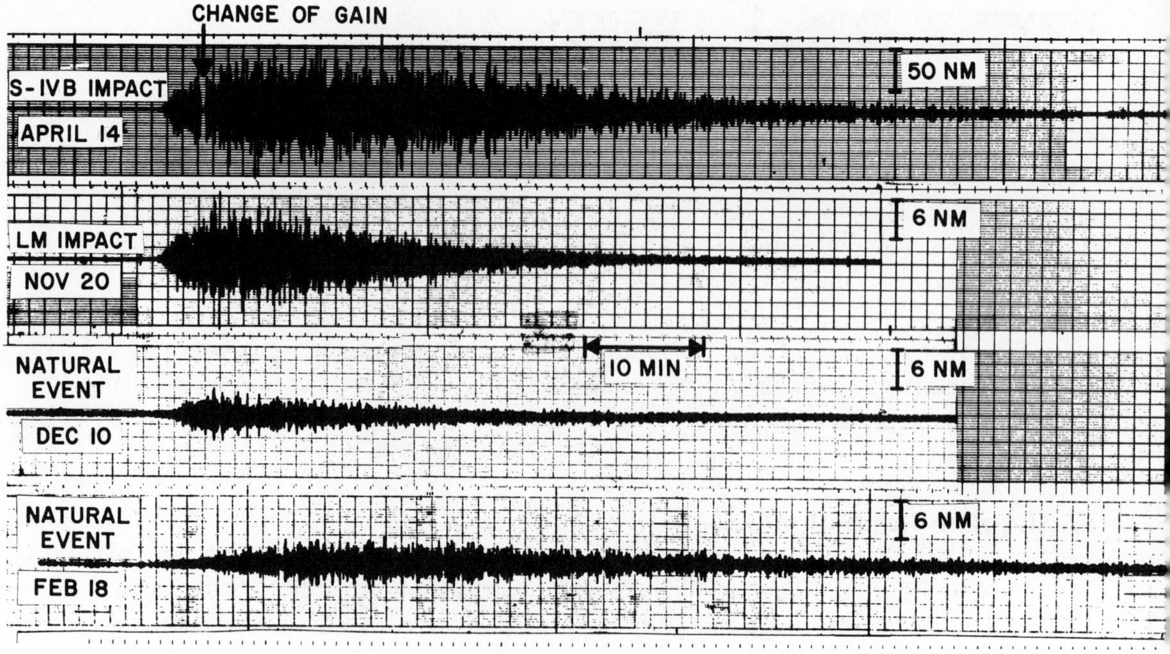

FIG. 10-33. Lunar seismograms from instruments left on the surface of the moon by the crew of Apollo 12. The top two records represent seismic signals from the impact of auxilliary parts of Apollo 12 and 13 expeditions. The lower two records represent lunar seismograms of natural events. Because of the compressed time scale, individual waves have fused into a single blurred image. The prolonged reverberations for about 1 hour are unusual for terrestrial earthquakes but common for lunar quakes and indicates a fundamental difference in structure between the moon and earth. (G. Latham and others, *Science*.)

performed (on April 14, 1970) with the third-stage booster rocket of the aborted Apollo 13 mission, which circled the moon and returned after a mishap aboard the spacecraft.

Both of these high-speed impacts generated strong seismic signals different from similar events on earth. As shown in the upper two records of Fig. 10-33, the seismic signals, which reached maximum intensity in a few minutes, persisted as seismic reverberations in the moon for about an hour. Seismograms from two of the largest natural events also shown in Fig. 15-33 have the same prolonged effect.

Seismic waves from terrestrial impacts similar to the artificial events persist for only tens of seconds. Although the interpretation of these seismic results is not yet complete, it is clear that the lunar subcrustal structure must be quite different, and less rigid, from that of the earth. As with the ages of lunar rock samples (see p. 244), the seismic experiment also paid off quickly, although not with such clear results. Future experiments will certainly solve the problem of lunar structure.

STUDY QUESTIONS

10.1. What is a "tsunami"?

10.2. Explain how one or more days of advance warning may be possible prior to the arrival of a tsunami.

10.3. Explain the principle of the seismograph.

10.4. Define (a) focus, (b) epicenter, (c) natural period, (d) microseism, and (e) microearthquake.

10.5. Determine the magnitude of an earthquake whose largest wave recorded by a standard seismograph is 1000 millimeters at a distance of 100 kilometers from the epicenter.

10.6. What is the energy in ergs of this earthquake?

10.7. Define "body waves" and describe the two main types.

10.8. Define surface waves and describe the two main types. Illustrate the ground motion involved with three-dimensional (block) diagrams.

10.9. What type of seismic wave motion is not transmitted by liquids?

10.10. Why are at least three seismograph stations required to locate earthquake epicenters?

10.11. How is the S-P time interval used to determine the distance to an earthquake?

10.12. What is the significance of observations of earthquakes whose foci are at a depth of hundreds of kilometers?

10.13. Define the "Mohorovicic discontinuity." How was it discovered?

10.14. What is the seismic "shadow zone"?

10.15. How did the observations of the shadow zone establish the presence of the earth's core?

10.16. How was the "liquid" nature of the outer core determined?

10.17. Illustrate the appearance of a train of waves that show wave dispersion. What causes this effect?

10.18. Briefly explain how surface wave dispersion is used to determine the nature of the layering within the earth's crust and deeper interior.

10.19. How are earthquakes related to the boundaries of the tectonic plates discussed in Chapter 8?

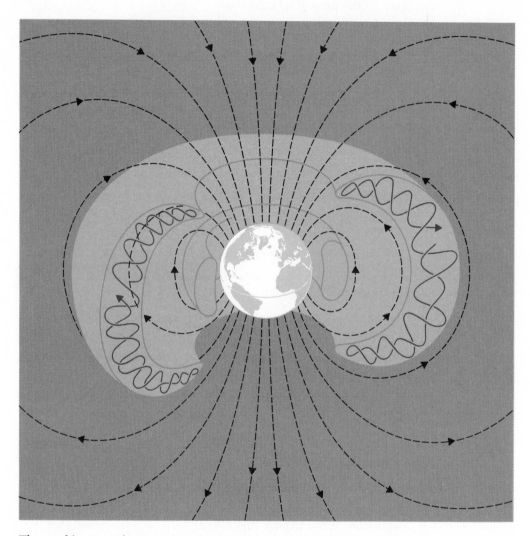

The earth's atmosphere, as detected by magnetic and radiation effects, extends thousands of miles away from the dense air near the surface. The magnetic field is shown by magnetic lines of force converging near the poles. The inner and outer Van Allen radiation belts girdle the globe. The actual paths of electrified particles are visible in the outer belt. Particles spiral endlessly between the polar regions.

The Atmosphere

The atmosphere is the thick gaseous envelope that surrounds the earth and permeates its crust and oceans. Although we breathe air and can feel it to a degree that varies from a zephyrlike breeze to a catastrophic wind, we cannot see it. Pure air is invisible; only natural impurities and man-made pollutants are visually apparent.

In addition to the geological importance of air in the weathering and erosion of rocks, it plays a dominant role in sustaining life on the earth. Nearly all organisms require oxygen in their respiration. Also, the weather and climate of the atmosphere exert a strong control over the affairs of man. From an aesthetic point of view much of the beauty of nature is attributable in one way or another to the behavior of the atmosphere. The many luminous and transparent shades of blue of the clear sky, or the dramatic tempestuous grandeur of a stormy, cloud-ridden sky, or the delicate, magical glow of shifting beams of auroras, are among the many phenomena that could be recounted.

NATURE OF THE ATMOSPHERE

The air about us seems to be a tenuous and almost weightless medium. But a simple, familiar experiment can quickly show that this is a deception by our senses. If a large flexible container—such as a rectangular five-gallon can—is evacuated of air, it seems to collapse quite mysteriously. Actually, the weight of the overlying atmosphere simply crushes the can when the air inside with its supporting pressure is removed. A slightly more quantitative experiment shows that the average weight of the air at sea level over each square inch is 14.7 pounds. This means that a column of air with a horizontal cross-sectional area of one square inch extending

TABLE 11-1 Composition of the Dry Atmosphere by Volume and Mass

Component	Volume (percent)	Mass (percent)
Nitrogen	78.084	75.51
Oxygen	20.946	23.15
Argon	0.934	1.28
Carbon dioxide	0.033	0.046
Trace components[a]	0.003	0.014

[a]Trace components consist of neon, helium, methane, krypton, nitrous oxide, hydrogen, ozone, xenon, nitic oxide, and radon. The present atmosphere also contains pollutants, which are referred to in detail in Chapter 16.

to the top of the atmosphere weighs 14.7 pounds. From the information about the surface area of the earth (see Chapter 2), we can determine the equivalent area in square inches. When this number is multiplied by 14.7, we can compute the total mass (weight) of the entire atmosphere. Expressed in tons rather than pounds, the mass of the atmosphere is found to be 560 million million tons (or 560×10^{12} tons).

Composition

Of this great mass of atmosphere, 99 percent is composed of only two gases—nitrogen and oxygen. In Table 11-1 a summary of the components of the atmosphere is given in both volume and mass measurements. Note that the table refers to dry air. Water vapor, a most important constituent, occurs in minor and variable amounts. It varies considerably from place to place and from time to time at the same place. Although the global water vapor content by volume is about 1 percent for moist air at sea level, the actual amount varies from near zero over some land areas (deserts) to about 4 percent in the humid tropics.

The composition in Table 11-1 is remarkably uniform even to great heights in the atmosphere. But the water vapor composition decreases drastically with elevation, as shown in Table 11-2. According to the values here, 90 percent of atmospheric moisture lies below 3 miles (about 5 km). This may seem strange when we consider that water vapor is the lightest of the common gas molecules in the atmosphere, having a molecular weight of only 18 ($2H^1 + O^{16}$) compared with oxygen, whose molecular weight is 32, ($2O^{16}$) and nitrogen, whose weight is 28 ($2N^{14}$).

The low-level concentration of water vapor stems primarily from the fact that atmospheric water vapor is close to its critical temperature—the point at which it changes state from a gas to a liquid. When water vapor is carried upward, it is soon cooled to its critical temperature, usually condensing to cloud particles. Precipitation in the forms of rain or snow then returns this *condensed* moisture to the earth's surface.

TABLE 11-2 Decrease of Water Vapor with Elevation

Height		Water Vapor (percent)
Km	Ft	
0	0	1.3
1	3,281	1.0
2	6,562	0.69
3	9,843	0.49
4	13,124	0.37
5	16,405	0.27
6	19,686	0.15
7	22,967	0.09
8	26,248	0.05

Despite its relatively small amount, water vapor has a very large effect on the atmosphere. It is the chief absorber of radiant (heat) energy in the atmosphere and thereby exerts a very strong influence on atmospheric processes. The nocturnal cold of the air in deserts and mountain regions is a direct result of their low water-vapor content. In humid regions, the atmosphere remains relatively warm at night because the water vapor present absorbs much of the heat transmitted by the surface after it has been warmed by the daytime sun. The absence of water vapor in the dry air of deserts and mountains permits rapid escape of the earth's heat with little absorption taking place in the air.

In addition to its thermal effect, atmospheric water as vapor, liquid, and solid constitutes the chief element in weather changes. The important weather changes of storms are clouds, rain, snow, fog, and wind. Of these five features, four are forms of atmospheric water resulting from condensation of water vapor. In view of this importance we will return to a more detailed study of atmospheric moisture in a later section.

It has been known for some time that the atmosphere, up to the heights reached by airplanes and balloons, is quite uniform in the mixture of gases present. With the earliest of the artificial earth satellites, we learned that this uniformity extends to about 50 miles. Above this region the air has a composition layering as shown in Fig. 11-1. In the lowermost layer one very important feature must be considered—the ozone layer. From about 20 to 30 miles in elevation, a large proportion of the oxygen present occurs in the form of ozone, whose molecular composition is given as O_3, compared to O_2 for normal oxygen. Because of the very rarified nature of the atmosphere at this height, all of the ozone present in the ozone-rich layer, if compressed to the density of air at sea level, would form a layer barely a third of an inch thick. Despite this small quantity, the ozone of the tenuous upper atmosphere is nearly opaque to the invisible ultraviolet rays from the sun. Without the shielding effect of the ozone layer, we would receive lethal sunburns after a relatively short

FIG. 11-1. (left) The compositional layering of the atmosphere based on information of the Explorer satellite program of the National Aeronautics and Space Administration (NASA).

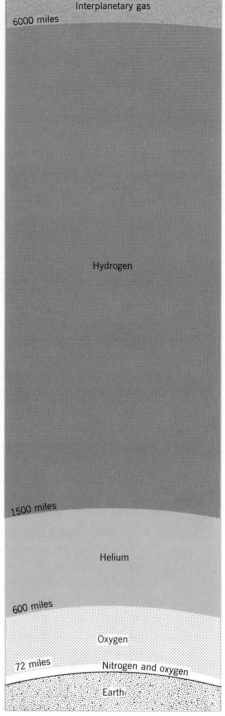

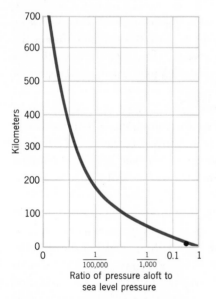

FIG. 11-2. (above) Graph showing the decrease of atmospheric pressure with elevation above sea level. The dot indicates the point on the curve where pressure is reduced to one-half the value at sea level.

exposure to sunlight. In fact, one theory of the evolution of land life (by L. V. Berkner and L. D. Marshall) involves the development of a significant ozone layer. Prior to this time, primitive life was mostly restricted to the earth's rivers and oceans, which provided a necessary radiation barrier. This theory then explains the rather rapid evolution of land life in late Precambrian time as occurring when ultraviolet energy reaching the earth's surface had been diminished below some threshold intensity by absorption in the ozone layer.

Thickness and Density

A gas, or a mixture of gases such as our atmosphere, has no fixed limits unless it is contained. Although the surface of the earth provides a sharp lower boundary for the atmosphere, there is no upper boundary to contain it. Hence, it has no specific thickness or height. Being a gas, it is compressible. The weight of the overlying atmosphere compresses the air beneath, the effect reaching a maximum at sea level, where the density of the air is greatest (1.2 kg per m^3 or 0.076 lb per cubic ft). A simple calculation reveals what may be a surprising fact, namely that a cubic yard of air contains two pounds of air.

Although it has no sharp upper limit, the bulk of the atmosphere lies quite close to the earth's surface because of the compressional effect. This is evident in the vertical distribution of the atmosphere, as in Fig. 11-2, which shows the decrease in air pressure with increasing elevation. The change in the lower atmosphere is very rapid. The arrow indicates the point on the curve where pressure is one half the value at sea level and the elevation is 18,000 ft (about $5\frac{1}{2}$ kilometers). One half of the entire atmosphere thus lies below 18,000 feet. Half of the remainder, or a total of three fourths, lies below 34,000 feet or about 10.3 kilometers. The amount of air present above 60 miles is less than the amount in the best artificial vacuum we can make at the earth's surface, but it is nevertheless readily detectable.

According to the curve in Fig. 11-2, the continued decrease of air pressure with increase in elevation in the upper, very tenuous, atmosphere takes place at a very slow rate. In other words, the air simply peters out more and more slowly in the upper atmosphere until the amount of gas present is equal to that in interplanetary space. Despite the small amount present at hundreds of miles above the earth's surface, the frictional effect of this "vacuum" is still felt by the artificial earth satellites. The accumulated frictional slowing ultimately causes them to plunge back toward the earth's surface.

It is interesting to realize that if the entire atmosphere were at the average temperature and pressure at sea level, it would occupy a layer only 5.5 miles (8.8 km) thick.

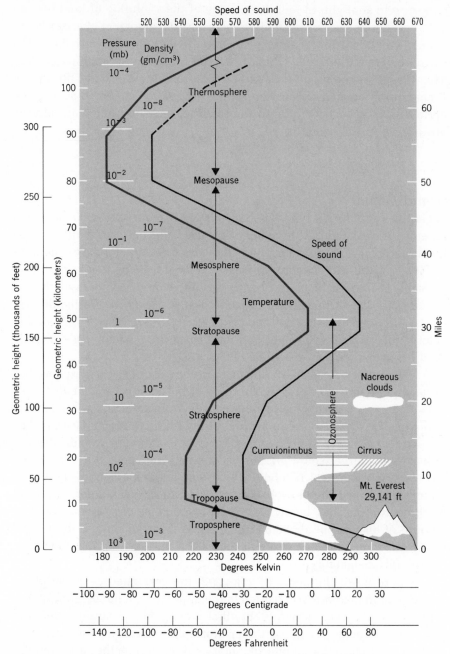

FIG. 11-3. The thermal structure of the atmosphere to an elevation of 70 miles. The elevation scale is given in both thousands of feet and kilometers on the left side and in miles on the right side. Temperature along the lower horizontal axis is given according to three scales: Kelvin K or Absolute, Celsius C (formerly Centigrade), and Fahrenheit F. The speed of sound, which varies with temperature, is shown by the thin line and read in knots (nautical miles per hour) along the top horizontal axis. (U. S. Navy Weather Research Facility.)

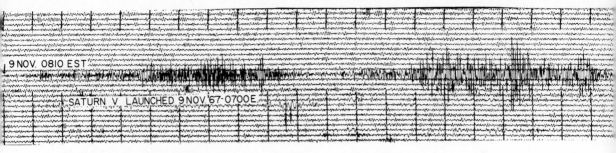

FIG. 11-4a. Record of the subaudible (low frequency) sound recorded at Palisades, N.Y. (Lamont Geological Observatory) from the launching of the first great Saturn V rocket used in the Apollo (lunar) space program at Cape Kennedy, Florida 920 miles to the south. Rocket ignition was at 7 A.M. and the sound arrived in two strong groups beginning at 8:11 A.M.

Thermal Structure of the Atmosphere

The atmosphere, even to hundreds of kilometers high, has a very definite structure produced by the vertical changes of temperature. This thermal structure is summarized in Fig. 11.3. In the *troposphere* or lowermost layer, temperature decreases with elevation until an average of 7 miles (11 km) is reached. Above this level (tropopause) the temperature remains fairly constant at about −70°F (−55°C). Toward the top of the *stratosphere*, characterized by nearly constant vertical temperatures, temperature increases again to a maximum average of 32°F. (0°C.) The *mesophere*, like the troposphere, is characterized by a decrease in temperature to a value below −140°F (90°C) between elevations of 50 to 55 miles. The *thermosphere* is characterized by temperatures that increase to about 1800°F (1000°C) as a result of the strong absorption of energetic but invisible short-wave solar radiation. It appears from this description that two layers of temperature minima exist in the atmosphere, with zones of higher temperature above and below.

Such a structure exerts an important influence on the propagation of sound over long distances. Since the speed of sound depends primarily on air temperature, its vertical variation will resemble that of the temperature, as evident from the appropriate curve in Fig. 11-3. Sound rays traveling upwards at different angles through the atmosphere will be refracted back toward the surface upon reaching the zones of temperature increase at 30 miles and above 55 miles. An example of how sound rays are guided by the atmospheres thermal structure is shown in Fig. 11-4b, which indicates regions of audibility where the rays from a distant sound source strike the ground and regions of silence between them. It is this channeling effect of sound rays that permits the detection of explosions and rocket launchings at long distances from the source (Fig. 11-4a and b).

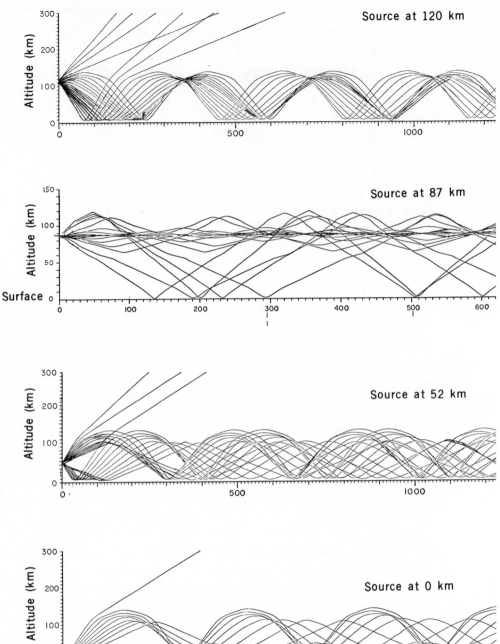

FIG. 11-4*b*. The paths of the sound rays coming from the rocket at different points as it rose above the surface. The paths at the ground show reflections from both lower (50 km) and upper (120 km) levels of high temperature. Sound from above the ground is affected primarily by the upper reflection level. Note the prominent zones of silence at certain distances, an effect long known from long-range explosions and gunfiring.

The troposphere is of particular interest because it is the layer where daily weather changes, often marked by violent storms, occur. The reason for this is the prominent decrease in temperature in the troposphere plus the concentration of atmospheric water vapor in the lowermost couple of miles. Storms and weather patterns will be discussed more completely in Chapter 13. Weather conditions are far less severe in the layers overlying the troposphere—hence, the advantage of high-level air travel.

During the first half of the twentieth century, most of the scientific study of the atmosphere was concentrated on the phenomena of the troposphere. As a result, the traditional term *meteorology,* meaning study of the air, is strictly applicable to this lowermost zone. However, with the development of tools and techniques for probing the outer atmosphere, the new term *atmospheric science* has arisen to include the study and knowledge of the entire atmosphere.

IONOSPHERE AND MAGNETOSPHERE

Very little of the intense, short-wavelength portion of the sun's radiation reaches the earth's surface. Nearly all of this form of solar energy is absorbed in the rarified upper atmosphere. As a consequence of this high-energy absorption, the atoms are stripped of one or more electrons, thus becoming ionized or electrified. Particularly strong concentrations of ions are present at several distinct elevations within the *ionosphere*—the zone between 30 and 200 miles. The more intensely ionized regions are the D region (50 miles), E region (75 miles), F_1 region (125 miles) and F_2 region (190 miles). The presence of the ionized layers is of great practical value in providing natural reflection surfaces for radio waves, thereby permitting long-distance radio communication.

In the outermost atmosphere (Fig. 11-1), hydrogen is the major constituent of the extremely thin remaining atmosphere. In this rarified outer layer, the hydrogen atoms have been ionized to proton particles. At this distance from the earth, the force of gravity becomes secondary to the earth's magnetic field, which exerts the more dominant control over the electrified gas present. For this reason, the outermost shell of atmospheric protons is often called the magnetosphere. Some of the common terminology of the high atmosphere is summarized in Fig. 11-5.

Extending through and beyond the atmosphere is the earth's magnetic field, which emanates from the hot, molten, iron core. The fact that the earth is a relatively strong magnet has long been known. But the distribution of this field in the space surrounding the earth's surface was not delineated until the advent of artificial satellites and space probes. From experiments conducted with these far-flung vehicles we have learned not only that the earth's magnetic field extends for thousands of kilometers out from the surface, but also that it is quite asymmetric, as shown in

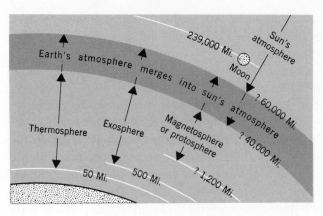

FIG. 11-5. Nomenclature scheme used to describe the layers of the earth's outer atmosphere. (U. S. Navy Weather Research Facility.)

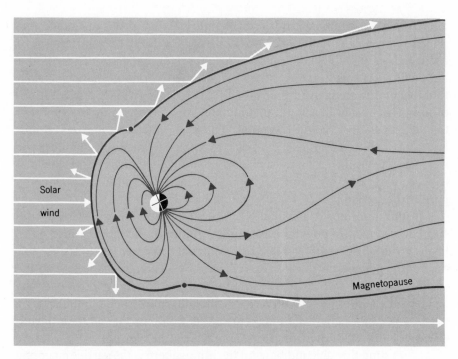

FIG. 11-6. Simplified view of the earth's magnetic field extending into space from the region of the earth's magnetic north and south poles. The entire field is compressed on the side of the earth facing the sun and extended on the opposite side from interaction with the solar wind. This stream of high speed electrified particles is reflected from the magnetopause or outer boundary of the magnetic field. (After N. Ness in *Science*.)

396   THE ATMOSPHERE

Fig. 11-6. Electrified particles issue from the sun at a high velocity in streams called the solar wind. This compresses the earth's magnetic field on the side facing the sun, as shown in Fig. 11-6, and extends the field on the opposite side.

One of the missions of the artificial space probes launched toward other planets is the determination of the nature of any magnetic field that may be present. The presence or absence of such a field, together with its intensity, if present, can then be used to interpret the planets' internal composition.

THE RADIATION BELTS

One of the most interesting and important aspects of the International Geophysical Year (IGY) of 1957–1958 was the American Explorer satellite program. In the interpretation of data from this program, J. Van Allen discovered the two radiation belts that now bear his name. Maximum radiation in these belts, which surround the earth, is centered at about 2000 and 10,000 miles respectively above the earth's surface. The radiation, in the form of very-high-energy protons and electrons, was detected originally by radiation-sensitive *Geiger* tubes carried aloft in a number of Explorer and later Pioneer satellites. The charged particles are trapped by the earth's magnetic field throughout the entire magnetosphere but have great concentrations in two zones, as illustrated in Fig. 11-7. Note how similar the distribution of contours of radiation intensity is to the magnetic-force lines of Fig. 11-6. The charged particles spiral back and forth from hemisphere to hemisphere, remaining within the confines of the lines of magnetic force.

The beautiful displays in the upper atmosphere known as auroras (*aurora borealis*—northern lights—in the Northern Hemisphere, and *aurora australis*—southern lights—in the Southern Hemisphere) have become better understood with the discovery of the high-energy ionized particles in the magnetic field. At times of high sunspot activity (discussed in Chapter 1), huge gaseous eruptions known as *solar flares* may occur in the sun's atmosphere. Concentrated streams of high-speed protons and electrons ejected by the flares may be directed toward the earth, partially disrupting the magnetic field. As a result, particles trapped in the field escape in large numbers where the horns of the crescent-shaped radiation belts (Fig. 11-7) are directed toward the earth. These high-speed particles, reinforced by those from the solar beam, strike the rarified upper atmosphere and energize the gaseous components, which are thus caused to glow with hues dependent on their compositions. Auroras show continuously changing, shimmering forms resembling beams, arcs, draperies, and the like, delicately tinted with a variety of colors in which a greenish hue usually predominates. (See Fig. 11-8).

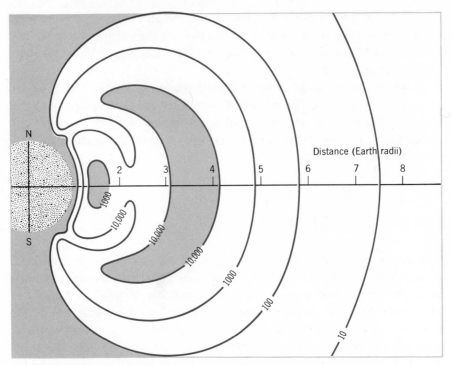

FIG. 11-7. Cross-sectional view showing the distribution of the corpuscular radiation field (high energy electrons and protons) that surrounds the earth. Contours of radiation intensity are labelled in actual count rates of the Geiger tubes carried by Explorer IV and Pioneer III satellites. The shaded zone shows the two regions of greatest radiation intensity known as the Van Allen Belts. Distances from the center of the earth C is shown on the horizontal scale in earth radii. (J. Van Allen, Journal of Geophysical Research.)

THE ENERGY OF THE ATMOSPHERE: HEAT AND GRAVITY

The atmosphere is often described as a huge heat engine powered by the external heat of the sun and the earth's internal gravity field. The primary physical behavior of the atmosphere is related to changes in density from unequal heating followed by the gravitational effect of the sinking of dense air and rising of light air. Except for the comparatively insignificant amount of heat escaping from the earth's interior and that due to radioactivity, the atmosphere receives all of its heat directly or indirectly, from the sun. Since the atmosphere is neither warming nor cooling, it must on the average, radiate away as much heat as it receives.

Because of the earth's curvature, the energy within a given solar radiation beam is spread over a larger area, with increase in latitude. On the average, the concentration of energy is greatest at the equator and least at the poles. Also, because of both the curvature and the seasonal effect (Fig. 1-12), the polar regions receive no sunlight during winter, when the earth's axis is tilted away from the sun. As a consequence of curvature and seasonal variations, low-latitude zones receive more heat than is lost through radiation, resulting in a heat surplus; high-latitude regions radiate away more than they receive, producing a radiation deficit. A balance of heat received from the sun and heat lost through radiation occurs at 35° latitude. The inequality of radiation leads to a flow of heat from the warm equatorial region to the cooler polar regions. *In the process of heat redistribution, the winds of the globe and the storms of the middle latitudes are generated.*

Insolation

The radiant energy received from the sun, called *insolation,* is spread over a band of wavelengths known as the solar *spectrum* (Fig. 11-9). Although the spectrum is so broad that the longest waves (radio waves) are about 10^{21} times the shortest (gamma rays), the energy peak and much of the total radiant energy is within the very narrow limits of the visible portion. The relationship of visible light to the rest of the solar spectrum is indicated in Fig. 11-9. Because the earth's orbit is elliptical, the amount of insolation varies very slightly during the year, but its average value, known as the *solar constant,* is 2.0 calories per square centimeter per minute. This number refers to the radiant energy falling on a surface perpendicular to the suns rays, before any absorption or loss in the atmosphere.

The gases in the atmosphere are quite transparent to the intense visible solar radiation. Short wavelength ultraviolet radiation, as noted previously, is absorbed in the very rarified upper atmosphere. Some infrared radiation is absorbed by carbon dioxide and water vapor in the lower atmosphere, but most of the sun's energy reaches the earth's surface without absorption in the air (see Fig. 11-10a and b). How then is the atmosphere heated? When we analyze the disposal of insolation by the earth, it becomes evident that most of our atmosphere is heated primarily by the surface beneath, after it is warmed by sunlight. Three different processes are involved in this heating.

1. When the earth's surface absorbs sunlight, it in turn becomes warmed. All warm objects radiate energy, the wavelength of which depends on the temperature of the radiating body. At the temperatures reached by the earth's surface, infrared wavelengths are radiated. Thus, the earth converts visible sunlight into invisible longer infrared wavelengths. Any object on earth can be photographed quite readily in the blackness of night if the camera used is equipped with infrared-sensitive film. Water vapor, and to a lesser extent, carbon dioxide, absorb infrared radiation—the more water vapor present, the greater the absorption. Nitrogen and oxygen absorb almost no outgoing infrared and no incoming

FIG. 11-8. Aurora as seen from Ogunquit, Maine. Painting by Howard Russel Butler. (Courtesy American Museum of Natural History.)

visible light—hence, the importance of atmospheric water vapor in the heating of the atmosphere. Liquid water droplets in a cloud are also very strong absorbers of infrared radiation, which is then reradiated to the earth's surface by the bottom of the cloud. As a result, cloudy nights are always warmer than clear nights. The prevention of infrared heat loss from the earth's surface by atmospheric moisture is known as the *greenhouse effect*, since it is analogous to the retention of heat by the glass of a greenhouse.

2. A second important heating process is the transfer of energy by direct *conduction* of heat from the earth's surface to the atmosphere in

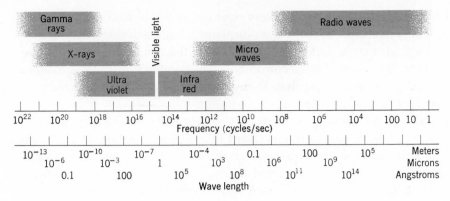

Frequency (cycles/sec)

Wave length

FIG. 11-9. The spectrum of solar radiation as received outside the earth's atmosphere. Wave lengths are given in several of the common units of science. (A meter is 3.28 feet; a micron is 1/1,000 of a millimeter, 1/25,000 of an inch; an angstrom is one-billionth of a millimeter.) The peak of solar energy is within the narrow visible band. Despite the great width of the spectrum, most of the suns radiant energy lies between 0.2 and 2 microns. (U. S. Navy Weather Facility.)

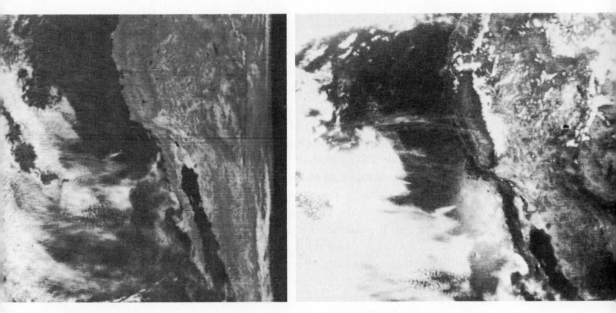

FIG. 11-10. Two views of western United States and northern Mexico transmitted by the weather satellite, Nimbus III, April 20, 1969. The left view is taken by means of infra-red light reflected from the clouds and surface, the right view in visible light. The surface can be seen with greater clarity in the infra-red view which does not, however, reveal the snow-capping of the Sierra Nevada and Rocky Mountains. (NASA Photo.)

The Energy of the Atmosphere: Heat and Gravity　　**401**

direct contact with it. Air is a very poor conductor of heat energy, so that only the lowermost part of the atmosphere is heated in this way. But once warmed, air expands and rises, transferring the conducted heat to higher levels by the process called *convection*. The combination of the processes of conduction and convection in heating the air is usually referred to as *turbulent heat exchange*.

3. A large quantity of heat, particularly over oceans or other large water bodies, is transferred as latent heat of vaporization. When water evaporates from a liquid to a gas, from 540 to 600 calories are absorbed by each gram of water converted to vapor. None of this heat is involved in a temperature change; it is used solely to provide the energy necessary for the water molecules to escape from the liquid. Recall that only 1 calorie is required to increase the temperature of a gram of water by 1°C. In comparison, the 540 to 600 calories required to evaporate a gram of water are enormous. Ultimately, the evaporated water vapor condenses to form the water droplets of clouds and then releases the heat gained in the evaporation process. After precipitation of rain or snow, the heat released upon condensation remains in the atmosphere and warms it. Notice that the heat transferred to the air in evaporation and later condensation is as much the result of sunlight as the heat transferred to the air by surface reradiation of infrared energy after absorption of visible sunlight.

Heat Balance of the Atmosphere

In order to understand fully the effects of sunlight in energizing all of the processes occurring in the atmosphere, it is necessary to evaluate methods by which the earth and atmosphere dispose of insolation. This evaluation, which balances heat loss against heat gain is known as the *heat budget* or *heat balance* of the atmosphere.

Recall the solar constant of 2.0 calories per square centimeter per minute at the outer edge of the atmosphere. When referred to the earth's surface, this value is decreased to one fourth or 0.5 calorie per square centimeter per minute. The reason for the decrease is evident in Fig. 11-11, which shows the surface perpendicular to the sun's rays or the surface to which the solar constant is referred. This surface has a radius r equal to the earth's radius, and therefore a surface area whose relationship to the earth's surface area is $\pi r^2/4\pi r^2$, or one fourth the area of the earth.

The disposal of insolation after entering the atmosphere is shown in Fig. 11-12, which summarizes the heat-balance factors. About 33 percent of the insolation is reflected back to space, primarily by clouds and partially by the varied features of the earth's surface so that 67 percent must be absorbed. A still smaller amount is reflected by particles within the atmosphere. The reflectivity of 0.33 is known as the earth's *albedo*. Reflected sunlight has little effect in heating the atmosphere.

Of the 67 percent of sunlight not reflected directly, about one fourth

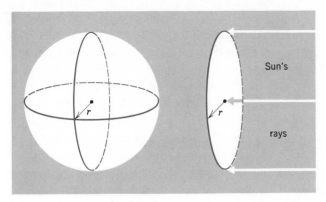

FIG. 11-11. The extra-terrestrial solar radiation perpendicular to a circular area of the earth's radius (r) would be reduced by an average of one-fourth when spread over the spherical earth.

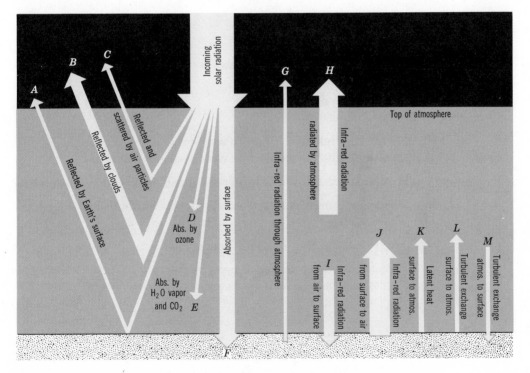

FIG. 11-12. Schematic diagram showing the disposal of insolation by the earth's atmosphere and surface. Processes on the left side indicate the direct disposal after solar radiation penetrates the atmosphere. Processes on the right side indicate the disposal of radiant energy after absorption at the earth's surface. Processes A, B, and C have no heating effect on the atmosphere, but constitute the earth's albedo (reflected sunlight). The sum of D, E, and F equal the sum of G and H and form the heat balance of the atmosphere. Note that a certain amount of heat is recycled between earth and atmosphere in the course of processes I, J, K, L, and M before final radiation to space (G and H).

The Energy of the Atmosphere: Heat and Gravity   403

is absorbed by the various gases in the atmosphere and three fourths is absorbed at the earth's surface. Of the original solar constant of 2.0, $\frac{3}{4}$ of $\frac{1}{4}$ or $\frac{3}{16}$ calorie per square centimeter per minute is the amount actually absorbed when averaged over the earth's surface. As noted previously, this energy is then utilized in heating the air by infrared radiation, turbulent exchange, and the transfer of latent heat of vaporization. The earth's surface gets a little bonus in the form of infrared radiated downwards by the atmosphere. But the atmosphere radiates most of its infrared energy to space. This radiation, when added to the smaller amount that escapes from the earth's surface, is just equal to the 67 percent of the insolation absorbed by the atmosphere and earth. If the atmosphere's heat gain and loss factors were not in balance, a net heating or cooling would occur. The fact that this in not now happening, nor, so far as we can judge, has it happened in the past, indicates that a state of balance of heat energy exists.

STUDY QUESTIONS

11.1. What is the average weight of air over each square foot of the earth's surface?

11.2. Give the percentage composition of the dry atmosphere by both mass and volume for its two main gaseous elements.

11.3. Explain why the water vapor composition of air is so variable.

11.4. What is the importance of water vapor from both the energy and weather standpoint?

11.5. What is ozone and why is it important?

11.6. Does the atmosphere have a recognizable top boundary? How does air density change with elevation?

11.7. Describe the thermal structure of the atmosphere.

11.8. Define the "troposphere" and give its special importance to man.

11.9. What is meant by the "radiation belts"?

11.10. Explain the two primary sources of atmospheric energy.

11.11. Define and give the value of the term "solar constant," and explain why its effect on the earth's surface varies strongly with latitude.

11.12. Describe the insolation spectrum.

11.13. What is meant by the "greenhouse effect"?

11.14. How is heat energy transferred by conduction and convection—or turbulent exchange?

11.15. How does evaporation of sea water transfer heat to the atmosphere?

11.16. Describe briefly the "heat balance" of the atmosphere.

11.17. A molecule of water, composed of 2 hydrogen atoms (total atomic weight of 2) and one oxygen atom (weight 16) has a molecular weight of 18. Nitrogen molecules (N_2) and oxygen molecules (O_2) have molecular weights of 28 and 32, respectively. As both are heavier than water vapor, explain why water vapor resides so close to the earth's surface (Table 11.2), whereas the heavier gases extend high into the atmosphere.

chapter 12

Sunset view of cumulus clouds of fair weather; undersurfaces of clouds masking the sun appear dark.

Elements of the Weather

Weather is the condition of the atmosphere for a short period of time—a day or so—and is described in terms of certain observational properties, the most important of which are temperature, humidity, clouds, precipitation, pressure, and wind. After a short discussion on how they are observed and how their changes are interrelated, we will see in the next chapter how the weather elements are related to storms and weather changes.

AIR TEMPERATURE

The irregular distribution of heat energy in the atmosphere leads directly to variations in global air temperatures. Gravitational effects on air having different temperatures then lead to further temperature changes of great significance in causing weather variations. Air temperature, including variations and distribution, is certainly the most fundamental of the weather elements. It is the direct measure of atmospheric heat energy whose variations are the ultimate driving force of all weather changes.

Thermometers and Related Instruments

A thermometer is a glass tube containing a fine capillary bore that extends from an enlarged reservoir chamber at the lower end of the instrument (Fig. 12-1). Mercury is the fluid used in standard weather thermometers; alcohol, being cheaper, is common in less-technical instruments. The liquid in the reservoir expands or contracts in proportion to changes in temperature, which can thus be determined from the calibrated scale etched into the outer surface of the glass tube.

The temperature scales in regular use are the Celsius (formerly Centigrade) and Fahrenheit. The former is arranged so that zero coincides with the temperature at which freshwater freezes and 100° matches that at which water boils at sea level; one degree C is $\frac{1}{100}$ of this range. The range from freezing to boiling is thus 100°. On the Fahrenheit scale, water freezes at 32° and boils at 212°, with a range of 180° from freezing to

Bore →

Mercury chamber

(b)

FIG. 12-1. (a) Principle of construction of a thermometer. (b) Standard exposed mercury thermometer.

(a)

boiling. (Thus the temperature range of one degree F is smaller than that of one degree C). The ranges for each scale are in the ratio of 5 to 9, respectively. This ratio must be used in converting from one scale to the other, together with the fact that freezing temperature is 0° on one and 32° on the other scale. To convert from Celsius to Fahrenheit and the reverse, we get the formulas:

$$F = \frac{9}{5}C + 32$$

$$C = (F - 32)\frac{5}{9}$$

A scale important in scientific use is the Kelvin or absolute scale. Zero in this scale is based on *absolute zero,* the temperature at which molecular motion ceases. This occurs at $-273°C$, so that $0°C$ is equal to $+273°K$. The Celsius temperature degree unit is used in the Kelvin scale.

Six's maximum-minimum thermometer (Fig. 12-2) is a U-shaped instrument containing fluids so arranged that a column of mercury rises in one arm and falls in the other as the temperature rises, and the reverse happens as the temperature falls. An index in each arm indicates maximum and minimum temperature over a given time period.

The thermograph, which makes a continuous record of temperature (Fig. 12-3), consists of a curved bimetallic element or liquid-filled tube whose shape changes with changes in temperature. A lever arrangement to which a pen arm is connected magnifies the small changes in the movement of the metallic element.

In the detection of the very small variations of temperature so important in scientific observations, a *thermistor* is commonly used. This is a metallic unit whose resistance to electric-current flow changes in proportion to the temperature change. By measuring the minute changes in current in an electrical circuit including a thermistor, the equivalent minute temperature changes can be determined. Changes of $0.01°C$ or less can easily be measured in this way. Instruments of this kind are in common use to measure temperatures in such remote regions as the sediment under the oceans and the near vacuum of the upper atmosphere. They are also involved in instruments designed to telemeter to earth, the temperatures of the surface of the moon and planets.

Time Variations in Temperature

To an observer remaining in the same place, the most obvious temperature variations are those of a diurnal and seasonal nature. Strangely, the warmest part of the day is not noon when the solar rays are most intense, but midafternoon, usually between 2 and 4 p.m. The reason for this lag lies in the fact, discussed in Chapter 11, that the earth heats the air. The sun heats the earth's surface beginning at sunrise and ending at sunset

(Fig. 12-4), with maximum intensity at noon. But even when decreasing, the early afternoon insolation is still strong. The optimum effect produces a peak in air temperature at some time in midafternoon, as indicated by the broken line in Fig. 12-4. Since the earth and atmosphere both radiate heat to space between sunset and sunrise without replenishment, the coolest air temperature occurs before sunrise. The diurnal effect is nicely shown on the thermogram in Fig. 12-5, which emphasizes the heating peak during insolation periods compared to the rather flat trough during continuous nocturnal cooling.

On an annual scale the same effect operates. The warmest part of the year is not at the time of the summer solstice, but one to two months later. The coldest season is generally one to two months after the winter solstice.

If a cloud cover blocks incoming solar radiation, lower peak temper-

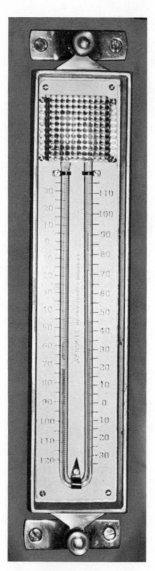

FIG. 12-2. Six's maximum-minimum thermometer. On the right side, indicating the maximum temperature, values increase from bottom to top. On the left side, indicating the minimum temperature, values increase from top to bottom.

FIG. 12-3. Thermograph. Changes in temperature cause changes in the curvature of the metallic element between the projections to the right of the case which in turn cause the pen resting on the rotating drum within the case to rise or fall as the drum rotates. (Friez Instruments.)

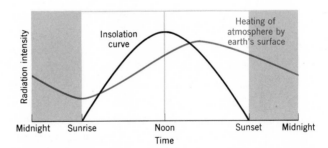

FIG. 12-4. Diagram showing the lag in heating of the atmosphere by the earth's surface. The peak of insolation is at noon, but the peak in air temperatures is in midafternoon.

atures occur during the daytime. Their greenhouse effect greatly reduces nocturnal radiation loss so that cloudy nights are invariably warmer than clear nights.

Variations over the Earth's Surface

The principal horizontal temperature variation is the decrease from equator to poles—the change being called the *meridional temperature gradient*. As noted earlier, the curvature of the earth is the principal cause

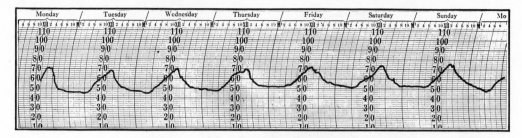

FIG. 12-5. Thermogram showing a continuous temperature record for one week. Note the time of peak temperature—3 to 4 P.M. and the time of minimum temperature—6 A.M.

of this strong gradient, whose mean annual value is about 80°F between the equator and the pole. The distribution of temperature over the surface of the earth is shown in Figs. 12-6 and 12-7 for midwinter and midsummer, respectively. Temperature distribution is shown by continuous lines of constant temperature called *isotherms*. The temperature gradient referred to above, being a change in temperature, always lies in a direction perpendicular to the isotherms. Hence, whereas the isotherms mostly trend east-west, matching the distribution of sunlight over the earth, the gradient is directed from equator to poles, and has a north-south orientation. In the Southern Hemisphere, which is the "water hemisphere," isotherms resemble the east-west parallel of latitude. But in the Northern Hemisphere they are much more irregular, showing a pronounced deflection when passing from land to water and the reverse.

The reasons for the pronounced differences in temperature distribution over land and water lie in the different ways that land and water dispose of insolation. On land, sunlight is absorbed in a very thin, immobile zone of soil or rock. The concentration of energy causes a rapid increase in temperature. Also, the specific heat (the heat necessary to raise one gram of a substance 1°C) of land is much lower than that of water, a further cause of the faster increase in land temperature compared to that of water. In the absence of sunlight, land cools rapidly in comparison with water because (1) its lower specific heat requires less radiation to cause a temperature drop, and (2) bodies with high temperature always radiate heat faster than those with lower temperatures. In midlatitudes, interiors of continents and the overlying atmosphere undergo rather extreme seasonal changes in temperature. Note that northeastern Asia experiences a seasonal temperature change from −50°F in January (Fig. 12-6) to 60°F in July (Fig. 12-7). Northwest Canada goes from −20°F to 60°F at the same time.

Compared to land, the seas are thermally conservative. They warm and cool slowly for a number of reasons.

1. Water has one of the highest known specific heats. It must gain or lose a relatively large amount of heat to change temperature.

FIG. 12-6. Isotherms showing mean sea level temperatures (°F) for January. Isotherms have a uniform east-west trend over the broad southern oceans compared to the irregular distribution in the Northern Hemisphere.

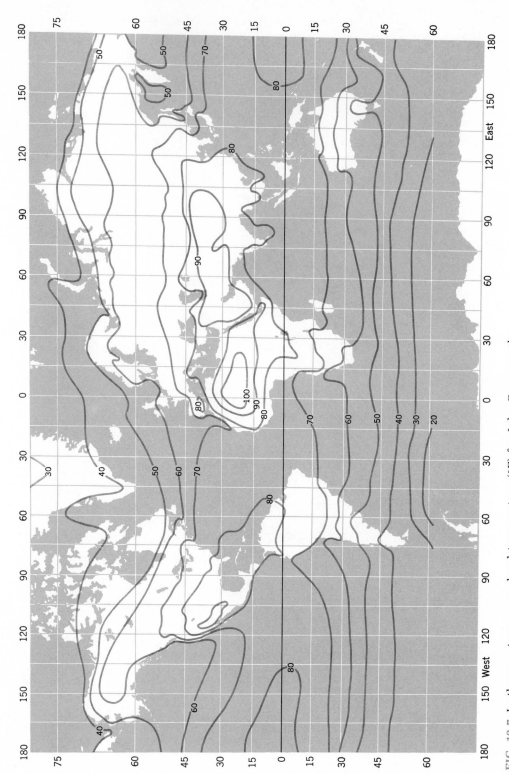

FIG. 12-7. Isotherms at mean sea level temperature (°F) for July. Compare the trend of the isotherms from land to water in the different seasons.

2. Water is transparent so that sunlight is not concentrated in a thin zone as with land, but penetrates to about 300 feet.

3. The seas are very mobile and undergo strong vertical mixing that spreads the heat received from above to a considerable depth.

4. Ocean currents redistribute surface heat excesses and deficiencies over thousands of miles; this warms higher-latitude seas beyond the expected insolation temperatures and cools low-latitude seas below expected values.

5. The moist atmosphere above oceans tends to reduce winter heat loss. Note in Fig. 12-6 how the air temperature of the North Atlantic for January (30°F) is the same as that in the United States at about the latitude of New York City (40°N). At the same time, northeast Asia at the same latitude as Iceland has a mean temperature of −50°F. These profound differences between land and water result in well-known differences between marine and continental climate. Peninsulas and islands commonly experience the moderate climate of the surrounding water body. In the middle latitudes (30° to 60°), the prevailing winds are from the west, so that western coasts of continents usually have pleasant marine climates with little of the temperature extremes of the interiors or eastern coasts. In the tropics where winds from the east prevail, east-facing coasts enjoy marine climates.

We will later reexamine the global temperature distribution because of its importance in the control of so many other atmospheric features to be studied.

Vertical Variations of Temperature

It is a matter of common experience that it grows colder as we increase our elevation. Two major reasons are (1) the source of heat is the surface and (2) water vapor, the chief radiant-heat absorber, decreases markedly with elevation. A third important reason will be examined later in this chapter.

The vertical temperature change or gradient is a very variable feature of the troposphere. The numerical value of this gradient, called the *lapse rate of temperature,* may vary from almost 0°F per 1000 feet to 10°F or more per 1000 feet. Although the lapse rate may be very different at different times, the average value is remarkably constant over the world. This average, known as the normal lapse rate, is 3.5°F per thousand feet (6.5°C per kilometer).

Frequently, the normal decrease in temperature with elevation is reversed so that an increase or *inversion* of temperature occurs in a relatively thin zone, followed again by a continued decrease. Examples of possible lapse rate curves, with and without inversions, are shown schematically in Fig. 12-8. The different slopes indicate some of the commonly observed distributions of temperature along the vertical. Inversions indicate that warm air overlies air immediately below that maybe cooler by a fraction of a degree to several degrees. The more pronounced the inversion, the greater is the temperature jog to the right in the curves. Curve D in Fig. 12-8 is typical of clear calm nights when strong surface radiation occurs causing the ground to cool rapidly until its temperature is less than that of the air above. The calm air in contact with the ground then cools by conduction of heat directly to the underlying cool surface. As the air cools, its temperature changes from its original condition shown by the dotted portion of curve D to the solid line to the left. Atmospheric temperature inversions are the chief meteorologic cause of smog, as explained in the section below on atmospheric stability.

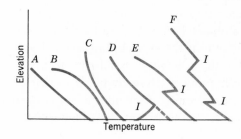

FIG. 12-8. Curves of temperature versus elevation showing different possible vertical lapse rates, all of which are common in the atmosphere. Curves D, E, and F also show inversions, indicated by "I."

Adiabatic Temperature Change

For a variety of reasons, summarized in Fig. 12-9, air can develop vertical motion. When it does so, temperature changes of the utmost meteorologic importance occur. As air rises, the pressure of the surrounding air through which it passes decreases so that the rising air expands just as does the gas in a rising balloon. The horizontal expansion of a rising column of air is shown schematically in Fig. 12-10, where the arrows indicate decreasing air pressure. When air expands, the work done by the expanding air uses energy from within itself, causing the temperature of the *rising air* to decrease. Recall (Fig. 11-2) that the pressure decreases at a constant rate in the lower atmosphere. Rising *dry* air thus expands, loses heat, and cools at a constant rate equal to 5.5°F per 1000 feet (10°C per kilometer). Because air is a poor conductor of heat, the vertical motion is rapid enough so that the rising air experiences the temperature decrease without any loss of heat to the surrounding, "motionless" atmosphere. The temperature change is thus called an *adiabatic* (*a* without, *diaba,* heat crossing over) change, since no actual heat exchange with the surrounding air occurs. This constant rate of change is called the *adiabatic lapse rate*. When air falls, it warms up at the same rate from compression due to increasing pressure.

To the two reasons why the atmosphere grows colder aloft (p. 415) we can now add the third—the cooling produced whenever air rises to higher levels from the surface. *Nearly all weather changes are directly*

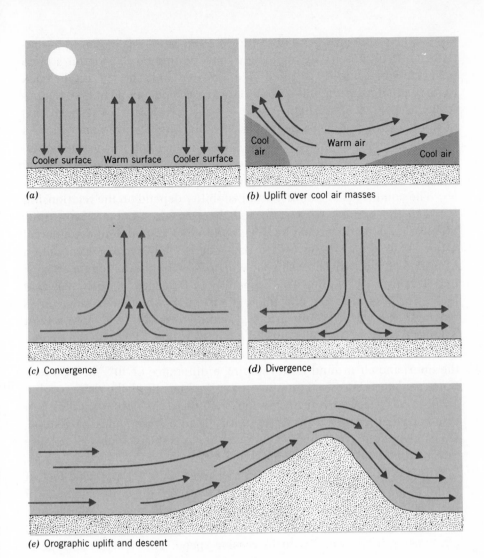

(a)

(b) Uplift over cool air masses

(c) Convergence

(d) Divergence

(e) Orographic uplift and descent

FIG. 12-9. Causes of rising and falling air motion.

related to the adiabatic cooling of rising air. Since this process does not occur consistently in the stratosphere because of its lack of vertical temperature change or lapse rate, weather changes are mostly limited to the troposphere or lowermost seven miles of the atmosphere.

Stable and Unstable Air

To properly understand the process of cloud formation—a process involving vertical air motion—we must consider first the very fundamental conditions of stability and instability in the atmosphere. Air is *stable* when it resists being displaced upwards or downwards. It is *unstable* when if displaced, it moves upwards or downwards with increasing speed. Stability and instability refer to the condition of the air, rather than its motion. Unstable air can be at rest until some cause—as in Fig. 12-9—initiates vertical motion. By analogy, a book resting on its side on a table is in stable equilibrium. If tilted up at one end and released, it falls back to its original position. But a book standing on end can be at rest, temporarily. If jostled or pushed, it will tilt and accelerate away from this position. After falling on its cover, the book will once again be in a position of stability. The atmosphere, depending on its vertical temperature structure, can be in a stable or unstable state. These states are simply less obvious than our book example.

The conditions of stability and instability depend on the relationship between the variable lapse rate and the constant adiabatic lapse rate. Stable air can be illustrated by the graph in Fig. 12-11a. Suppose a parcel of air at the ground is forced to rise—say be flowing over a mountain or wedge of cold air. The rising air will cool at the adiabatic rate of 5.5°F per 1000 feet, developing the temperatures indicated at each 1000-foot level. Suppose also that the quiet air through which the rising air is moving has a vertical temperature gradient expressed by a lapse rate of 3°F per 1000 feet. The quiet air will have the values shown on the steeper lapse rate curve. Note that the rising air becomes progressively cooler than the nonrising air around it, developing a difference of 10°F at 4000 feet. Since cool air is heavier than warm air at the same level, the rising, cooler, air becomes progressively heavier and will resist the uplift or will return to its original level as soon as the uplifting force terminates. In general, we can conclude from this example that air is stable whenever the lapse rate is numerically less than the adiabatic rate.

In Fig. 12-11b, the lapse rate in the quiet air is 7°F per 1000 feet. If a unit of air now rises, it will still cool at the adiabatic lapse rate of 5.5°F per 1000 feet. But the surrounding air temperature decreases faster; hence, the rising air becomes progressively warmer and lighter than the adjacent air, causing it to ascend with increasing speed. This it will continue to

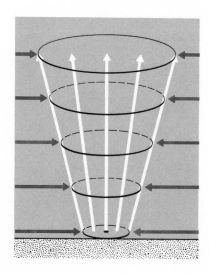

FIG. 12-10. Expansion of rising air caused by a decrease in surrounding air pressure.

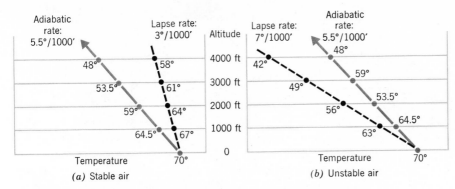

FIG. 12-11. (*a*) Relationship of the lapse rate to the adiabatic rate in the case of stable air. Rising air which cools at the adiabatic rate would always be cooler than the surrounding air cooling at the lapse rate and would tend to sink to the original level. (*b*) Relationship of the lapse rate to the adiabatic rate for unstable air. Rising air would in this case become increasingly warmer than the surrounding air and would tend to accelerate its uplift.

do until the lapse rate decreases, whereupon the rising air temperatures will catch up with surrounding air temperature and the air will come to rest. This must always happen at the tropopause (if not at a lower elevation), where the lapse rate becomes zero as described in Chapter 11. From this example, we can conclude that an unstable air condition exists whenever the lapse rate exceeds the adiabatic rate. Vertical air motion tends to dissipate very quickly in the stratosphere because of the constant vertical temperature gradient prevailing there.

We can now appreciate why temperature inversions are conducive to smog formation. The broken line in Fig. 12-12 represents a lapse rate with a pronounced temperature inversion overlying an unstable lapse rate zone as indicated by the lower segment of the curve having a slope gentler than the adiabatic rate slope. Ordinarily this represents an unstable condition. If air is forced aloft without being differentially heated, its

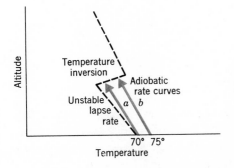

FIG. 12-12. The effect of inversions in promoting stability in the atmosphere. Rising air would tend to come to rest at the inversion level where its temperature becomes equal to that of the surrounding air.

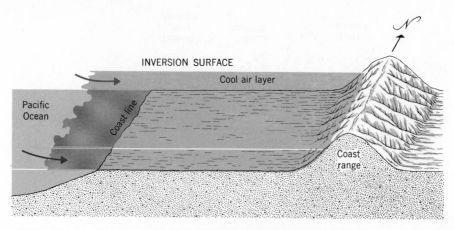

FIG. 12-13. The development of a strong inversion between the
California coastal mountains and the Pacific Ocean from the flow
of cool air off the ocean. Smog is concentrated in the stable cool
air layer below the inversion.

temperature will change (decrease) along the solid adiabatic curve, a. The
rising air will at first become progressively warmer (relative to the surrounding air) as it ascends, but on encountering the inversion will come
to rest because the temperature difference disappears abruptly. Even if
the air at the gound in the situation shown becomes 5°F warmer, it, too,
will only rise a short distance (following curve b) and then come to rest.

We can see from this example that a temperature inversion acts as
a physical lid, preventing air near the ground from rising very high. Fumes,
noxious gases, and pollutants of all sorts are thus trapped in a rather
shallow layer. The stronger the temperature inversion and the greater the
source of pollution, the more pronounced is the smog, until at times a
dangerous level of irritants develops.

The now-infamous southern California smog owes its origin to just
such a meteorological condition compounded by the addition of a geographic trap. Especially in summer, cool air from the Pacific is carried
over the coastal region and becomes confined as a shallow layer by the
coast range to the east (Fig. 12-13). The strong inversion that exists between
the cool lower air and overlying warmer air traps fumes and smoke,
causing dense smog accumulation.

MOISTURE AND WATER VAPOR

The vertical distribution and oceanic source of the water vapor in the
atmosphere is considered in Chapter 14. Variations in this distribution,
which occur continuously in the course of the hydrologic cycle, are
responsible for a large part of the phenomena of weather.

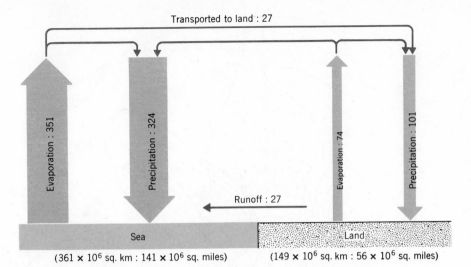

FIG. 12-14. The earth's water balance or hydrologic cycle. The numbers are centimeters of water per year. Total evaporation from land and sea of 425 cm is balanced by the total precipitation of 425 cm.

The Hydrologic Cycle

The water-vapor content of the atmosphere is on the average quite constant despite local changes due to seasonal and weather variations. Continuous evaporation from all of the oceans, lakes, rivers, and moist soil replenishes the moisture lost by rain, snow, and the like; the entire process of maintaining the constant water-vapor content of the atmosphere is known as the *hydrologic cycle*. It involves evaporation directly from surface waters, particularly the oceans, and moist lands. Moisture loss from vegetation, known as *transpiration,* may be quite important locally. *Evapotranspiration* includes the combined effects of evaporation and transpiration from vegetation-covered land areas. The next stage in the hydrologic cycle involves the conversion of airborne water vapor to a precipitable form. Precipitation of rain or snow may be directly into the ocean, completing the cycle, or it may occur over land, requiring surface runoff or groundwater seepage to complete the cycle. The elements of the hydrologic cycle can be treated quantitatively, resulting in an atmospheric water budget analagous to the heat budget described previously. Figure 12-14 contains a quantitative evaluation of the hydrologic cycle expressed in a form leading to a balanced water budget. Another useful way to regard the water budget is to consider the actual distribution of evaporation and precipitation over the globe as given in Fig. 12-15 for the Northern Hemisphere. Climatic deductions of great importance can be made from an illustration of this kind.

Humidity

Humidity can be described in a number of ways, all of them important and all related to the significant fact that the water vapor of the atmos-

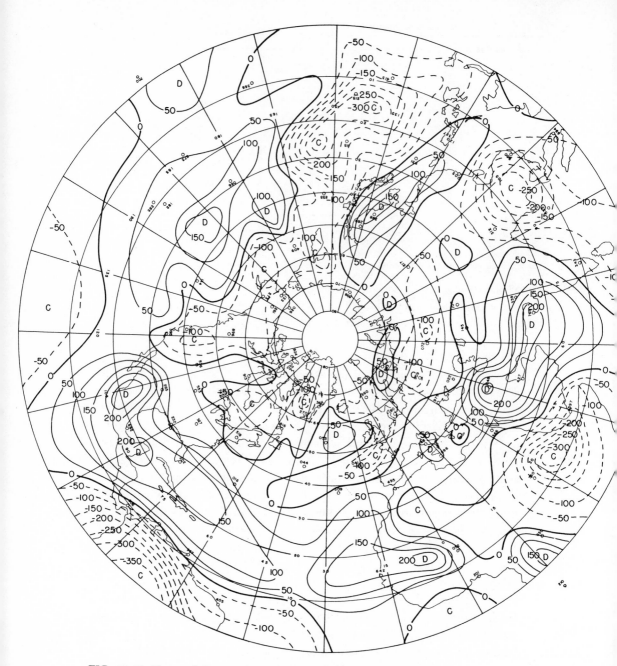

FIG. 12-15. Chart of the Northern Hemisphere showing the earth's
water budget in terms of evaporation minus precipitation. Solid
lines indicate evaporation exceeds precipitation by the amount
given. Broken lines indicate precipitation exceeds evaporation.
Units are in grams of total precipitable water over one square centimeter. (H. Lufkin, Sci. Rep. 4, Mass. Inst. Tech. General Circulation Project.)

phere varies directly with the temperature. At a particular temperature the maximum possible amount of water vapor in the air is called the *capacity* and the air is said to be *saturated* when this quantity is present. Saturation is achieved naturally through two processes: (1) by a decrease in temperature until the capacity just reaches the amount present or (2) by addition of water vapor through evaporation. Of the two processes, saturation from a temperature decrease is more common. The temperature at which saturation occurs is called the *dew point.* At and below the dew-point temperature water vapor is converted to liquid water by the process of *condensation.* (When the dew point is below freezing temperature, the change is from vapor directly to ice and the process is called *sublimation.*)

The absolute amount of water vapor in the air is described by the specific humidity, which refers to the weight of water vapor per *weight* of air (usually as grams per kilogram). Specific humidity, which depends on the weight of air regardless of volume changes, is independent of temperature and therefore of much meteorologic value in tracing air as it moves over the earth's surface.

Perhaps the most familiar of humidity terms is *relative humidity,* the ratio (expressed in percent) of the actual amount of water vapor present in the air to the air's capacity, or

$$\text{relative humidity} = \frac{\text{specific humidity (gms/kg)}}{\text{capacity (gms/kg)}} \times 100\%$$

Since the capacity of the air varies hourly with changes in air temperature, relative humidity is equally variable. It is generally lowest in midafternoon, when the temperature is maximum, and highest at night, when air temperature is minimum. This effect is shown strikingly by the comparison of simultaneous records of temperature and humidity in Fig. 12-16.

Measuring Humidity

Relative humidity is measured by using the *sling psychrometer* (Fig. 12-17), which consists of two standard thermometers, one of which is covered with muslin that is moistened when in use. Evaporation from the "wet bulb" lowers its temperature below that of the "dry bulb." The "depression" of the wet bulb depends on the relative humidity. Standard meteorological tables give the relative humidity and the dew point when the dry-bulb temperature and the depression of the wet bulb are known. Further humidity properties can then be calculated as explained above.

Permanent records of relative humidity are obtained from the *hygrograph* (Fig. 12-18). This recorder magnifies the changes in length of a sheaf of blond human hair, which is particularly sensitive to atmospheric water vapor.

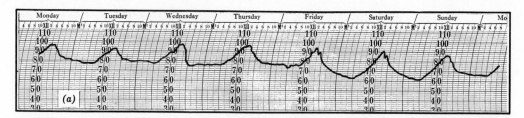

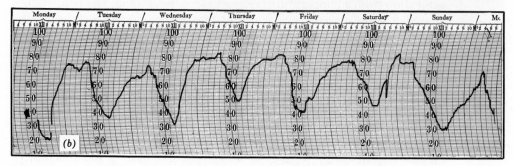

FIG. 12-16. Comparison of simultaneous thermograph (a) and hygrograph (b) records showing the inverse relationship of temperature and relative humidity. The latter is recorded by a hygrograph (Fig. 15-19). Temperature is shown in °F and relative humidity in percent.

Condensation

The varied forms of condensation of atmospheric moisture constitute some of the most important and most beautiful of the phenomena of weather. Condensation (sublimation, when below freezing), in the atmosphere requires that three conditions be met: (1) an adequate amount of water vapor, (2) cooling to and below the original dew-point temperature, and (3) nuclei of condensation. The reasons for the first two conditions have already been considered. The third is necessary because the vapor pressure (evaporation pressure) of a water droplet is inversely proportional to its radius:

$$\text{vapor pressure } \alpha \frac{1}{\text{radius}}$$

As a result of this relationship, the high vapor pressure of small droplets causes them to evaporate rapidly. Hence, some threshold size is necessary for droplet condensation to begin, otherwise evaporation will destroy the embryonic droplet as fast as condensation occurs. Nuclei of condensation must be somewhat water-absorbent or *hygroscopic*. Thoroughly insoluble mineral dust is of little natural importance in water condensation compared to particles of salt, products of combustion, and organic particles. Dust does serve as a nucleus for ice sublimation, however.

When condensation takes place directly on objects on or close to the earth's surface, fine beads of moisture, or *dew*, coat the objects. Dew forms mainly on clear, calm nights when the ground surface, or other materials, cools rapidly by radiation. Nocturnal deposits of dew are especially heavy in summer, when the absolute amount of water vapor in the air is high. If the dew point is below the freezing temperature, *frost* is deposited rather than dew.

When a thick layer of air near the land or water surface is cooled below the dew point, condensation occurs throughout the layer, forming a blanket of *fog*. *Radiation fog* forms under the same conditions as dew. To spread condensation through a thicker layer of air, there is an added requirement of just enough wind stirring to lift the air that cools by contact with the ground. As air cools, it becomes heavier and drains into hollows or low areas that are thus the first to collect radiation fog in characteristic "pockets." Despite the opacity of the densest of fogs, the droplets are so fine that little actual water is present. For example, a

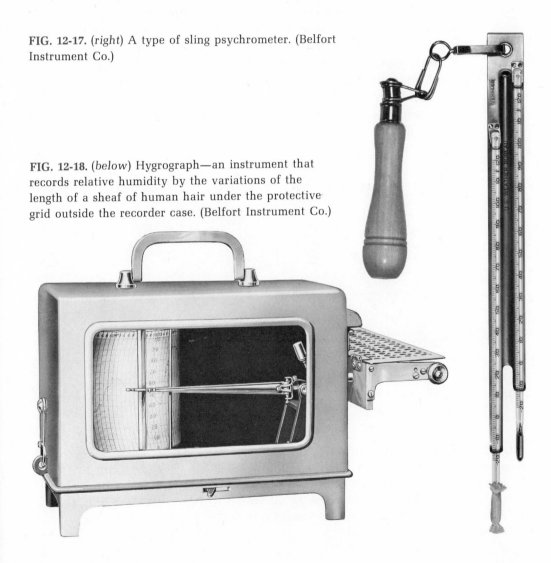

FIG. 12-17. (*right*) A type of sling psychrometer. (Belfort Instrument Co.)

FIG. 12-18. (*below*) Hygrograph—an instrument that records relative humidity by the variations of the length of a sheaf of human hair under the protective grid outside the recorder case. (Belfort Instrument Co.)

volume of dense fog the size of an average room through which objects would not be visible beyond 50 yards would contain less than half a glass of water.

When moist, warm air is blown over a cooler surface, cooling often results in a thick fog blanket known as advection fog. Famous examples are the fogs of the Grand Banks south of Newfoundland and the English "pea soup" fogs. The former occurs when warm air over the Gulf Stream is cooled by northward passage over the cooler Labrador current on the continental shelf (Fig. 12-19). The British fog occurs during the cold season when westerly winds transport relatively warm, moist air from over the Atlantic Ocean across the colder British Isles. Fog conditions are intensified over London owing to air pollution involving a high density of combustion-formed nuclei of condensation.

One of the greatest weather hazards in aviation is the form of fog known as *precipitation fog*, which often forms in the air beneath a cloud system from which rain or snow has been falling for a protracted period of time. Evaporation from falling droplets slowly increases the water-vapor content of the air, thereby raising the dew-point temperature until it nearly coincides with the true air temperature. The slightest cooling may then cause dense fog formation from the ground all the way up to the cloud level. Whereas other forms of fog are usually restricted to a fairly thin layer above the ground, precipitation fog can be thousands of feet thick.

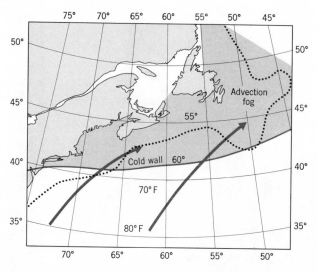

FIG. 12-19. The formation of advection fog over the Grand Banks. Warm humid air from the south is cooled by passage over the cool waters of the Labrador Current which is separated from the warmer Gulf Stream by a strong temperature discontinuity known as the "cold wall."

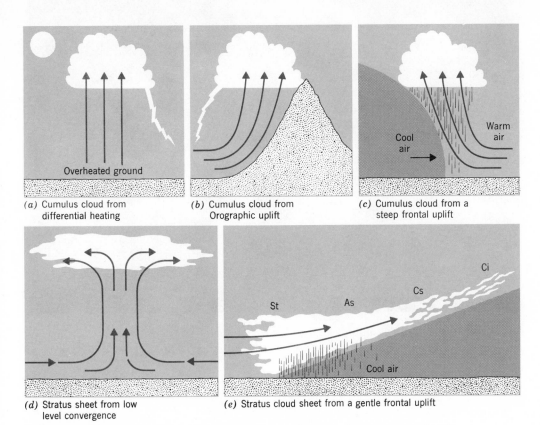

(a) Cumulus cloud from differential heating

(b) Cumulus cloud from Orographic uplift

(c) Cumulus cloud from a steep frontal uplift

(d) Stratus sheet from low level convergence

(e) Stratus cloud sheet from a gentle frontal uplift

FIG. 12-20. The generation of clouds from the cooling air caused to rise by the different processes shown.

CLOUDS AND PRECIPITATION

Clouds are by far the most important form of condensation, because they are the direct source of precipitation. The great diversity of shape and thickness of clouds is related to the quantity of water available for condensation and the cooling process that produces condensation at the cloud level. Although all clouds develop from adiabatic cooling of rising air, the mechanism of uplift varies with a consequent variation in the nature of the clouds. Cooling may occur from any of the causes shown previously in Fig. 12-9.

Cloud Types

The relationship of cloud type to uplift mechanism is indicated in Fig. 12-20. Isolated *cumulus* type clouds form from air having vertical or near-vertical uplift. These clouds are typically dense white with a castellated or cauliflowerlike side structure and a gray flat bottom (Fig. 12-21). The flat lower surface marks the level at which the rising air cools to

FIG. 12-21. Fair weather cumulus clouds. These clouds are typical of good weather in the warm seasons and are produced by rising air columns over ground areas that are differentially warmed by the sun (Fig. 12-20a). (U. S. Weather Bureau.)

the dew-point temperature. Continuous cooling and condensation occurs with further uplift above the dew-point level. Cumulus clouds are typical of fair-weather convection caused by the sun's heating the ground surface, which often becomes warmed differentially. The satellite photograph in Fig. 12-22 is a most striking example of the results of the difference in heating of land and water.

When air spreads laterally aloft after uplift from horizontal convergence (Fig. 12-20d) near the surface, cloud condensation is spread into a cloud sheet known as *stratus*. Stratiform clouds are produced even more commonly from the structure shown in Fig. 12-20c, where warm air is lifted over a sloping wedge of cool air.

In the case shown in Fig. 12-20c, thin wispy clouds known as cirrus often develop at high elevations beyond the deck of stratus clouds. These highest of common clouds are composed of thin crystals of ice, since temperatures are always subfreezing at cirrus-cloud levels. Even the upper part of the stratus sheet, called *cirrostratus*, is formed of ice particles. The familiar halos about the sun or moon are produced when their light is transmitted through high cirrostratus sheets. The thicker, lower-level part of the stratiform deck is called altostratus; this is characterized by a uniform pale-gray color and a "watery" appearance of the sun or moon. The thickest, lowest portion of the cloud sheet, referred to as *stratus*, has a dense, gray appearance. All-day rains or snows commonly fall from clouds of *altostratus* or *stratus* type. These clouds then take on a "wet" appearance and are called *nimbostratus* (*nimbus* meaning water). Precipitation fog, referred to earlier, often forms below the stratus sheet from the evaporation of continuous rain in the cool air layer.

Examples of standard cloud types are illustrated in Figures 12.23 to 12.26.

Precipitation: Rain and Snow

Because all clouds are composed of water droplets or ice crystals or a combination of both, a natural question is, why doesn't rain or snow always fall from clouds? The answer lies in both the size of the cloud particles and the strength of the rising air currents that are required for cloud formation in the first place. Most cloud particles are so small (0.00016 to 0.008 inches) that they are easily buoyed up by the gentlest

FIG. 12-22. Tiros IV photograph showing cumulus cloud development over England—relatively warm compared to the surrounding water. The dark strip just south of England is the English Channel. To the south of the channel, the coast of France can be recognized. (U. S. Weather Bureau.)

FIG. 12-23. Cumulus type clouds forming ring patterns within the eye of Hurricane Esther, September 1961. (U. S. Weather Bureau.)

of air currents. Only when conditions within clouds permit particles to grow to a size sufficient to overcome the natural air flotation will precipitation reach the ground.

Often we can see vertical gray streamers of rain or snow hanging below a cloud but not reaching the ground. On these occasions the moisture is evaporated completely in falling through dry air. Raindrops that do reach the ground are invariably smaller than they were when they left the parent cloud, owing to evaporation and even fragmentation during their descent.

When cloud condensation occurs below freezing temperatures, the water vapor is converted directly into ice crystals rather than water droplets. Snow consists of single or clumps of such crystals that are large enough to reach the ground. When water condenses into snow crystals, a tremendous and beautiful variety of six-rayed star-shaped units develop. Each one is slightly different from the other, as seen in the photograph of snow crystals in Fig. 12-27.

Occasionally rain droplets freeze to form *sleet* in falling through a layer of air colder than that in which the cloud particles condensed; or falling snow may partially melt and refreeze to form a softer kind of sleet. Sleet should not be confused with hail which is really a summertime feature of thunderstorms, to be described in the next chapter.

FIG. 12-24a. Cirrostratus clouds formed as in Fig. 12-20e.

FIG. 12-24b. Cirrus clouds illuminated by the rising sun while the surface is still dark.

FIG. 12-25. Altostratus clouds formed as in Fig. 12-20e.

FIG. 12-26. Stratus clouds formed as in Fig. 12-20e.

PRESSURE AND WIND

Cloud type, air pressure, and wind direction are the three most important indexes of the existing state and probable changes in weather conditions. These variables, of which clouds have already been discussed, will be related to storms and other weather patterns in the following chapter.

Air Pressure: Its Meaning and Measurement

Air pressure is often stated in terms of length (inches or centimeters) or weight—really mass (pounds or kilograms)—or in terms of actual pressure (millibars). The reason for the different nomenclature is historical, being related to the development of measuring procedures and understanding. Although changes in pressure can be described adequately by length or weight units, true pressure units are required in mathematical solutions in meteorology.

The units of length and mass are derived from the mercurial barometer—the traditionally standard instrument for measuring air pressure. The principle of this instrument is illustrated in Fig. 12-28a. If an evacuated long glass tube were placed with its open end in a well of mercury, air pressure on the surface of the mercury would drive the dense liquid into the tube. The mercury rises to a height such that its weight balances the weight of air on an area of the well surface equal to the area of the tube. On the average at sea level, this height is 29.92 inches or 76 centimeters. If the tube has an area of one square inch, the weight of the mercury column on the average is 14.7 pounds. (Mercury is used because of its high density. A column of water one square inch in cross section weighing

FIG. 12-27. The diversity of snow crystals shown in a photograph in which all the instruments must be kept chilled to prevent melting of the snow flakes. (U. S. Weather Bureau.)

14.7 pounds would occupy about 34 feet—a most inconvenient length.) As the pressure changes with changes in weather, the height of the mercury column in the barometer tube changes proportionally—falling pressure gives a falling column, rising pressure a rising column. Variations of the length of the mercury column thus describe variations in pressure.

Since air is gaseous, the property of pressure is much more meaningful than the weight or length of the equivalent mercury column. Technically, pressure is a force per square inch or square centimeter. The pressure of a gas on a surface is determined by the number and speed of the molecules striking the surface. The speed of the molecules depends on the temperature. The number of molecules of air in a given volume at any level—sea level, for example—depends on the compressional effect of the overlying atmosphere. If the amount of air from the surface to the top of the atmosphere increases, compression increases and the pressure—determined by the number of molecules striking a given surface—increases, and the reverse. To further illustrate the concept of pressure, suppose we completely enclose a given volume of air—say a cubic foot—at sea level and isolate this volume within a vacuum. A barometer within this isolated air would read exactly the same pressure as a barom-

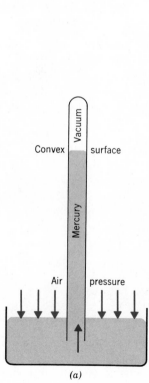

FIG. 12-28. (a) Principle of the mercurial barometer, (b) Fortin type standard Weather Bureau or laboratory barometer.

(a)

(b)

eter exposed to the free air at sea level. It would do so because the same number of molecules would strike the walls of the enclosed and isolated volume of air as would strike a surface of equivalent area in the unenclosed atmosphere. Recall from Chapter 11 that a cubic foot of air contains only 0.076 pounds, not 14.7 pounds, yet the pressure of this small quantity of air is the same as that under a column of air 14.7 pounds in weight.

The common unit used to describe pressure is the *bar* (one million dynes per square centimeter, the approximate sea-level pressure of the atmosphere—actually it is 1.0132 bars). Changes in air pressure associated with weather are very small. To avoid using long decimal values, meteorologists use the unit of the *millibar* (mb)—$\frac{1}{1000}$ of a bar. Average air pressure at sea level is thus 1013.2 millibars. Hence, 1013.2 millibars is equivalent to the length units 29.92 inches and 76 centimeters. These units, in both the English and metric systems, can be converted from the relationships:

$$1 \text{ in.} = \frac{1013.2 \text{ mb}}{29.92 \text{ in.}} = 33.86 \text{ mb}$$

$$1 \text{ mb} = \frac{29.92 \text{ in.}}{1013.2 \text{ mb}} = 0.03 \text{ in.}$$

$$1 \text{ cm} = \frac{1013.2 \text{ mb}}{76 \text{ cm}} = 13.33 \text{ mb}$$

$$1 \text{ mb} = \frac{76 \text{ cm}}{1013.2 \text{ mb}} = 0.075 \text{ cm}$$

Although the mercurial barometer is the standard and accurate instrument for measuring atmospheric pressure, the aneroid barometer (Fig. 12-29) is more portable and more readily read. The aneroid barometer contains an evacuated disc-shaped chamber with appropriate springs that prevent the walls from collapsing. Small movements of the disc faces produced by variations in pressure are transmitted to a dial indicator by a mechanical coupling that magnifies the movements. The face of the dial is calibrated in appropriate units of pressure—often both millibars and inches or centimeters. A recording aneroid, known as a *barograph*, provides a continuous record of pressure; when more sensitive, the instrument is called a *microbarograph* (Fig. 12-30).

Isobars and the Pressure Gradient

The horizontal distribution of pressure is shown by means of *isobars*, or lines of equal pressure. Although methods of weather analysis and forecasting have become more and more sophisticated over the years, the pressure distribution as revealed by isobars is still the primary characteristic of weather systems migrating over the earth's surface.

This relationship is exactly analogous to topographic contours whose spacing indicates the steepness of land slope. Isobars can thus be con-

FIG. 12-29. Aneroid barometer. (Friez Instruments Division of Bendix Aviation Company.)

sidered to be contours of pressure. Also, the direction of the pressure gradient must be perpendicular to the isobars (arrows in Fig. 12-31), since they are lines of no pressure change. (When isobars are close together, the horizontal change in pressure, or pressure gradient, is great; when relatively far apart, the pressure gradient is weak). Roughly circular isobar patterns enclosing regions of low or high pressure are among the most important and familiar features of a weather map. These are illustrated in Fig. 12-31, together with other typical isobar configurations.

Wind

Wind is air in *horizontal* motion. Troposphere winds are defined by the direction from which they blow; north winds come from the north, south winds come from the south, and the like. Wind direction is observed by a *vane* that points into the wind; speed is measured by a variety of instruments known as *anemometers*. Both speed and direction can be read directly from dial indicators or from permanent recordings. Examples of wind instruments in common use are illustrated in Fig. 12-32. A fair estimate of wind speed on land and sea can be made by observing the effects of wind on surrounding objects. Criteria for judging speed are given in Tables 12-1 and 12-2.

Wind speed and direction are depicted on weather charts by means

FIG. 12-30. Microbarograph—A sensitive recording aneroid barometer.

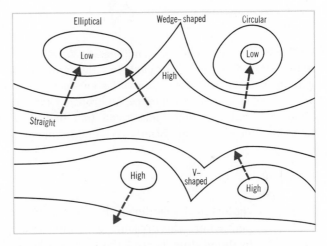

FIG. 12-31. Common isobar patterns: the arrows indicate the direction of the pressure gradient (change in pressure).

of arrows that fly with the wind. Speed is designated by barbs on the shaft of the arrow; each half barb represents 5 knots and each full barb is 10 knots. Pressure patterns and associated winds extending from western North America to Europe are shown in Fig. 12-33. Many low- and high-pressure systems are evident on this simplified weather chart. Notice the difference in intensity of the pressure gradient between the eastern and western parts of the prominent low-pressure area (labeled *L* at its

FIG. 12-32. Types of wind transmitters. (a) Wind vane or wind direction transmitter—transmits electrical signals to either a dial or chart recorder, (b) three-cup anemometer whose rotation varies an electrical current that drives a wind speed indicator in the office, (c) propeller type anemometer combining wind speed and direction transmission. (a,b, Belfort Instrument Company, c, Friez Instrument Division, Bendix Aviation Company.)

center) in the center of the chart. Note also that the winds in the western part where the pressure gradient is steep, are much stronger than those in the eastern part of the system where the pressure gradient is weaker. Winds in the western half are northerly and reach 45 knots; in the eastern half they are southerly to easterly and are as low as 15 knots.

Forces Affecting Wind Speed

These observations reveal a fundamental fact about winds, namely that their speed is determined by the strength of the horizontal pressure gradient. The steeper the pressure gradient, the higher is the wind speed, and the reverse. Friction between air and the land or ocean surface slows the wind in the lowest layers of air; the effect is greater the rougher the surface. Above the level of friction, the wind moves with a speed controlled only by the pressure gradient.

TABLE 12-1 Criteria for Estimating Wind Speed on Land

Wind Speed		Description
Mph	Knots	
0–1	0–1	Calm; smoke rises vertically
1–3	1–3	Direction of wind shown by smoke drift but not by wind vanes
4–7	4–6	Wind felt on face; leaves rustle; ordinary vane moved by wind
8–12	7–10	Leaves and small twigs in constant motion; wind extends light flag
13–18	11–16	Wind raises dust and loose paper; small branches are moved
19–24	17–21	Small trees in leaf begin to sway; crested wavelets form on inland waters
25–31	22–27	Large branches in motion; whistling heard in telegraph wires; umbrellas used with difficulty
32–38	28–33	Whole trees in motion; resistance felt in walking against wind
39–46	34–40	Wind breaks twigs off trees and generally impedes progress
47–54	41–47	Slight structural damage occurs (chimney pots and slate removed)
55–63	48–55	Seldom experienced inland; trees uprooted; considerable structural damage occurs
64–72	56–63	Very rarely experienced; accompanied by widespread damage
73	64	Maximum wind damage

The driving force is gravity, as illustrated in the schematic diagram in Fig. 12-34. Air pressure is greater above the high-pressure region than above the low, resulting in a downward pressure differential (heavy arrow). A force is thus directed from the high- to the low-pressure region, tending to set up the air motion shown by the broken lines. The greater the pressure differential, the higher is the wind speed between the "high" and the "low." A quantitative expression of this differential is given by the values and spacing of the isobars between the high- and low-pressure centers—the pressure gradient.

Forces Controlling Wind Direction

When a pressure gradient exists between two points, the wind initially tends to blow directly "down" the gradient, or across the isobars from high to low pressure. But note that in Fig. 12-33, the wind arrows about the low pressure center already referred to are more nearly parallel to than perpendicular to the pressure gradient, resulting in circular rather than radial inward wind motion. Hence, an additional effect, the *Coriolis force,* must be introduced to explain the deflection of wind from the direction of the pressure gradient.

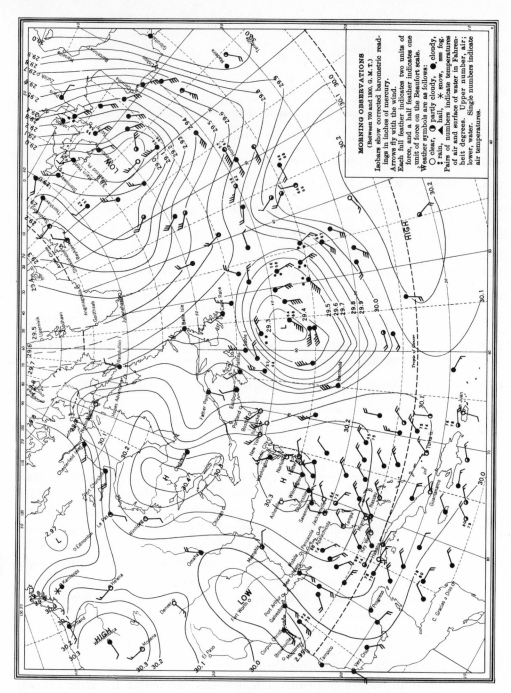

FIG. 12-33. Weather map of the North Atlantic Ocean and adjacent continents, January 17, 1939. Isobars (lines of equal pressure) are drawn for pressure intervals of 0.1 inches on this old map. Arrows fly with the wind; each half barb represent a 5 knot wind speed, each full barb, 10 knots. The circles at the head of each arrow locates the point of observation. The map is dominated by a series of low and high pressure systems, (cyclones and anticyclones), the former being more intense.

440   ELEMENTS OF THE WEATHER

TABLE 12-2 Criteria for Estimating Wind Speed at Sea

Wind Speed		Description
Mph	Knots	
0–1	0–1	Sea like mirror
1–3	1–3	Ripples with scaly appearance; no foam crests
4–7	4–6	Small wavelets, crests of glassy appearance and not breaking
8–12	7–10	Large wavelets with crests beginning to break, scattered whitecaps
13–18	11–16	Small waves growing larger, numerous whitecaps
19–24	17–21	Moderate waves with greater length, many whitecaps with some spray
25–31	22–27	Larger waves, whitecaps very numerous, more spray
32–38	38–33	Sea tends to heap up, streaks of foam blown from breaking waves
39–46	34–40	Fairly high waves of greater length, well-marked streaks of foam
47–54	41–47	High waves with sea beginning to roll, dense streaks of foam with spray blown higher into air—may cut visibility
55–63	48–55	Very high waves with overhanging crests, sea is white with foam, heavy rolling and reduced visibility
64–72	56–63	Waves exceptionally high, sea covered with foam, visibility further reduced
73	64	Sea completely covered with spray, air filled with foam, greatly reducing visibility

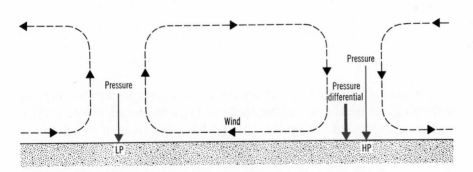

FIG. 12-34. Schematic relationship of wind to regions of low and high pressure. The pressure excess in the high is indicated by the heavy arrow. Air has a falling component in high pressure, and a rising component in low pressure areas.

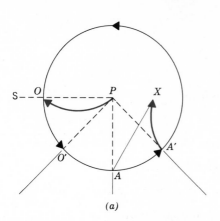 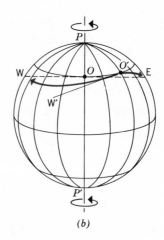

(a)   (b)

FIG. 12-35. (a) Coriolis deflection as viewed at the north pole. An object (bullet or wind) moving freely away from the North Pole would follow the straight line PO as viewed by an observer above and not rotating with the earth. An observer on the rotating earth would see the apparent movement indicated by the curved line, PO as he is carried by rotation to O'. A wind or bullet travelling toward the pole from A to P would follow the straight line AX as seen by an observer above the earth. To an observer on the earth, the path would appear to be the curved line A'X as he is carried by rotation to A'. (b) Coriolis deflection as viewed at the middle latitudes. An object (wind, bullet) travelling from O due east would follow the straight (dashed) line OE as seen by an observer in space. With reference to the rotating earth the apparent path is the curved line O'E. An object moving due west would follow the straight line OW as seen by an observer in space, and the curved line O'W by an observer on the earth as his west point shifts to W'.

To understand the Coriolis force we must refer to the explanation of the Foucault pendulum in Chapter 2 and Fig. 2-15. Recall that the earth's surface at any point but on the equator has a component of rotational motion about a radial axis through that point. The amount of rotation is a maximum at the poles and decreases to zero at the equator. If an object is moving freely (unattached) over the surface or through the atmosphere, the surface tends to rotate beneath the object. There are different ways to illustrate this effect.

In Fig. 12-35a we see a portion of the earth as viewed from a point above the North Pole (P). The counterclockwise rotation visible in this view is shown by the arrows. Now imagine that an observer at the pole fires a bullet toward the star S along the meridian PO (broken line). By the time the bullet reaches O, the meridian has rotated to PO'. To an observer above the earth, the bullet travels the straight path *POS*. To an observer on the rotating surface, the bullet seems to depart more and more from the meridian as it rotates from *PO* to *PO'* and *appears* to follow the curved path *PO*. Note that the deflection of the bullet is to the *right* of *PO'*, the new position of the observer's meridian.

If the bullet is fired toward the pole along the meridian *AP*, the latter will rotate to *A'P* as the bullet travels. Since the bullet partakes of the earth's rotational speed from west to east plus its own muzzle velocity, to an observer above the earth it will travel to X along the straight line *AX* as *A* rotates to *A'*. But to an observer on the rotating earth, the bullet appears to leave the meridian behind as it travels the curved path *A'X*. Again it is deflected to the right of the initial direction of motion. The reader should develop the case for a bullet fired in a more general path— say from *O* to *A'*—and show that for an observer on the earth, deflection to the right will still occur.

If the bullet is deflected as indicated in these examples, a force must exist to cause the deflection. Named after the French mathematician, G. Coriolis, who first explained the effect in detail, *Coriolis force* is an apparent or fictitious force in that it exists only for an observer whose reference is the earth's surface; it does not exist for an observer viewing the earth's surface from a position in space.

Another view of the effect is given in Fig. 12-35*b*, where the earth is viewed from the side. Imagine an observer on a parallel of latitude at point *O*. If he fires a bullet due east, it will travel a path toward *E* as the observer rotates to *O'* and will appear to follow the solid arrow *O'E*. If the bullet is fired westward toward *W*, when the observer is rotated to *O'*, his west will be along the line *O'W'*. The bullet will appear to travel the curved path *O'W*. In both cases the deflection is to the right of the initial direction—or line of sight.

In general, all freely moving bodies in the *Northern Hemisphere* are deflected to the *right* of their initial direction of motion. The reader should reverse the direction of rotation in Fig. 12-35*a* to show that with clockwise rotation in the Southern Hemisphere, deflection will be to the left.

Because the amount of rotation of the earth beneath a moving body varies as the sine of the latitude, as shown in Chapter 2 for the Foucault pendulum, the resulting Coriolis force varies in the same way. Coriolis force can be expressed in terms of the wind speed and the earth's angular velocity as

$$F = 2\ mv\omega \sin \phi$$

where *m* is the mass of the moving object, be it wind or bullet, *v* is its speed, ω is the earth's angular velocity expressed in radian measure, and ϕ is the latitude. *At any instant, the Coriolis or deflecting force is always perpendicular to the motion of the object.* As this force causes changes in direction of movement, the direction of the deflecting force is also changed.

If we simply substitute air moving in response to a pressure gradient in place of a bullet, the same Coriolis effect can explain the deflection of the wind. This is illustrated for the case of straight isobars in Fig. 12-36, where the direction and magnitude of the pressure-gradient force from higher to lower pressure is indicated by the solid arrows, and the resulting wind motion, which increases steadily in speed between points 1 and 6, is shown by the heavy curved arrow. As soon as the wind starts to blow as at point 1, the Coriolis force shown by the dashed arrows begins to deflect the wind to the right (in the Northern Hemisphere). Although the pressure gradient force maintains a constant direction, the Coriolis force, always perpendicular to the wind at any instant, constantly changes its direction as the wind suffers continuous deflection. Finally, at 6 the pressure gradient and Coriolis force exactly oppose each other, at which point the wind motion is now parallel to the isobars (perpendicular to the pressure gradient or initial direction of motion). This wind, which is in equilibrium with the gradient and the Coriolis force, is called the *geostrophic wind*.

Frictional effects prevent the full wind speed from being realized near the earth's surface so that the low-level winds always blow slightly inward—across the isobars. At about 2000 feet in altitude, frictional effects disappear and the true geostrophic wind is present.

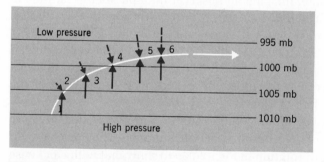

FIG. 12-36. Coriolis deflection in the case of straight isobars. The wind, starting from rest at point 1, is deflected to a path shown by the curved arrow until it blows parallel to the isobars at point 6. Wind speed increases steadily until parallel to isobars. The direction of the pressure gradient is constant as shown by the solid arrows. The Coriolis force (dashed arrows) increases in magnitude as the wind velocity increases. Because it it always perpendicular to the wind direction, the Coriolis Force changes direction continuously until it is equal and opposite to the wind direction (point 6). The wind continuing beyond point 6 is called the *geostrophic* wind. Wind in the layer below 2000 feet never achieves true geostrophic balance because of frictional forces.

Cyclonic and Anticyclonic Winds

In the case of circular pressure systems, such as the low- and high-pressure systems discussed previously, the pressure gradient is directed radially inward or outward, as indicated by the dashed arrows in Fig. 12-37a-d. The Coriolis force produces a deflection to the right in the Northern Hemisphere resulting in a counterclockwise inward wind spiral (Fig. 12-36a); in the Southern Hemisphere the deflection, being to the left, causes a clockwise spiral (Fig. 12-37b).

At about 2000 feet above the ground, the wind flow is essentially parallel to the isobars, as explained for the geostrophic wind.

The circular wind motion (counterclockwise in the Northern Hemisphere) about low-pressure areas is known as *cyclonic* motion and the entire pressure-wind system is called a cyclone. As shown in Fig. 12-37c and d, wind motion in high-pressure areas is deflected into a clockwise spiral in the Northern Hemisphere and a counterclockwise one in the

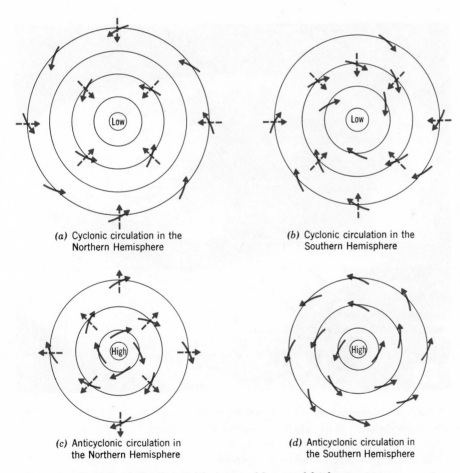

(a) Cyclonic circulation in the Northern Hemisphere

(b) Cyclonic circulation in the Southern Hemisphere

(c) Anticyclonic circulation in the Northern Hemisphere

(d) Anticyclonic circulation in the Southern Hemisphere

FIG. 12-37. Coriolis deflection in the case of low and high pressure systems. The direction of the pressure gradient is shown by dashed arrows and the wind by solid arrows.

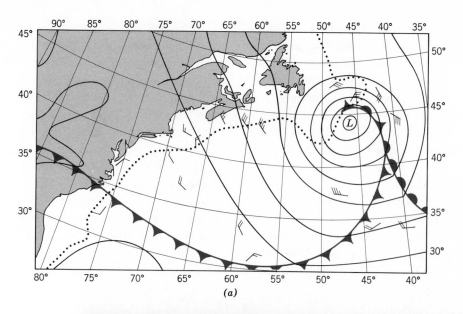

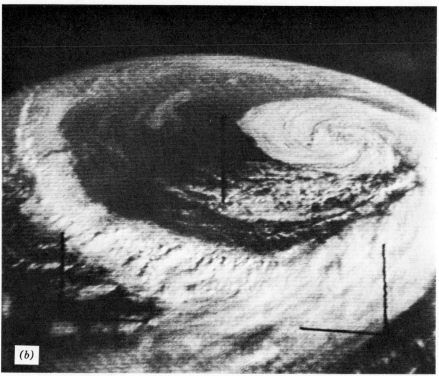

FIG. 12-38. (a) Cyclonic storm system in the western North Atlantic Ocean with characteristic counterclockwise spiral wind pattern. (b) Tiros satellite photograph of the storm system in (a) showing counterclockwise spiral structure of the cloud pattern corresponding to the intense part of the storm (upper right). (NASA Photos.)

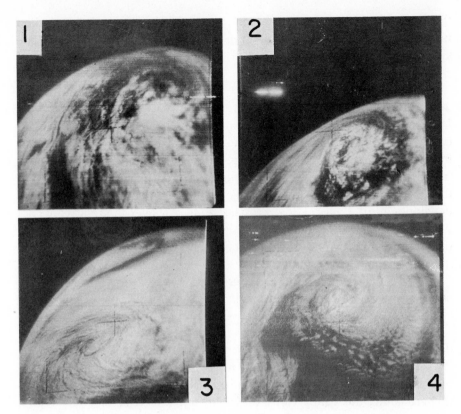

FIG. 12-39. Satellite photographs of cyclone systems in the Southern Hemisphere. (NASA Photos.)

Southern Hemisphere. High-pressure system winds are known as *anticyclonic* winds and the combined system is called an anticyclone. The terms cyclones and anticyclones do not imply any disastrous or intense storm systems, as is commonly inferred. In the middle latitudes (30° to 60°), the daily weather is nearly always under the influence of one or the other such systems. The weather chart in Fig. 12-33 shows several cyclonic and anticyclonic systems with associated counterclockwise and clockwise motion, respectively.

The airflow patterns within cyclones of both hemispheres are beautifully illustrated in Figs. 12-38 and 12-39. A simplified weather chart of an intense cyclonic storm over the western north Atlantic Ocean (Fig. 12-38a) shows the typical counterclockwise wind pattern with a slight inward drift. The Tiros satellite photograph of the same region contains a striking cloud pattern organized into a counterclockwise spiral in the region corresponding to the central part of the storm (upper right in photo). The long curving band of clouds extending from the center of the storm marks the position of the cold front, a feature to be studied in Chapter 13. Four satellite photographs of Southern Hemisphere cyclone systems (Fig. 12-39) illustrate typical clockwise spiral air flow.

Local Wind Systems

Land and sea breezes and mountain and valley breezes are common wind types, related to local geography. The former occur along coastal regions from the differential diurnal heating of land and water. On sunny days in the summer, the general winds may be weak or calm. At such times the air over the heated land warms rapidly and rises, resulting in the development of a narrow low pressure zone. Air over the relatively cool water subsides creating a zone of high pressure. The pressure gradient between land and water usually becomes strong enough to start the sea breeze by about 10 a.m. local time.

After sunset, when the pressure gradient reverses direction as the land cools by radiation, the land breeze develops. Land and sea breezes rarely extend far inland or over the water.

Mountain and valley breezes are also diurnal convective effects that develop where fairly steep slopes exist. On sunny days, particularly during the warm season, valleys become warmer than the adjacent upland with the creation of a current of air (known as the valley breeze) that rises upward along the valley sides. At night, upland areas cool faster than valleys as the land cools by radiation. The air near the land slope cools faster than the adjacent air over the central part of the valley and subsides to form the mountain breeze. This nocturnal breeze is much steadier and stronger than the daytime valley breeze.

STUDY QUESTIONS

12.1. Explain the principle of the thermometer and the thermograph.

12.2. Convert (a) $-40°F$ to degrees C.
      (b) $37°C$ to degrees F.

12.3. Draw two curves showing the variation trends of temperature during a day and during the year. Label temperature on the vertical axes but do not give actual values and label time of day (year) along the horizontal axes.

12.4. Why does the heating of land and water differ so much?

12.5. Plot a graph with two curves that show the change in temperature from summer to winter, respectively, across the United States in a south-north direction (from about New Orleans to the Canadian border). Use Figures 12.6 and 12.7.

12.6. Which curve shows a greater horizontal temperature change, that for summer or winter? Can you explain why they are different?

12.7. Distinguish between the lapse rate and the adiabatic lapse rate. Give values related to each.

12.8. Define stable and unstable air. What conditions are necessary for each condition?

12.9. What is the meteorological condition responsible for most smog?

12.10. What is meant by the "hydrologic cycle"? How does it (on the average) keep an exact balance of water between the atmosphere and the earth's surface?

12.11. Why does the relative himidity nearly always rise at night?

12.12. Define condensation and give the conditions in the atmosphere necessary for this process.

12.13. How does the cooling necessary for cloud formation occur in the atmosphere?

12.14. Describe the main types of clouds.

12.15. Distinguish between condensation and precipitation and give the forms of precipitation.

12.16. Explain air pressure in terms of both (a) weight of air and (b) molecular motion of air particles.

12.17. What are the common units in air pressure measurement? Give the average value of normal pressure in each of these units.

12.18. How is wind force related to the pressure gradient?

12.19. Show two examples of pressure gradients by means of isobars and indicate which of the two would be related to the stronger wind. Use a pressure difference of 3 millibars for each isobar and label them in values you select.

12.20. What is the physical cause of wind motion?

12.21. Briefly explain "coriolis force." Is it a real or apparent force?

12.22. How does the coriolis force cause cyclonic and anticyclonic wind motion? Explain the difference in this motion in the Northern and Southern Hemispheres.

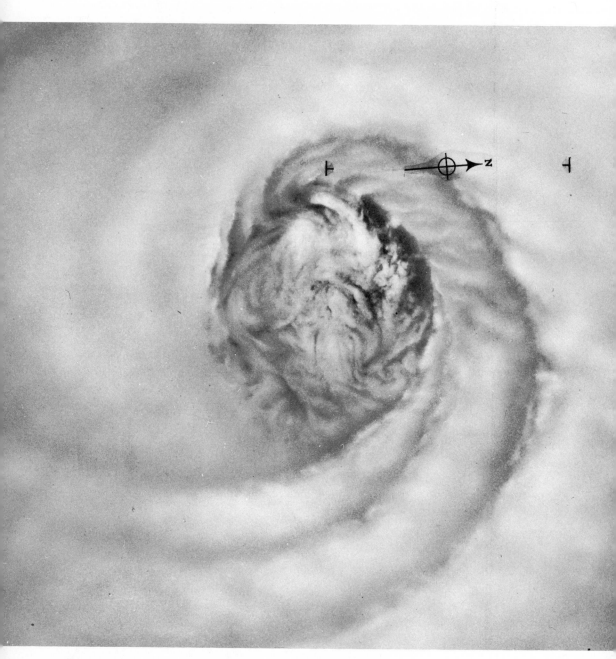

chapter 13

Photograph of Typhoon *Ida,* 1958 from high elevation (U-2) aircraft. (Air Weather Service, U. S. Air Force.)

Planetary Winds, Weather, and Storms

Our local weather and storms are a closely woven part of the global air circulation. Local storms affect the wind systems of the world and at the same time the winds steer and direct the movement of local weather patterns. So close is the relationship that one cannot say precisely which causes which. Let us start, rather arbitrarily with a review of the large-scale winds and then review current knowledge of storms and their connections to the prevailing winds.

THE GLOBAL WIND PATTERN

To the two sources of atmospheric energy, heat and gravity, we must now add a third—the energy of the earth's rotation—in order to understand the behavior of the global or planetary winds and related weather movement. Remember that the atmosphere is rotating at essentially the same

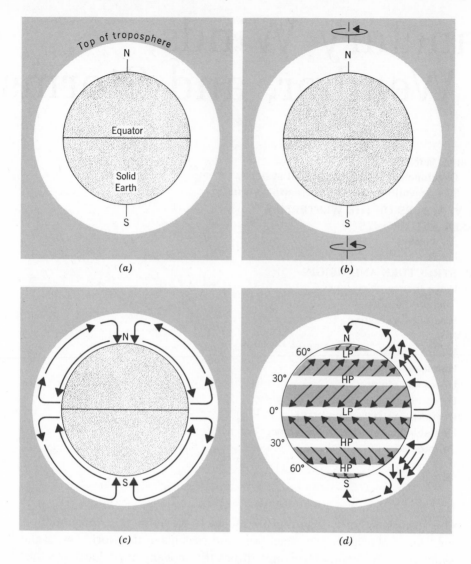

FIG. 13-1. Wind systems that would exist on an earth with (*a*) the same temperature over the surface and no rotation, (*b*) the same temperature over the surface and rotation, (*c*) a warm equator and cold polar regions but with no rotation, and (*d*) warm equator and cold poles plus rotation.

speed as the solid earth. If it were not, we would experience a wind, depending on latitude, of up to a thousand miles an hour as the earth turned through the stationary atmosphere. But the atmosphere differs from the solid earth in being able to move about relative to the surface.

Development of the Ideal Wind System

The development, in principle, of the planetary wind systems can be seen from the several views in Fig. 13-1. Imagine the earth to be fixed and nonrotating, a la Ptolemy, with the atmosphere having the same horizontal temperature at all levels—the same at the equator and at the poles. Both atmosphere and earth would be at rest as in Fig. 13-1a. Now imagine the earth to rotate realistically in b. If the atmosphere is still uniform in temperature from equator to poles, the earth and atmosphere would rotate as one unit with no differential motion between them. And now let us once again keep the earth fixed but consider a more realistic heat distribution, in which the tropical belt is strongly heated and the polar regions cooled throughout the year. Two simple, giant convection cells would develop. Warm equatorial air would rise and flow northward at high levels; after cooling and sinking at polar latitudes, it would return to the tropics as in c.

Finally, let us combine the effects in b and c in which rotational, thermal, and gravitational energies are all active. The Coriolis deflection discussed in the last chapter would cause a wind deflection to the right in the Northern and to the left in the Southern Hemisphere. In the Northern Hemisphere, northerly winds would be deflected to become northeast to east winds. The resulting pattern can be generalized in d. A calm, low-pressure belt (the doldrums) develops from the heated rising air of the equatorial region. The descent of this air at 30° north and 30° south produces the high-pressure belt of subtropical calms known as the "horse latitudes" ever since the days of sailing ships when horses were thrown overboard in these calm regions in order to conserve water as well as lighten the vessel.

From the horse latitudes to the doldrums blow the *northeast* and *southeast trades*, the most constant winds of the earth. From the horse latitudes toward the low pressure belts of the subpolar regions blow the prevailing westerlies, the wind zone in which most of the civilized world resides.

The Prevailing Winds at the Earth's Surface

The ideal planetary wind systems in Fig. 13-1d are modified by the seasonal heating and cooling of the earth's nonuniform surface. A more realistic but still generalized view of the distribution of pressure and winds at the surface is shown for the months of July and January in Figs. 13-2 and 13-3, respectively.

In the Southern Hemisphere, where the surface is mostly water, no striking seasonal difference is evident. But in the Northern Hemisphere, vast changes do occur over rather large areas. For example, Asia becomes very warm in summer relative to the oceans and very cold in winter. Since warm air rises, creating a surface low-pressure region, while cold air subsides, creating a surface high-pressure area, atmospheric pressure over Asia undergoes a marked seasonal transformation from low to high as is evident in Figs. 13-2 and 13-3.

Note that winds over the Indian Ocean blow toward the low-pressure

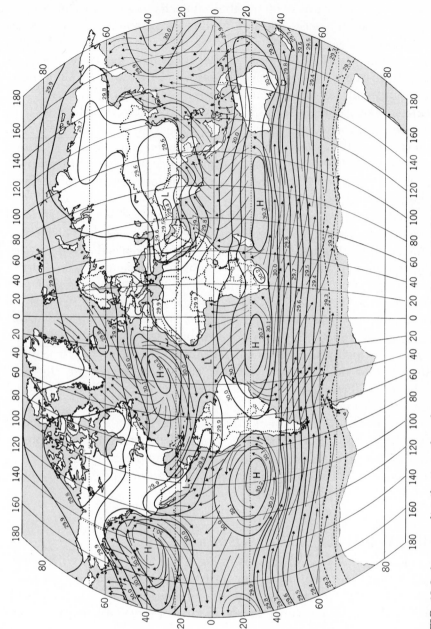

FIG. 13-2. Average distribution of surface pressure and winds for July. (After McNight & McNight.)

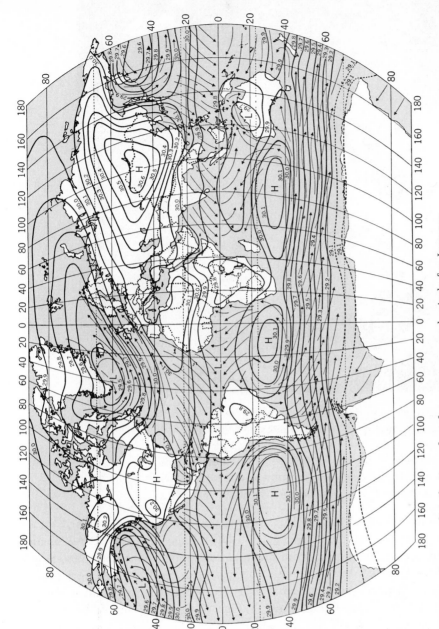

FIG. 13-3. Average distribution of surface pressure and winds for January. (After McNight & McNight.)

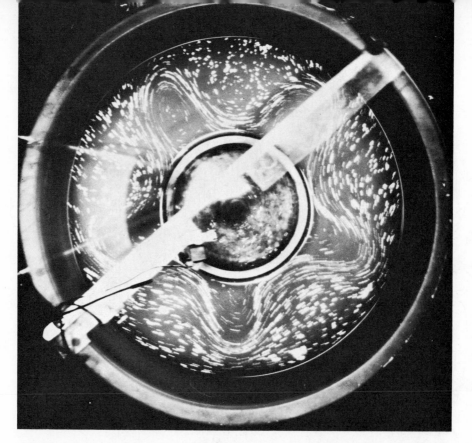

FIG. 13-4. Rotating dishpan-model of the atmosphere in which the perimeter is warmed analogous to the equator and the center is cooled, analogous to the poles. The combined effect of the thermal gradient from the edge to the center and the rotation of the fluid causes development of the wave pattern shown. The same effect is clearly shown by the earth's atmosphere. (Courtesy, D. Fultz.)

area over Asia in summer (Fig. 13-2), giving the well-known warm, moist Indian *monsoon;* in winter (Fig. 13-3) the winds are reversed and become cool and dry in coming from the cold continent. Other seasonal changes are evident over the continents and oceans of the Northern Hemisphere, all due to pronounced thermal differences between land and water.

The prevailing winds shown in Figs. 13-2 and 13-3 should be regarded as average conditions for the seasons shown. Secondary wind and storm effects, to be described later in this chapter, may completely alter the pattern of prevailing winds at a particular time. This is particularily true in the prevailing westerlies zone.

THE GREAT WAVES OF THE WESTERLIES

According to the cross-sectional portion of Fig. 13-1c, the winds in the entire troposphere between about 30° and 60° are southerly. But this side view indicates only the component in the north-south plane. Actually, the Coriolis force deflects these winds into westerlies from the surface to the tropopause. This we must picture as a broad current of air covering some 30 degrees of latitude and about 35,000 to 40,000 feet thick.

This current is the boundary between cold air on the poleward side and warm air on the equatorial side. Under these conditions it is now well known from both theory and experiment that such a zone is unstable—it cannot maintain a straight boundary region. Rather, it is thrown into great horizontal waves, a few of which girdle the entire earth. This effect has been shown experimentally in a dishpan model of the earth (Fig. 13-4) in which the center of the pan is cooled (analogous to a pole) and the perimeter is warmed (analogous to the equator). When the model is rotated at the proper speed (equivalent to the earth's rotation), a number of waves develop, giving the fluid a wave motion superimposed on its rotational motion. A very similar effect is shown by the winds of the atmospheric westerlies, particularily above 15,000 feet.

The development of such waves in the westerlies, known as *Rossby waves* after G. C. Rossby, who predicted them mathematically, is shown schematically in Fig. 13-5. Note how these waves form alternate low- and high-pressure areas, with associated cyclonic and anticyclonic air motion. Near the ground a number of factors may distort or mask this effect, but

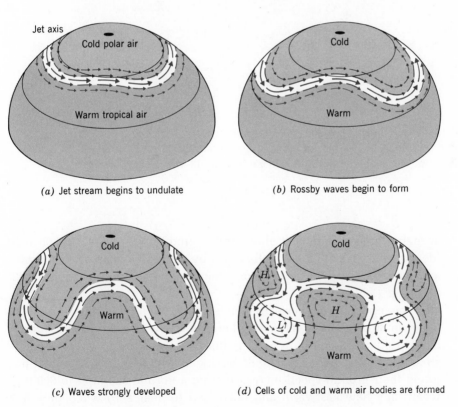

(a) Jet stream begins to undulate

(b) Rossby waves begin to form

(c) Waves strongly developed

(d) Cells of cold and warm air bodies are formed

FIG. 13-5. The development of Rossby waves in the upper westerlies. (After A. N. Strahler, *The Earth Sciences*, Harper Bros.)

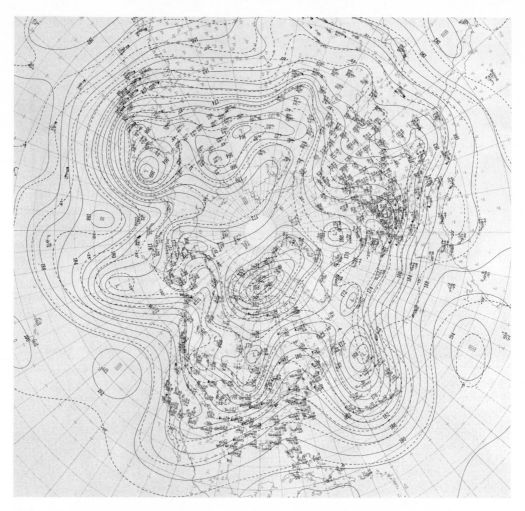

FIG. 13-6. Distribution of atmospheric pressure and winds over the Northern Hemisphere (February 5, 1955) at an elevation of about 18,000 feet. Compare this actual situation to the model and scheme in Figs. 13-4 and 13-5, respectively.

at many thousands of feet above the ground, Rossby waves reach their ideal development. This is well-shown by the typical map of air pressure and winds for the Northern Hemisphere at an elevation of about 18,000 feet (Fig. 13-6).

The upper-level westerly winds reach their maximum velocities between 30,000 and 40,000 feet, where they flow as a serpentine tube of high-speed air that follows the pattern of the Rossby waves. This *jet stream* was discovered by pilots during World War II, who sometimes found their ground speed increased from a normal 300 mph to 600 mph when they flew eastward, and decreased to near zero when they flew westward.

AIR MASSES AND FRONTS

In the last chapter cyclonic and anticyclonic wind motion was explained for areas of low and high pressure. We now see that these low- and high-pressure areas, which for the most part make our everyday weather, are really parts of the global pattern of Rossby waves in the westerlies.

To understand the full picture of the surface weather conditions associated with cyclones and anticyclones, we must include in their behavior the nature and effects of air masses and fronts.

Air Masses

In size, air masses cover hundreds of thousands of square miles; vertically, they extend upward for thousands and tens of thousands of feet. There is no difficulty in conceiving of uniform ocean currents within the main ocean body. We can see them; ocean water is visible. The Gulf Stream is readily apparent by its movement, color, temperature, seaweed content, and so on.

Uniform bodies of air, or air masses, are not so obvious, but their presence is adequately shown by meteorological observations, particularly of their temperature and humidity. Although air masses are identified and their motion traced through instrumental rather than visual observation, their presence is often detected very noticeably by our senses. We are all aware of the oppressively hot, sticky, summer heat waves. We are also aware of the dramatic end of such a hot-weather spell, when, following a violent thunderstorm, a wave of cool dry air is experienced for several days. A large, hot, humid air body responsible for the heat wave was simply replaced by a cool, dry air mass with its consequent relief for the heat sufferers.

Air masses derive their original properties from the surface over which they form. The temperature and humidity characteristics of an air mass are determined directly by the nature of the surface beneath.

In considering the relatively large volume of air masses and the poor powers of heat conduction that air possesses, it is apparent that bodies of air with uniform temperature and humidity will not form too rapidly. In the development of an air mass, a large volume of air must remain stagnant or circulate for some time over a particular portion of the earth, gradually acquiring its distinguishing temperature and humidity characteristics.

Air masses develop more commonly in some regions than in others, the areas of formation being known as *source regions*. We may note, for example, that the common source regions for air masses affecting American weather are the Northern Pacific west of Canada; the northern interior of Canada; the North Atlantic east of Canada; the Pacific west of southern California; the desert areas of southwestern United States and northern Mexico; and the Gulf of Mexico and the Caribbean Sea.

It is noticed that the source regions tend to bound the belt of prevailing westerlies. One set of source regions exists along the northern boundary in the vicinity of the subpolar low-pressure circle, while the other set exists along and to the south of the horse latitudes. The basic difference between the air originating in the northern source regions and that in the southern is therefore one of temperature.

Cold northern air masses are called *polar air masses* and designated "P" on weather maps, while the warm air bodies originating in low latitudes are called *tropical air masses,* "T". Then, depending on whether they form over land or water, the air masses will be dry or humid, respectively. This leads to two subdivisions for the above air types. Dry polar air of continental origin is known as *polar continental air,* designated cP, and when of oceanic origin, *polar maritime air,* mP. Similarly, tropical air is known as *tropical continental* (cT) and *tropical maritime* air (mT). Arctic air masses (A) are the coldest and driest of those that invade the midlatitudes of the Northern Hemisphere.

When air masses break loose from their source regions they are steered by the strong wind currents of the upper troposphere and move in a generally eastward direction. In the midlatitudes we are usually within one or the other of the above types of air masses and experience the temperature and humidity related to them: A and cP airs are cold and dry, mP air is cold and moist, mT air is warm and moist, and cT air is warm and dry. Important but rarely encountered air masses are the warm, very humid equatorial air (E) and the very cold and dry antarctic air (AA).

Fronts

Major changes of weather characterize the transition from one air mass to another. The border regions or boundaries of air masses are known as *fronts.* Very different weather is developed in the frontal zone at the leading edge of an air mass from that at the trailing edge.

In Fig. 13-7, the cold (P) air mass is shown moving from west to east, through a surrounding warm (T) air mass. The polar air mass is asymmetric in shape because the leading surface is greatly steepened by friction with the surface. The ground contact of the steep forward edge of the cold air is called a *cold front* because cold air replaces warm air along this line. And because warm air replaces cold at the trailing contact of the cold air, this line is known as a *warm front.*

The cold-front region is characterized by a rather narrow zone of cumulus-type clouds produced by the steeply rising warm air just ahead of the front. Precipitation from these conditions is in the form of showers or thundershowers. In contrast, the warm air rising over the gentle slope of the warm frontal zone spreads out over a broad area, thus forming sheet- or stratus-type clouds. Rain or snow may fall over a broad zone

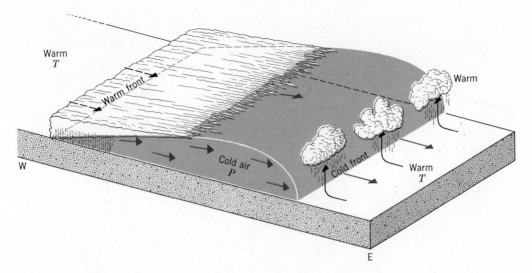

FIG. 13-7. Characteristics of cold and warm fronts. The former mark the leading edge of advancing cold air masses and the latter, the trailing edge.

hundreds of miles ahead of the warm front. Regions within the cold air mass between the two bad weather areas of the frontal zones usually experience fine sunny-day weather.

CYCLONE STRUCTURE AND ORIGIN

We have now examined (in this and the preceding chapter) the nature of low- and high-pressure areas, the cyclonic and anticyclonic winds that circle these areas, and the characteristics of air masses and fronts. Let us now see how all of these features are combined into the weather patterns of the middle latitudes.

Wave Theory of Cyclones

Recall that the movement of air masses is controlled by the winds of the upper troposphere. Prior to the generation of a new cyclonic or low-pressure system, the wavelike motion of the upper westerlies is so disposed that a large volume of air becomes stagnant in the polar source region. As the polar air mass develops, it becomes a great cell of high pressure because it is cold and dense. Clockwise or anticyclonic wind motion thus develops (in the Northern Hemisphere). With a change in the upper Rossby wave pattern, the great cold cell may begin to drift southward, and as it does so the air at the southern portion will encounter

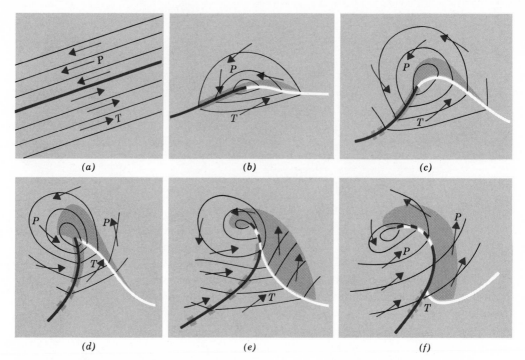

FIG. 13-8. Stages in the life history of a typical wave cyclone. The initial polar front is modified into a growing wave that moves eastward along the front causing the front of the wave to develop warm front features, and the rear portion cold front characteristics. (Courtesy, W. Donn, *Meteorology,* McGraw-Hill Book Co.)

the warmer air of the prevailing westerlies that originate in the tropics. Winds in the southern part of the cold, high-pressure cell will move westward about the central region while the warmer air from the south will move eastward. The contact region will look somewhat like that in Fig. 13-8a, where the wind patterns are shown by arrows and the continuous lines are isobars.

The heavy black line represents the margin of the polar air, often called the *polar front*. A great contrast in thermal energy exists between the moist, warm tropical air on one side of the polar front and the cold, dry air on the other side. Now in nature, such zones of energy contrast are very unstable and rarely maintain a straight boundary. Great horizontal waves commonly develop along unstable boundaries of this type. Sequential stages of wave development appear in Fig. 13-8, which shows the deformation of the pressure and wind pattern into a roughly circular counterclockwise or cyclonic system. The life history of the wave is really the life history of a cyclonic wind system in the Northern Hemisphere.

The significant points in the development of this wave cyclone are (1) the wave grows in size, (2) the wave moves eastward at 20 to 30 miles per hour along the frontal zone, (3) a cyclonic wind system develops, (4) as the wave moves the western portion becomes a cold front as it pushes

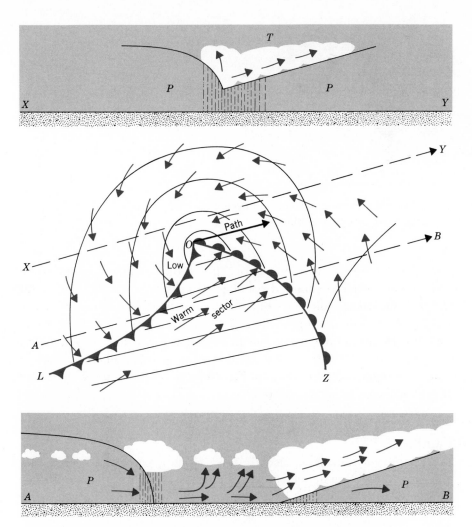

FIG. 13-9. The mature stage of a wave cyclone with vertical sections showing conditions above the ground along the lines *AB* and XY. The warm front is depicted by the conventional half-round symbols always drawn on the side toward which the front is moving. The cold front is shown by conventional wedge-shaped symbols pointing in the direction of motion. (Courtesy, W. Donn, *Meteorology,* McGraw-Hill book Co.)

into the warm air and the eastern portion becomes a warm front, (5) at the ground most of the cyclone consists of polar air that surrounds a wedge-shaped region of tropical air, (6) a broad (shaded) area of warm-front precipitation precedes the warm front while a narrow band of precipitation is associated with the cold front, and (7) the cold front moves eastward faster than the warm front and finally catches up with it to form an *occluded cyclone*—usually the most intense stage.

The views in Fig. 13-8 are all two-dimensional, surface views of cyclone development. But weather processes are primarily related to

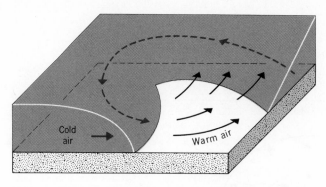

FIG. 13-10. Three dimensional view of a mature cyclone showing the air mass structure and motion.

vertical behavior of the atmosphere. The vertical structure of the wave cyclone must also be examined as in Fig. 13-9, which shows the horizontal and two vertical section views of a storm in the mature stage of the wave, *d* of Fig. 13-8. Note that the section south of the center shows both fronts reaching the ground with the warm air between. To the north of the center, the warm air, although present, is above the ground, where it rises to produce clouds and precipitation. Fig. 13-10 is a three-dimensional view of a typical mature wave cyclone.

Cyclone Families

A series of wave cyclones usually form along a particular outbreak of polar air. When this happens there is a progression in age of the storms from youngest in the west to the oldest in the eastern part of the polar front. Fig. 13-11 shows a fairly typical cyclone family stretching across the North Pacific Ocean. The easternmost, at the Alaskan Coast, is dying out followed to the west by two storms in different stages of occlusion. A newly developing wave brings up the last of the storms.

Fig. 12-38 shows a strong, mature wave cyclone centered just south of Newfoundland in the Atlantic Ocean. The polar front on which this storm developed makes a strong arcuate loop to the south and then west from the storm center. The cloud- and wind-flow patterns of this weather situation are depicted very strikingly in the Tiros weather satellite photograph of Fig. 12-38 where the storm center is marked by a cloud vortex. From this extends a broad loop of clouds related to the polar front.

Weather Forecasting

Let us now connect lower-level with upper-level weather. We have noted that air-mass movement related to the development of cyclones is controlled by the upper level wind system. In fact, the entire surface pattern of lows and highs is so closely related to the behavior of the upper pressure and wind pattern that simultaneous maps of sea-level and upper-level weather have much in common. Compare, for example, the 18,000-foot pressure-wind map of the Northern Hemisphere in Fig. 13-6 with the sea-level weather map in Fig. 13-12. Note how each southern extension of the Rossby waves in the wind pattern of Fig. 13-6 corresponds to a large outbreak of polar air (bordered by polar fronts and resulting high-pressure cells in Fig. 13-12). Northward flow patterns in the upper westerlies match the surface intrusions of warm into polar air and mark cyclone positions.

Modern weather forecasting is concerned primarily with the prediction of the day-to-day changes in the upper westerlies and the consequences of these changes to surface weather patterns. The broad features of these patterns are now predictable to a degree that gives 70 to 80 percent forecast accuracy. The details of prediction necessary to give complete reliability are still in the research stage.

HURRICANES

Imagine a violent, circular marine storm, 200 to 300 miles across, that releases enough energy in a day or two to provide all of the electrical power consumed in the United States in a year! Such is a hurricane. As critical as these storms are in providing an important element in the

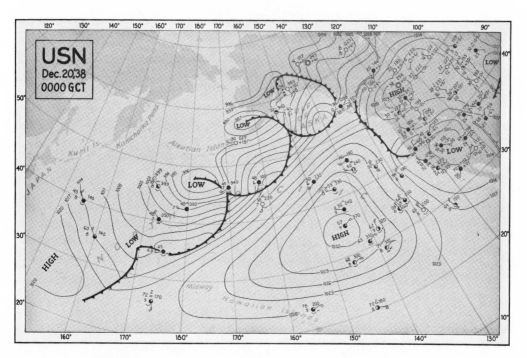

FIG. 13-11. A family of cyclones distributed along a polar front over the North Pacific Ocean.

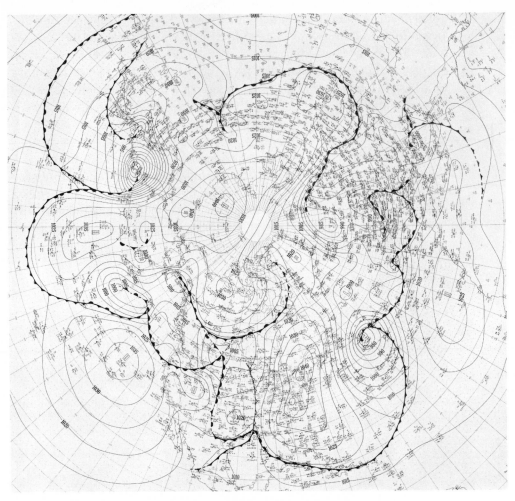

FIG. 13-12. Weather map of the entire Northern Hemisphere corresponding to the time of the upper westerlies wind map of February 5, 1955 in Fig. 13-6.

atmosphere heat and water balance, they are at the same time a great hazard to all who encounter them and are one of the greatest causes of natural catastrophes. So great is the potential peril of hurricanes to maritime areas in their paths that huge investments have been made by many countries in the attempt to forecast their origin and behavior. Some of the study is actually concerned with the modification of these storms sufficiently to "take the sting out of them" if not to destroy them completely.

Where and When

"Hurricane" is the regional name given originally to *tropical cyclones* in the West Indies region. Tropical cyclones are small, intense low-pressure

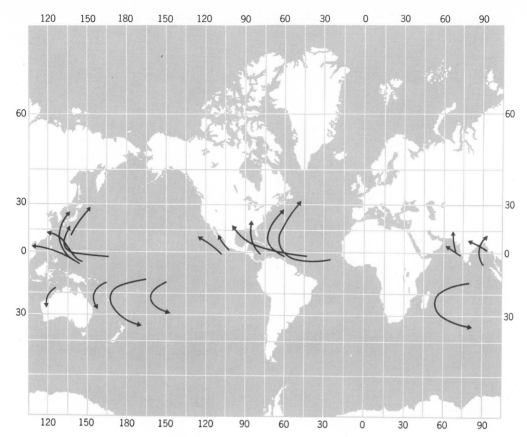

FIG. 13-13. Average hurricane tracks of the world-strong preference exists for the western sides of the oceans.

areas that originate in the low latitudes and move toward higher latitudes, drifting with the prevailing wind circulation. Their normal motion carries them toward the northwest (southwest in the Southern Hemisphere); after crossing the horse latitudes (30°) and encountering the prevailing wester-lies, they move eastward. However, the details of both the upper- and lower-level air circulation determine individual paths so that the motion of any particular hurricane may not be easily predictable. Compare, for example, the average world storm tracks of hurricanes (Fig. 13-13) with the individual paths of a large number of western North Atlantic storms (Fig. 13-14).

With the exception of the storms in the Pacific Ocean off the coast of Mexico, they are mostly restricted to the western sides of the world's oceans and are most common in tropical marine latitudes. But considerable coastal devastation has been caused by storms reaching the higher latitudes of the eastern coast of the United States and Japan.

Most hurricanes develop between midsummer and midautumn, but many have been known both prior to and following this period. Since

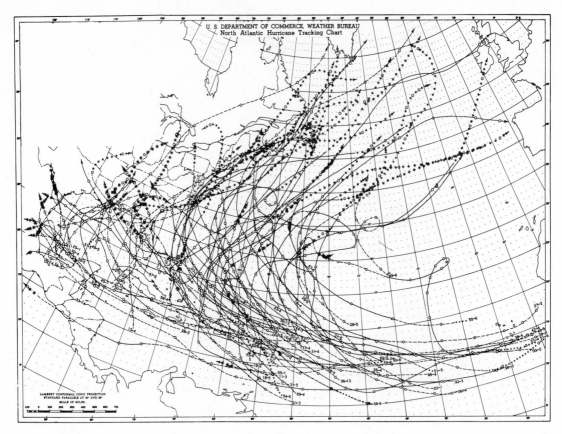

FIG. 13-14. Paths of individual North Atlantic hurricanes during the first ten days of each September from 1886 to 1958.

individual storms may never have "heard" of the averages, people who live in exposed portions of hurricane country have learned to be on the alert all through the hurricane season.

Hurricane Weather

The wave cyclones of the midlatitudes extend from 500 to 1000 miles across, and the change in pressure is commonly a uniform one of about 20 millibars from the edge to the center. A typical horizontal pressure gradient for the midlatitude storm would be about 1 millibar per 35 miles. A tropical cyclone, in contrast, has a pressure gradient of some 40 millibars in a distance of about 125 to 150 miles. If the gradient were uniform, it would thus be 1 millibar per 3 to 4 miles. Since the pressure gradient in a hurricane is not uniform but is more like that shown in Fig. 13-15a, where it increases continuously toward the center, the cyclonic winds pulled inwards increase in velocity and reach maximum intensity at speeds measured between 75 and 200 miles per hour. (A storm is not called

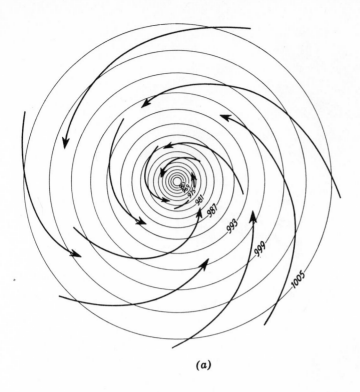

(a)

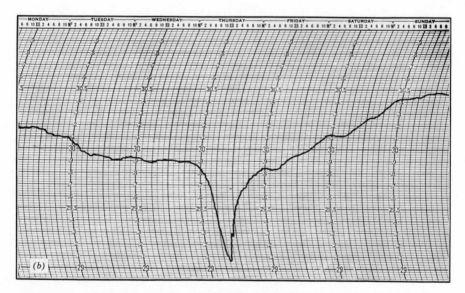

FIG. 13-15. (a) Pressure and wind pattern typical of a hurricane. Note the increasing pressure gradient toward the center. It is the steep pressure gradient of hurricanes that causes the devastatingly high winds. (b) The rapid fall of pressure from the passage of a hurricane near Kings Point, N.Y., September 14, 1944.

FIG. 13-16. Photograph of the eye of Hurricane Betsy, September 2, 1965 from an elevation of 11 miles by a high-flying Air Force reconnaissance plane shows why the center of a hurricane is called the "eye." (U. S. Air Force photo.)

a hurricane until the winds reach 75 mph). As the winds spiral into the storm with ever-increasing speed, they become more and more circular and never get quite into the center, which thus remains surprisingly calm.

Because the central part of the hurricane is the warmest portion, the air also rises more and more steeply as it speeds around the calm center. This vertical motion produces clouds that release the torrential showers so characteristic of hurricanes. Some of the heaviest rainfalls on record were generated by this process. Single storms have dumped up to 88 inches of rain in one locality (Baguio, Phillipines, July 1911), while many, many others have yielded tens to scores of inches of precipitation.

The central calm region that is often cloudless or nearly so due to the absence of rising air has become known as the *eye of the storm*. This clear shaft penetrating the storm clouds is shown strikingly in many high-level airphotos (Fig. 13-16) and radar-scope views (Fig. 13-17) of hurricanes.

Hurricane Energy

Heat of condensation released in connection with the rains just referred to is essential to the life of the storm. The maintenance of the very steep pressure gradient and resulting high winds requires a storm whose central region is continuously warmer than the marginal area. Latent heat of vaporization released to the atmosphere is the mechanism that provides this heat. The heat was originally picked up from the warm tropical seas when moisture evaporated elsewhere over the nearby ocean. As the storm moves, it continuously entraps fresh moisture-laden winds into its center to feed its demanding heat engine. When the storm crosses onto land, this source of energy disappears and the storm weakens rapidly. Should it curve back to the sea, almost immediate rejuvenation occurs.

Despite its tempestuous winds and drenching rains, the greatest damage from these marine storms has come from severe ocean floods that spread rapidly over shallow coastal regions under the onslaught of the storm winds. Many cities have been destroyed from the secondary impact of rising sea waters rather than from the direct atmospheric effects. In East Pakistan on November 12, 1970, 300,000 persons were lost, mostly by drowning from the high seas associated with what was perhaps the deadliest hurricane in history.

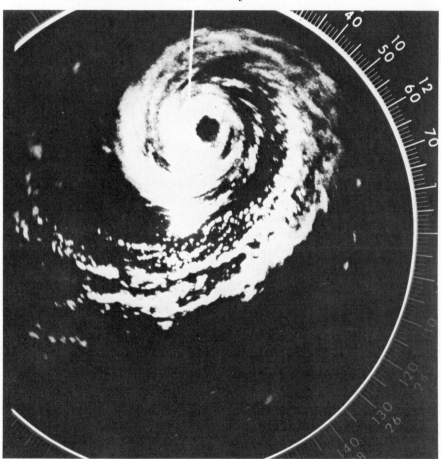

FIG. 13-17. Eye of Hurricane Donna, September 10, 1960 and spiral cloud associated with the characteristic spiral wind structure. (Courtesy U. S. Weather Bureau.)

Detection of Hurricanes

Most hurricanes originate in the low latitudes of the midocean tropics, far from the sight of land and often far from discovery by transocean trade vessels. Until the advent of weather satellites, the discovery of tropical storms depended on frequent aircraft reconnaissance. Many storms were missed, particularly those forming in the remote tropical Pacific Ocean. But no regions are out of sight of modern weather satellites, one of the greatest values of which is the positive detection of all hurricanes that originate anywhere in the oceans. Once they approach within the limits of coastal radar surveillance, these violent storms of the sea can be monitored continuously by precise radar observations.

THUNDERSTORMS AND TORNADOES

To see the ominous graylike cloud of a thunderstorm approach out of the west is to witness one of the violent displays of nature. A thunderstorm is really a huge, overgrown cumulus cloud, called a *cumulonimbus* cloud, which has grown vertically from the cooling of moist air that has risen through tens of thousands of feet after being triggered by any of the vertical uplift processes (Fig. 12-9).

Unstable Air—A Prime Requirement

Although cumulus and cumulonimbus (thunderstorm clouds) are both direct results of steeply rising moist air, thunderstorms certainly do not develop from all of the very commonly occurring cumulus clouds. Most cumulus clouds occur with stable air conditions, whereas thunderstorms require that the air be unstable. At such times the conditions of Fig. 12-9 are simply triggering mechanisms to get the air started, whereupon the instability (recall Fig. 12-11b) does the rest.

A consideration of the adiabatic lapse rate in slightly more detail is required to appreciate the contribution of instability to thunderstorm development. When clouds form in rising air, the cloud base represents the level at which condensation begins. Above this level latent heat of condensation is released continuously. The amount released depends on the amount of water condensing—a variable factor. As an average value, rising saturated air in the lower troposphere of the middle latitudes releases latent heat at a rate that would warm the air by 2.3°F per 1000 feet. But rising air is cooling by adiabatic expansion at 5.5°F per 1000 feet, resulting in a net cooling of rising saturated air of 3.2°F per 1000 feet. This is known as the pseudo- or saturated adiabatic rate and is applicable to air rising within clouds.

A comparison of cumulus and cumulonimbus (thunderstorm) cloud development is made in Fig. 13-18a and b, where uplift is triggered by a local warming of the ground. In *a*, the warmed air rises and cools to the cloud base according to the dry adiabatic lapse rate (solid line). Within the cloud, the air cools at the lower, saturated adiabatic lapse rate. Since the lapse rate (broken line) in the surrounding air is stable, the rising air comes to rest where its temperature equals that of the surrounding air— the altitude where the two rate lines intersect. Further cooling, further condensation, and cloud development cease at this level.

In Fig. 13-18*b*, two possible lapse-rate curves are shown. Lapse rate is greater than the dry adiabatic rate so that the air is unstable from the ground upwards; and once warmed, the rising air, even after the condensation level is reached, continues ascending to form the towering cumulonimbus. Lapse rate 2 is less than the dry adiabatic rate, so that the two curves approach each other below the cloud base. But once in the cloud, the rising air cools at a rate even less than the prevailing lapse rate, as indicated by the divergence of the two curves. Under the conditions of lapse rate, the air is stable below the cloud base but unstable within the cloud. One of the lapse-rate conditions in Fig. 13-18*b* is invariably present during the generation of a thunderstorm. The continuous growth of the storm can be described by three distinct stages (Fig. 13-19). The first, early stage is characterized by updrafts throughout the growing cumulus cloud.

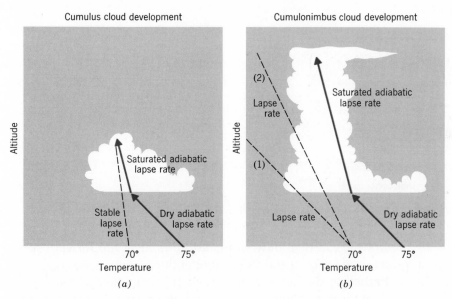

FIG. 13-18. (*a*) The development of a cumulus cloud under stable air conditions. Below the cloud the air rises and cools at the dry adiabatic rate. Within the cloud the rising air cools at the pseudo-adiabatic rate. (*b*) The development of a cumulonimbus cloud when unstable air conditions prevail. Lapse rate 1 exceeds the dry adiabatic lapse rate and 2 exceeds the pseudo-adiabatic rate within the cloud.

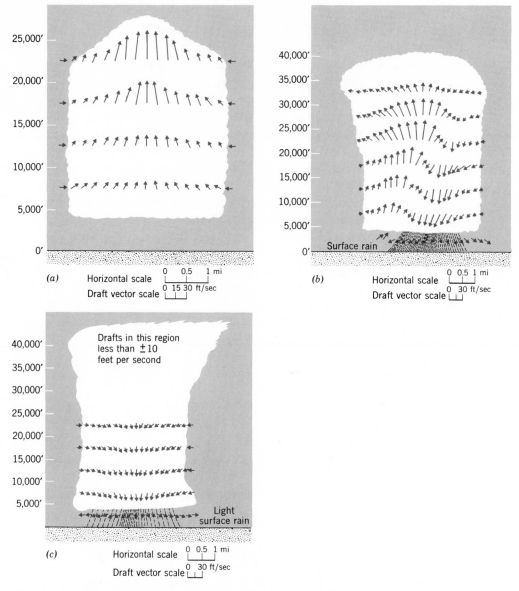

FIG. 13-19. Three stages in the development of a thunderstorm.

In the mature stage, strong downdrafts are present in the forward part and updrafts in the rear of the cloud. By the third, weakening stage, air motion is limited to downdrafts that result in dissipation of the storm through lack of the continuous-condensation necessary for storm maintenance.

Updrafts within the cloud are usually so strong that raindrops grow to large, sometimes unusual size before falling to the ground. Also, the details of wind motion within the cloud show the air to be engaged in strong vertical eddy motion so that drops may be whirled continuously

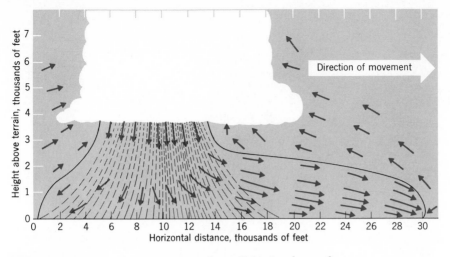

FIG. 13-20. Details of air motion and precipitation beneath a thunderstorm.

between higher and lower elevations. If the circuits carry the drop above and below the freezing level, alternate freezing and water accumulation take place on the drop until it is large enough to fall as a concentrically layered ice particle called *hail*. In especially violent storms, hailstones can grow to egg- or fist-sized particles. Hail should not be confused with *sleet*, which is simply rain frozen in passing through a cool air layer beneath a stratus-type cloud, or partially melted and refrozen snow.

Thunderstorms drift with the prevailing winds. In the middle latitudes they move from a generally west to east direction. The phenomena of a thunderstorm experienced by an observer at the ground can be predicted from conditions shown in Fig. 13-20 for a mature storm in which the motion of the cloud would be from left to right (west to east). A very strong downdraft is commonly experienced as the leading edge of the storm approaches. This is followed by very intense rain, which lessens toward the rear of the cloud as the downdraft weakens. The schematic conditions of Fig. 13-20 appear in more realistic form in the storm photograph of Fig. 13-21.

Friction between the air and the rapidly whirling water and ice particles develops electrical charges within the cloud. Positive and negative charges become segregated in different parts of the turbulent cumulonimbus cloud. A very strong electrical gradient soon develops between the high charge concentrations within different parts of the cloud or between the cloud and the earth's surface. *Lightning* is the electrical discharge between these regions (Fig. 13-22). *Thunder* is the explosive sound caused by the rapid expansion of air along the lightning stroke. When masked by clouds, the lightning stroke produces a general illumination called *sheet lightning*, which is not in itself a special type of discharge.

FIG. 13-21. View of a thunderstorm east of Pensacola, Florida. Heavy precipitation is evident beneath the central portion of the cloud. (U. S. Weather Bureau.)

Tornadoes

Tornadoes, by far the most violent of atmospheric phenomena, are offsprings of very severe thunderstorms. They are produced only in those storms whose air uplift is generated by cold fronts in which a cold mass of air pushes vigorously into a warm, moist mass of air. The greater the contrast in temperature and moisture between the two air masses, the greater is the chance of tornadoes developing from the thunderstorms along the zone of contrast. Although tornadoes have occurred in every one of the original 48 states, they are by far most common in central United States in the spring.

Tornadoes appear as gigantic dark funnel clouds that extend to the earth's surface from the most active part of the parent cumulonimbus (Fig. 13-23). The lower the funnel cloud dips, the larger is the area of coverage. This extent may vary from almost nothing to about a quarter of a mile. The damage from a tornado is twofold. (1) The violent winds composing

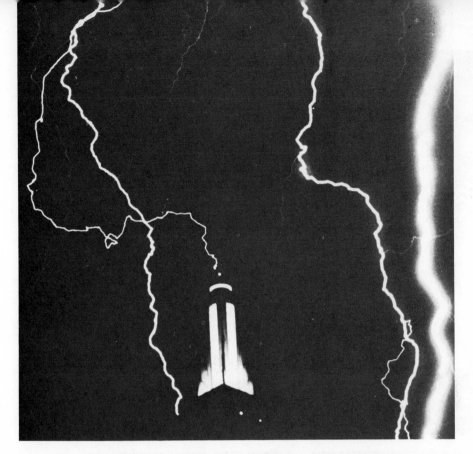

FIG. 13-22. Lightning striking the tower of the Empire State Building in New York City. The high frequency of occurrence of this event belies the old adage about lightning not striking twice. (U. S. Weather Bureau.)

the funnel whirl about the vertical axis of the storm with speeds that have been estimated up to 500 mph. These estimates are based on observations of objects that are driven through other objects by the force of tornado winds as in Fig. 13-24. (2) The second major cause of damage is the greatly decreased air pressure within the tornado funnel cloud. So reduced is the pressure that when the tornado passes over a building, especially one with closed doors and windows, the structure tends to "explode." Although we tend to close doors and windows for most storms, they should be open when a tornado approaches to permit easier air exit.

Although the meteorological conditions that give rise to a tornado are known as described, the details of their origin are still unsettled. Only when such details are understood will it be possible to forecast the development of these most destructive of atmospheric storms. In fact, when tornado origin is understood, the means to prevent tornado formation may be developed by meteorological scientists.

FIG. 13-23. Stages in the development of a tornado. (U. S. Weather Bureau.)

FIG. 13-24. Winds in the tornado of April 18, 1970 that struck Clarendon, Texas, drove a stick of wood through the 1½ inch pipe shown. (NOAA Photo.)

STUDY QUESTIONS

13.1. Draw a diagram of the earth's surface that summarizes the mean pressure and wind belts.

13.2. Describe Rossby waves. How are they related to the Jet stream?

13.3. Explain the relationship of fronts to air masses.

13.4. What conditions define an air mass?

13.5. Explain with a diagram how fronts cause the development of clouds and rain.

13.6. Describe the origin and movement of a cold arctic air mass that may invade central United States all the way to the Gulf of Mexico.

13.7. Explain with diagrams the development of a wave cyclone along the polar front.

13.8. How does a hurricane differ from the cyclones of the middle latitudes (extratropical cyclone)?

13.9. Why do hurricanes normally dissipate when they move over the land?

13.10. What is the form of the great damage caused by hurricanes?

13.11. Describe conditions in a typical thunderstorm.

13.12. What is the difference between the dry adiabatic lapse rate and the saturated adiabatic lapse rate?

13.13. Why is unstable air necessary for thunderstorm development?

13.14. Explain lightning, thunder, and hail.

13.15. Describe tornadoes and conditions associated with them.

chapter 14

"Paul Bunyon's foot," an example of nature's sculpting of a large iceberg photographed in summer in the North Atlantic Ocean. The Coast Guard icebreaker *Eastwind* is steaming northward passing the iceberg heading south. Icebergs are most common in the North Atlantic Ocean in spring and summer after they break from the Greenland ice cap during warm weather thawing. (U. S. Coast Guard photo.)

The Oceans: Our Marine Environment

The oceans cover some 70 percent of the earth's surface. The fascination of the sea lies in its mood, which is as everchanging as the clouds in the sky. At times it is placid with an almost mirrorlike quality. Then, within a very short time, the waters can be whipped into a devastating fury soon to return to a calm that belies the threat of the sea.

The sea is more than a subject for poets and artists. During most of historical time it has been the universal means of transportation. Until the advent of international air travel, the sea was the only useful medium that connected all lands and continents. Of still greater importance and significance, the ocean serves the primary function of providing water—water to the atmosphere for the nourishment of life on the continents, water to support the myriad life in the sea, and water to act as the agent of erosion and the medium for sediment deposition whose importance in geological processes we have examined throughout this book. And back when the earth was still young, life arose in the sea. Despite all of the vital elements necessary for the organic chemistry of life, water is still the major component of living cells. A world without the oceans would be as desolate as the barren surface of the moon.

Also, the sea is playing an increasing vital role in the direct livelihood of man. The sea abounds with life—life extremely rich in proteins, minerals, and vitamins. Observations indicate that the world ocean, on the average, supports many tons of dried organic matter per acre per year. This is an almost staggering amount of food when the total acreage of the oceans is included. With the ever-increasing population, the harvesting of sea life in its many forms and fish and seafood farming have been turned to as a means of increasing our food supply.

HISTORY OF OCEAN STUDY

From the Greeks to the Moderns

In a sense, the science of oceanography began when man first began to travel on the seas. Only some of the milestones marking the development of this fascinating study can be touched on here. The influence of the oceans, their bays, and smaller seas on the behavior of sailing vessels was, and is, so great that an accumulation of ocean knowledge was vital to continued navigation. Earlier, in Chapter 2, we gave some attention to the importance of voyages of the ancient Greeks on the growth of knowledge about the earth's shape and the making of maps (Fig. 14-1). Recall also that early and accurate knowledge of the size and shape of the earth was contributed by Erotosthenes of Alexandria in the third century B.C.

During the "Dark Ages" following the fall of Rome, culture in general and travel and map-making in particular sank to a low ebb in the Mediterranean countries. But the thread of exploration continued in the north as the Vikings established settlements in Iceland and Greenland (in the ninth and tenth centuries A.D.), which were then used as jumping-off points to Canada. The voyages of Eric the Red and Leif Ericson into Canada and Newfoundland, respectively, are well-known mileposts in the history of the New World.

The journeys of the Vikings were favored by a warming of climate that made the northern oceans less harsh and the northern lands more habitable. A deterioration of climate set in during the latter part of the Middle Ages that cut off further exploration of the Norsemen and caused

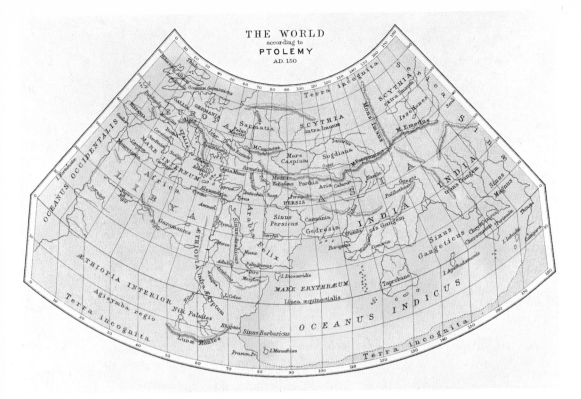

FIG. 14-1. An early map of the world according to Ptolemy, A.D. 150. (From Challenger Office Report, Great Britain, 1895, Courtesy P. Weyl, *Introduction to Oceanography*, John Wiley & Sons.)

retreat from the settlements other than Iceland. The "Little Ice Age," as this climatic interlude is called, resulted in the spread of Alpine glaciers over mountain roads constructed by the Romans centuries earlier.

With the beginning of the Renaissance came the voyages of Columbus. Vasco de Gama, and the first circumnavigation of the globe by Magellan (Fig. 14-2). These and other journeys of discovery led to a rebirth of map-making and a renewed need to know and understand the oceans.

After the Renaissance

All of the travels from the Greeks through the Renaissance were primarily voyages of trade and discovery in which understanding of the oceans was a necessary "part of the job." Real oceanographic exploration began in the latter half of the eighteenth century when James Cook, who rose from a poor beginning in England to become a master mariner, conducted three epic oceanographic expeditions to nearly all parts of the world. He charted new islands, mapped details of the "newer" continents, and touched on all of them except Antarctica, which awaited discovery by Nathaniel Palmer in 1820.

The art of measuring the depths of the seas and of dredging from

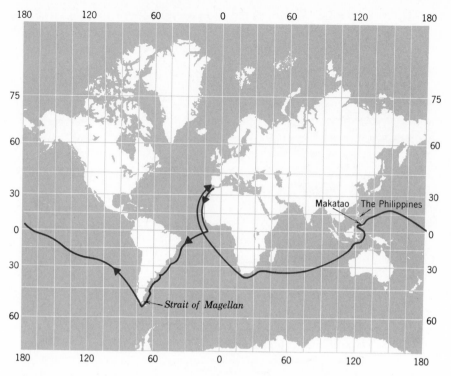

FIG. 14-2. The voyage of circumnavigation of Ferdinand Magellan. Five vessels departed from southern Spain on September 20, 1519. Only one, the *Victoria*, returned on September 8, 1522. Magellan himself was killed on the small island of Makatan in the Central Philippines on April 27, 1521. The success of the voyage is a testimony to Magellan's navigational skill as well as his foresight and imagination in sailing the unknown waters. (Based on *Seas, Maps and Men* Ed. by G. E. R. Deacon, Doubleday & Company, Inc.)

the sea bottom started with Sir John and Sir James Ross. Sir John, the uncle, explored the high latitudes of the Atlantic Ocean and the fringes of the icy Arctic, while Sir James journeyed to the southern extremities of the ocean to investigate the great depths of the seas surrounding Antarctica. The Ross shelf ice, a sheet of thick ice fringing part of Antarctica, today bears his name.

The true modern science of the seas probably began with the ocean-current charts of Matthew Fontaine Maury published in 1855 in his now-classic *The Physical Geography of the Sea*. Maury synthesized his ocean-current charts from the wealth of observational data buried in the logbooks of vessels from years gone by. The way that ocean currents are calculated from a ship's navigation log is quite simple as illustrated in Fig. 14-3. On a chart of the type shown, the navigator plots the vessel's course from "dead reckoning," a procedure in which the estimated positions of the vessel based on observed speed are plotted along its compass course. A vessel leaving the coast at *C* might arrive at dead-reckoning position *DR* in a given time. But careful observations of latitude and longitude might establish a "fix" at *F*. Although not apparent to the navigator, the vessel would really have traveled the course *C-F* rather than *C-DR*, owing to an ocean current moving parallel to the broken line *DR-F*. If the ship traveled for 10 hours over the distance shown and the difference in position between *DR* and *F* is *20* nautical miles, the current strength (drift) is 2 knots (a knot being a nautical mile per hour). In addition to information related to navigation, the deck logs of past ships also include periodic observations of winds at least once during each watch. Maury averaged these observations and produced the first wind chart of the oceans, charts not unlike those in Figs. 13-2 and 13-3.

Much of our knowledge of the currents of the world ocean are based on such observations and calculations, although far more sophisticated current-measuring devices are now in use.

The first truly scientific exploration of the seas was undertaken by the British Corvette H.M.S. *Challenger* under the direction of the British Royal Society. During its monumental cruise from December 1872 to May 1876, the *Challenger* travelled 69,000 miles and accumulated the first massive scientific data on the physical, chemical, and biologic and geo-

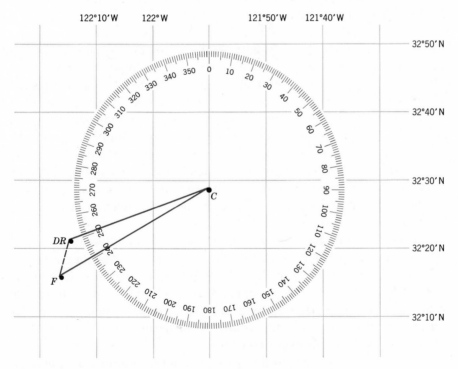

FIG. 14-3. The method of determining average ocean currents between given points in a ship's voyage. (Explanation in text.)

FIG. 14-4a. The Research Vessel *Vema* that has cruised hundreds of thousands of miles since 1950 in the course of carrying out pioneering geophysical studies of the sea and the crust beneath the sea in all the open oceans of the world. It is operated by the Lamont-Doherty Observatory of Columbia University.

logic aspects of the world oceans. The 50 volumes of technical data resulting from this 3½ year expedition provided the real basis for our modern oceanographic knowledge and is still a source of information used in the analysis of conditions in the seas.

Twenty years later the Norwegian research vessel, *Fram,* directed by the polar explorer F. Nansen, entered the Arctic Ocean for the first exploration of that frozen sea. Among many of the important discoveries of Nansen and Fram was the knowledge that the Arctic Ocean, despite its ice cover, reached true oceanic depths. Nansen also observed that floating pack ice drifted to the right of the wind rather than traveling with the wind. This important discovery, now known to apply to all surface ocean currents in the Northern Hemisphere, was then explained by V. W. Ekman on the basis of effects related to the Coriolis deflection (See p. 442, explaining wind motion). Nansen also developed the famous Nansen bottle, a special container for collecting water samples at any desired depth for later laboratory anaylsis.

The oceanographic sequel to the *Challenger* expedition was that of the German research vessel *Meteor* during 1925–1927. *Meteor* continued the detailed observations of the kind begun by *Challenger* but had the advantage of more modern equipment for measuring temperatures, the Nansen bottle for water sampling, and the electronic echo sounder for recording ocean depths as the vessel steamed along (See Chapter 3 on bottom and subbottom depth recorders).

Following the very fruitful voyage of *Meteor,* a number of American oceanographic institutions, in particular the Woods Hole Oceanographic Institution, the Scripps Institution of Oceanography, and the Lamont-

FIG. 14-4*b*. The Research Vessel *Conrad* owned by the U. S. Navy, but also operated by the Lamont-Doherty Geological Observatory. It has also carried out major oceanographic and geophysical studies of the seas.

FIG. 14-4*c*. The Research Vessel *Researcher* of the National Oceanic and Atmospheric Administration. This is one of the most advanced research vessels afloat. (NOAA Photo.)

Doherty Geological Observatory of Columbia University entered into very active global exploration programs with the most modern equipment. Woods Hole's *Atlantis I* and *Atlantis II* and Scripps' and Lamont's *Vema* and *Conrad* (Figs. 14-4 *a* and *b*) have added tremendously to our knowledge of all aspects of the seas and have truly revolutionized our knowledge of the land beneath the sea and with that our knowledge of crustal processes (Chapters 8 and 9).

Some of these vessels participated in the international program of exploration and discovery of 1957–1958 known as the International Geophysical year (IGY). An example of IGY ship tracks for the Atlantic Ocean appears in Fig. 14-5, and detailed resulting oceanographic sections of temperature and salinity for particular east-west sections are reproduced in Figs. 14-6 and 14-7, respectively. Despite all of the seeming emphasis on ocean study during the past century, we still have far to go, because the oceans are a huge three-dimensional volume of water practically impenetrable to the eye.

THE NATURE OF INNER SPACE

It is almost paradoxical that regions of the earth like the oceans, often called "inner space," which are rather readily accessible to exploration, have lagged well behind the much more difficult adventures into space beyond the earth. In the 1960's, more complete exploration of the earth's liquid shell or hydrosphere has been undertaken by deep manned and unmanned *submersibles*. Men have been lowered into some of the greatest depths in the seas. Others have drifted along for weeks while embedded in the riverlike flow of the deep ocean currents. Many surprises about composition and deep-sea life have already been encountered by this direct observation of the murky depths of the oceans.

Light in the Sea

Prior to this epoch of direct observation, our knowledge of conditions below the surface had mostly been gained through indirect study. Although the atmosphere above the oceans and the astronomical world beyond has long been studied by visual means, this has not been possible with the oceans because sunlight penetrates only a relatively skin-deep layer. For example, of the intense sunlight that strikes the ocean surface, only 40 percent penetrates below 3 feet and only about 2 percent is left at a depth of 100 feet. The sea is thus a dark world requiring artificial illumination whenever direct exploration is carried out below the surface layer. Shallow as is the illuminated layer, it is, as we shall see in the section on sea life, all-important in the maintenance of marine life.

The sunlight absorbed in the ocean is also all important in maintaining the earth's heat balance. Earlier, in Chapter 11, we examined in some detail the distribution of sunlight after it entered the atmosphere. We noticed that the heating of the lower atmosphere comes primarily from the earth's surface after sunlight is absorbed there. Because some 70 percent of this surface is marine, the sea plays a major part in the heating of the atmosphere and in controlling its weather and climate. Just as the

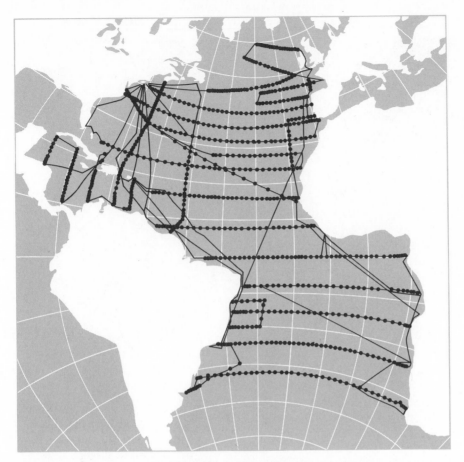

FIG. 14-5. Courses of oceanographic vessels during the International Geophysical Year (IGY) 1957–1958. Note the systematic pattern of cruise spacings. The dots are locations where a series of observations were taken form the surface to the sea bottom.

irregular heating of the atmosphere is the motivator of winds and storms, so the irregular heating of the oceans is a fundamental factor in driving its circulation.

Also, as noted in Chapter 11, the absorption of sunlight by land may not only be quite different from that absorbed in the sea, it is also quite variable from place to place over the land because of irregular topography and variable types of ground cover and composition. Light absorption in the sea is much more uniform and varies mainly with latitude (from the earth's curvature) and the length of the day. In tropical regions where the altitude of the sun is highest, the greatest amount of surface heating takes place. In the polar regions where the altitude of the sun is least and where the sun is absent for many months during winter, solar heating is least.

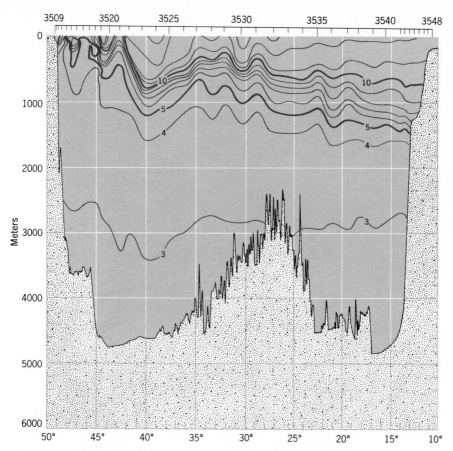

FIG. 14-6. A vertical section of ocean temperature made along latitude 48°N. See Fig. 14-5 for specific IGY ship tracks. The sections
extend from the submerged edge of North America just off New
Foundland to the edge of Europe just south of Ireland and crosses
the mid-Atlantic Ridge. Great vertical exaggeration is used. Note
that most of the ocean temperature is below 4°C (39°F). This is
true for all of the seas regardless of latitude. (After Atlantic Ocean
Atlas, F. C. Fuglister, Woods Hole Oceanographic Institution.)

Since the earth's surface is always losing heat through reradiation,
evaporation, and conduction beyond latitude 35°, the amount of heat
lost exceeds the amount of heat gained. This heat loss is particularly
large over the Arctic Ocean and Antarctica. The ice-covered surface of
the north polar ocean and the thick glacial ice sheet of Antarctica reflect from 60 to 70 percent of the sunlight compared to about 10 percent
reflection for regular sea water. As we shall see in Chapter 15, the presence of these huge ice regions in polar areas has a strong effect on world
climate. Profound changes would occur with the disappearance of polar
ice.

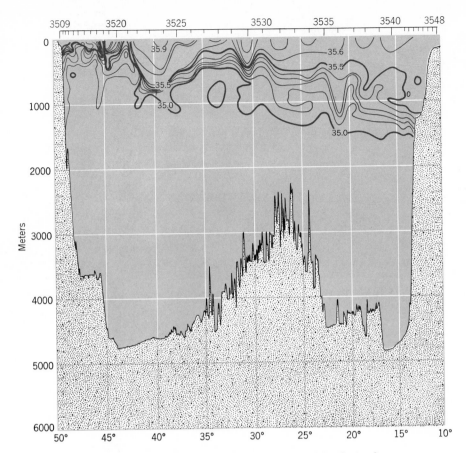

FIG. 14-7. A vertical section of ocean salinity made along latitude 48°N. Salinity values are in parts of salt content per thousand parts of water. Although salinity also decreases with depth, the change is very small compared to the temperature effect. (After Atlantic Ocean Atlas, F. C. Fuglister, Woods Hole Oceanographic Institution.)

Heat and Temperature of the Ocean Surface

The temperature of the sea surface is a consequence of the balance between the solar radiation absorbed and the energy lost by radiation, evaporation, and conduction to the air plus the amount of heat distributed below the surface by the natural stirring of the waters. In low latitudes, the amount of heat absorbed exceeds the amount lost; in high latitudes the amount lost exceeds the amount absorbed. Surface temperature is therefore high in low latitudes (about 80°F in equatorial waters) and low in high latitudes (about 0°F in the polar region). A horizontal transport of heat must occur from low to high latitudes to keep the tropical seas from becoming progressively warmer and the polar seas from becoming progressively colder. This transport is accomplished both by surface ocean

  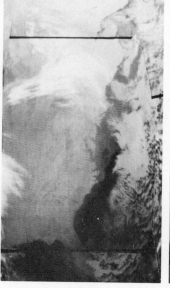 

| NIMBUS III | NIMBUS III | ITOS I | NIMBUS III |
| 4 FEBRUARY 1970 | 22 FEBRUARY 1970 | 22 FEBRUARY 1970 | 26 FEBRUARY |

FIG. 14-8a. A nocturnal infra-red photograph showing the sharp boundary between the warm Gulf Stream and the cooler waters just along the east coast of United States. (Dark tones are warmest and white coldest.) The mottled white area to the east of the Gulf Stream represents clouds covering part of the sea. White cloudy regions are also evident over the interior. (NASA photos.)

currents and by the atmosphere. Just as the circulation of the atmosphere results primarily from the attempt to equalize the temperature difference between low and high latitudes, so also does the surface and deep circulation of the oceans. This will be explained further in the following section on ocean currents.

Over the broad central portions of the oceans, the distribution of surface temperature follows parallels of latitude. Marine isotherms (lines of equal temperature) are nearly east-west lines. Only near continents, where strong northerly or southerly ocean currents exist, do the temperatures depart from the latitudinal trend. For example, off the east coast of the United States, the Gulf Stream carries a considerable quantity of warm water northward in the Atlantic Ocean; off the west coast, the California current transports cold water southward.

To measure sea-surface temperatures, the water can be sampled directly either by scooping up a bucket of water on a moving vessel or by measuring the temperature of the water drawn in to cool the ship's engine. Another very interesting and useful way to observe and study marine temperatures has been developed for use with artificial weather satellites. Recall (Chapter 1) that radiant energy is proportional to the fourth power of the temperature. Warmer water emits more infrared radiation than does cool water. Weather satellites equipped with appropriate cameras loaded with infrared-sensitive film show differences in ocean shading that are directly related to the sea-surface temperature. Water currents of different temperature can quickly be delineated by this procedure, as illustrated in Fig. 14-8. Measuring the temperature of the water beneath the surface has mostly been accomplished by use of the Nansen bottle and reversing thermometer described in Fig. 14-9a, the bathythermograph (Fig. 14-9b), or the more modern STD (Fig. 14-9c).

Temperatures of the ocean surface, as we noticed earlier in Chapter

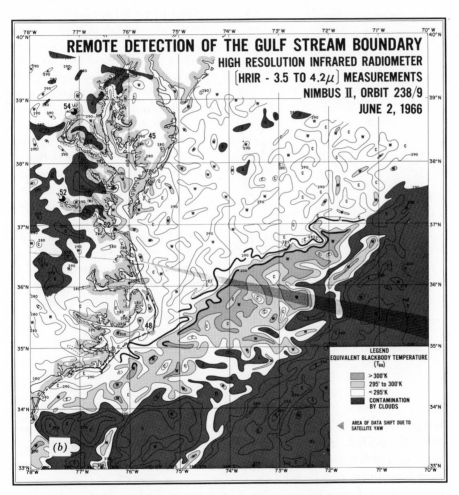

FIG. 14-8b. An enlarged portion of the coast and waters off Cape Hatteras, North Carolina showing sea surface temperatures by means of both contours and shading. This temperature chart was made from the analysis of the tone density on infra-red pictures like those in Fig. 14-8a. The deeper the black tone the warmer the water. (Courtesy NASA.)

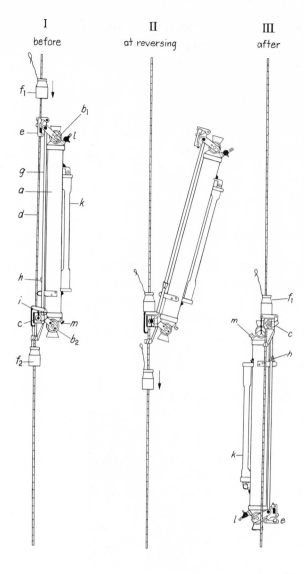

FIG. 14-9a. Diagram of the Nansen. collecting bottle (*a*) and attached reversing thermometer (*b*). A number of these units are attached at particular intervals on a cable which lowers them to predetermined depths in the sea. The sampling bottle (*a*) remains open permitting water to pass through from bottom to top. After the units are lowered a weighted messenger (*f₁*) is dropped down the cable. On striking the release mechanism (*e*), the upper clamp is opened so that the instrument can rotate as in II about the fixed clamp (*e*). As the instrument rotates, the guide rod (9) causes the valves (*b₁, b₂*) to close trapping the water. Also, as the unit begins to rotate, the lower messenger (*f₂*) is released from its clamp (*i*) and it slides down the cable to strike the second Nansen bottle where the process is repeated. After reversal, as in III, the unit is locked by the device at (*h*). Also on reversal, the thread of mercury in the thermometer (*k*) is separated from the main mercury reservoir and its length indicates the water temperature at depth. Water salinity and composition are from the sample in the bottle. (After G. Dietrich, *General Oceanography,* John Wiley & Sons, Inc.)

11, are quite conservative, particularly when compared with land-surface temperatures. The high specific heat of water, the loss of heat in evaporation, the partial transparency of water to sunlight, and the effects of vertical and horizontal motions all contribute to the small time changes in temperature experienced by surface sea water. On the average, the temperature change between day and night is less than 0.5°F and between winter and summer the average temperature variation is only 10°F. Changes of 100°F are common over land. In the next chapter we will return to the important climatic consequences of the small marine temperature variations.

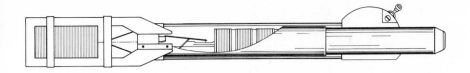

Temperature element   Pressure element

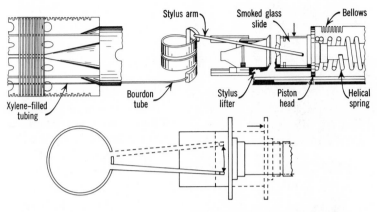

FIG. 14-9b. The bathythermograph. A temperature recording instrument that is lowered on a wire cable to a maximum depth of 1000 feet. A long fine copper tube containing the liquid zylene (whose volume changes rapidly with temperature change) is coiled about the upper section of the instrument. Movements of the coil move the point of a stylus attached to its lower portion across a small plate of smoked glass. As the instrument descends in the water, increased pressure causes the entire slide, which is attached to a pressure-sensitive bellows, to move at right angles to the movement of the stylus. A continuous curve of temperature is thus recorded on the calibrated glass slide. (ESSA photograph.)

The Thermocline

Perhaps most important in understanding the physical nature of the sea is the vertical distribution of temperature. Despite the relatively high temperature of much of the world ocean surface, this warmth is only skin deep, for the sea is a tremendous reservoir of cold water. This is readily apparent in the temperature section across the North Atlantic Ocean in Fig. 14-10. In fact, below a depth of one mile the temperature everywhere in the oceans is less than 39°F. Since the average depth of the oceans is about 2.5 miles, the temperature of most of the seawater is not very much above freezing. Even in tropical latitudes including the equator, the bottom water is very nearly freezing.

For much, if not most, of the oceans, a characteristic vertical temper-

FIG. 14-9c. The salinity-temperature-depth instrument (STD) gives more continuous but less reliable values than the Nansen bottle and reversing thermometer. The STD transmits information electrically to the research vessel as it is lowered. Sampling bottles attached to the periphery can be sealed at desired depths. These water samples are then used to check the STD values for the same depths.

FIG. 14-10. The vertical distribution of temperature in most of the oceans resembles that in this graph. The upper layer of constant temperature, known as the mixed layer, is caused by both vertical mixing from wind stirring and sinking water which cools by radiation from the surface. This layer can be up to 700 feet thick. The zone of rapid decrease of temperature to a depth of 3500 or 4000 feet is called the *thermocline*. Below this level, temperatures are low (Fig. 14-6) and decrease but slightly to the bottom.

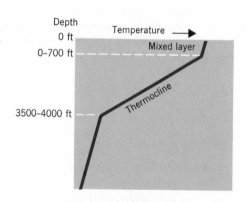

ature distribution exists (Fig. 14-10) that is absolutely basic to the understanding of many of the physical and biological processes in the sea. Although the shape of this temperature-depth curve is quite typical, actual temperature and depth values vary with location, so that only the form of this temperature curve is discussed here.

The upper, vertical portion of the curve that shows no change of temperature with depth is called the *mixed layer*. Ocean waves and vertical turbulence from currents thoroughly mix this surface layer. Also, in winter when the air is colder than the sea, cooling of the surface water causes sinking of the heavier cold water, thereby increasing the thickness of this zone.

Below the mixed layer, temperature falls rapidly in the *thermocline* until a depth commonly about 4000 feet, at which level the temperature is close to 40°F. From here to the bottom of the ocean, temperature falls by only a few more degrees.

A good view of just how steady these temperature-depth relations are can be seen in actual data taken in the deep sea near Bermuda in the western North Atlantic Ocean (Fig. 14-11). This diagram shows the distribution of temperature with depth during a period of nine years. Depth is on a logarithmic scale in order to keep the vertical dimension small enough for illustration. Time runs horizontally as shown. Lines of constant temperature are shown as labeled. The upper 100 to 200 meters (325 to 650 feet), where the temperature contours are essentially vertical, showing little or no change with depth, is the mixed layer that increases in thickness from summer to winter. Between 200 and 1000 meters (3300 feet) a rapid change of temperature occurs, as shown by the close packing of the contours. The bottom of this thermocline at Bermuda is at 6°C (43°F). Note that it is only 2°C (3.6°F) colder at the lowest line, 800 meters (2600 feet) below the thermocline. Below this, the temperature decreases even more slowly.

Sources of the Deep Cold Water

Where does all of the cold water of the ocean come from? Why is the vertical temperature distribution of the sea so like a mirror image of that in the atmosphere? In the latter, the temperature is highest near the sea or land surface and decreases with increased elevation. In the sea the temperature is also highest at the surface (at the air-sea contact) and then decreases downward.

The answer lies in the basic heat budget of the ocean. The amount of sunlight received is only enough to warm the oceanic surface layer in the low and middle latitudes. Remember, at latitudes above 35° the ocean surface loses more heat than it receives from the sun, but some warming is maintained by the poleward transport of warmer flows of air and water. But in the very high latitudes, particularly around icy Antarctica in the southern Hemisphere and ice-capped Greenland in the North Atlantic, ocean-surface temperatures fall to and below 32°F. This cold, dense water sinks to the bottom and then flows equatorward, making up most of the cold waters of the ocean.

In the Southern Hemisphere, water flowing from the Weddell Sea off Antarctica sinks to the bottom and then flows northward as the great cold water mass known as Antarctic bottom water, which reaches as far as 40°N in the North Atlantic Ocean. Cold water originating at the surface near Greenland can be found in the deep sea of the South Atlantic just

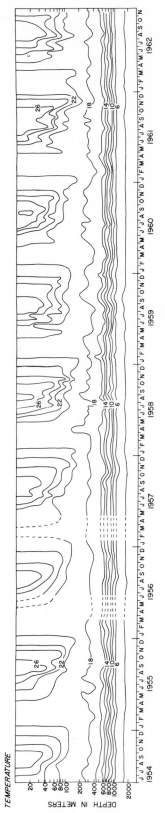

FIG. 14-11. The variation of temperature with depth over a nine-year period in the deep water near Bermuda. Temperature is shown by contours for every 2°C (ca 3.6°F). The upper 200 meters (656 feet) where the contours are vertical is the mixed layer. The zone of closely spaced contours is the thermocline. Apart from obvious, repetitive seasonal changes, the pattern is quite constant for the nine-year interval.

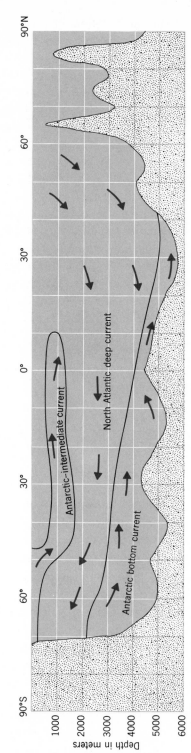

FIG. 14-12. The deep ocean circulation between Antarctica (left) and the Arctic (right).

above the Antarctic bottom water. A view of the deep-ocean circulation from pole to pole in the Atlantic can be seen in Fig. 14-12, where the importance of the cold polar water masses is evident.

Although the basic circulation of the deeper ocean water below the surface layer is caused by the sinking of cold heavy water and the rising of warm light water, the effect of salt composition is an important secondary factor in the deep circulation.

Salt in the Sea

The sea was "born" pure. But for billions of years rivers have been concentrating the oceans with chemical matter dissolved from the lands. Apart from calcium carbonate, much of which has been deposited as limestones, most of the salts are still in the sea. From every 100 pounds of sea water that is evaporated, 3.5 pounds of a white salt mixture remain. If all of the oceans could be evaporated into space, enough salt would remain to cover the earth with a layer close to 200 feet in thickness. Most of this salt (78%) is common sodium chloride (table salt). Of the remaining 22%, only five other compounds are important quantitatively, as indicated in Table 14-1. Despite these numbers, at least 35 other elements have been detected in sea water and despite their relatively minute amounts it is these *trace* components that are all-important in providing the basic mineral nourishment of the life that, in places, teems in the sea.

Throughout most of the oceans, the total salt content, or salinity, usually expressed in parts of salt per thousand parts of water (*parts per mille*), varies from about 33 to 37. Where precipitation is high, as in the equatorial and stormy midlatitudes, salinity is somewhat lower from dilution by rain or snow. Where evaporation is high, as in the dry subtropical calms and trade-wind zones, salinity is relatively high. The interesting variations of many of these interrelated factors with latitude is shown in Fig. 14-13. In some restricted waters where precipitation and stream runoff are very high, the water is surprisingly fresh, as in the Baltic Sea where salinity is only about 7. However, places like the Red Sea, where evaporation greatly exceeds precipitation, reach a salinity of 41.

Despite the variations that may occur in total salinity from place to place, a very important fact has been discovered about the composition, namely that the relative composition of the salt components is quite constant. Once the amount of any component is determined, the total salinity is calculated very quickly. In practice, the amount of chlorine, as chloride ion (Cl^-), is determined in a few minutes by chemical analysis of the water collected in Nansen bottles. Since Cl^- is 55.26 percent of all of sea salt,

$$\text{salinity} = \text{amount of } Cl^- \times \frac{100}{55.26}$$

$$\text{salinity} = 1.8 \times Cl^-$$

TABLE 14-1 Main Components of Dissolved Solids in Sea Water
and River Water

Molecule or	Ion	Oceans g/kg (Percent)	Percent	Average Rivers (Percent)
Chloride	Cl^-	18.97	55.06	5.68
Sodium	Na^+	10.56	30.58	5.77
Sulfate	SO_4^-	2.64	7.68	12.14
Magnesium	Mg^{++}	1.29	3.74	3.41
Calcium	Ca^{++}	0.41	1.20	20.39
Potassium	K^+	0.38	1.12	2.12
Bicarbonate	HCO_3	0.14	0.41	35.15
Bromide	Br	0.07	0.19	
Silica	SiO_2			11.67
Iron aluminum oxide	Fe,Al_2O_3			2.75
Nitrate	NO_3			0.90
Total		34.46	99.98	99.98

Although this chemical procedure has for years been the only really reliable way of determining this property, it is quite time-consuming for a vessel to stop and then lower and retrieve collecting bottle samples from the depths. A new instrument, the *salinometer,* has been developed that enables salinity to be determined with great precision by measuring the electrical conductivity of sea water, a property that depends on salt composition. Pure water is almost nonconducting, but as salinity increases so also does electrical conductivity. In this way salinity of a water sample can be measured on the deck of a vessel, or by lowering the sensitive element of the salinometer into the depths while the meter itself is read aboard the ship. Frequent checks against chemically determined values are still necessary.

Freezing of Seawater

Do lakes freeze to the bottom? Can the oceans freeze to the bottom? Where do the fish go in winter when lakes freeze? These are familiar and age old questions that we can answer here. To understand how seawater freezes, it may be easier to see what happens in freshwater first.

As freshwater cools, it contracts, becoming denser and heavier. If the cooling occurs from radiation and conduction to the air, as in a pond, lake, or pool in winter, the cooling water sinks and is replaced by warmer,

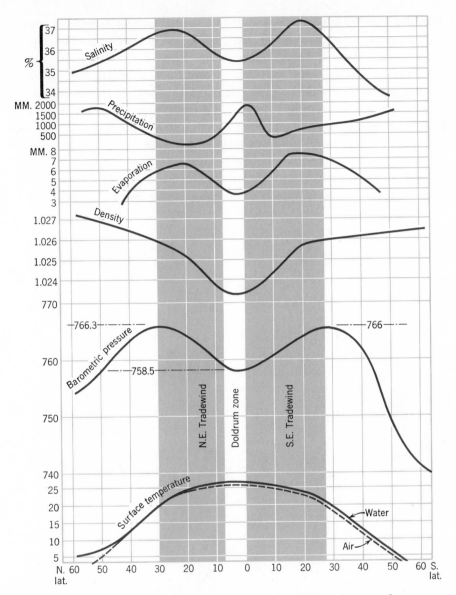

FIG. 14-13. The latitudinal relationships of the variables of sea surface temperature, salinity, and density and atmospheric precipitation and evaporation. These in turn are controlled by the primary air pressure and wind patterns described in Figs. 13-2 and 13-3. In zones where the atmosphere is dry, evaporation from the seas causes increased salinity and water density; where precipitation is high, dilution decreases ocean salinity and density.

less dense water from below. When freshwater cools to 39°F (4°C), it reaches its greatest density (see Fig. 14-14). Thus, if cooling continues it begins to expand, becoming less dense and lighter in weight. The more it cools, the lighter it becomes and the greater is its tendency to remain at the surface, leaving the warmer 39°F water below. On reaching 32°F the freshwater freezes, leaving the denser, warmer water beneath. Hence freshwater ponds and lakes do not freeze to the bottom, to the good fortune of any fish residents, who usually survive beneath the ice until the following spring, when they once again have access to the surface. In summary, the temperature at which maximum density occurs in freshwater is above the freezing point, so that freezing water remains at the surface, preventing freshwater from freezing to the bottom.

Now as the salinity increases, the temperature at which maximum density occurs is lowered progressively below 39°F, and the temperature at which freezing occurs also decreases. Once the salinity exceeds 24.7 ppm, the temperature at which maximum density occurs is below the

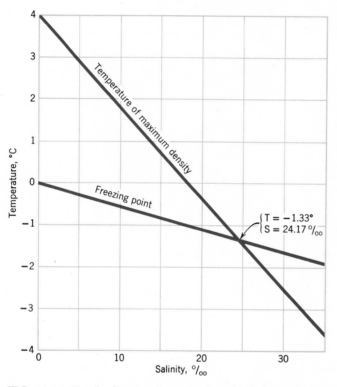

FIG. 14-14. Graph showing how the freezing of water varies with salinity. Two curves are shown: one, the temperature at which water reaches maximum density, and the other the temperature at which water freezes. Both curves depend on the salinity. The freezing point of water is seen to decrease from 0°C for fresh water to −1.8°C for water of average ocean salinity (35‰). The two curves cross at a salinity of 24.7‰ and a temperature of −1.33°C. At lower salinities, water reaches its maximum density at temperatures above the freezing point. Because natural water bodies cool at the top, deeper water will be heavier preventing the cold water from sinking so that freezing is restricted to the surface layer. At salinities above 24.7, water will continue to sink as it cools so that the entire layer can be brought to the freezing point.

freezing point. Since normal ocean salinity is about 35 ppm, seawater could continue to fall as it cools even after the surface freezes. Thus, seawater could theoretically freeze all the way to the bottom. Despite this, seawater rarely freezes at all, much less to the bottom. Oceans are kept from freezing by the mixing due to winds that bring warmer waters to the surface. Even when the sea surface does freeze in winter, as in the Arctic Ocean and around Antarctica, freezing is limited to a fairly thin (about 10 feet) layer called *pack ice* (Fig. 14-15a) by the reappearance of the spring and summer sun before much of the ocean, with its high specific heat, can become chilled too deeply.

Icebergs, the bane of shipping in the North Atlantic Ocean (Fig. 14-15b), are not frozen ocean at all. They are really huge masses of glacial ice that tear loose from the Greenland ice cap where it meets the sea. Currents then carry the bergs southward into the shipping lanes. Because water expands when it crystallizes to form ice, ice is less dense and lighter than water, having from 80 to 90 percent of the original water density. Icebergs thus float in the heavier water but have 80 to 90 percent of their mass beneath the surface, and therein lies the great danger to shipping.

Maritime history is replete with records of ships lost through collision with icebergs, but no occurrence was more tragic than the sinking of the "unsinkable" *Titanic* on her maiden voyage to America. A total of 1517 men, women, and children were lost when the vessel collided with a berg in midocean at about the latitude of New York City.

Following this sea disaster, the international ice patrol, in which the U. S. Coast Guard plays a large role, was organized to keep track of bergs in the North Atlantic and provide warning to shipping. Some 15,000 icebergs break or *calve* from numerous ice tongues of the main Greenland ice sheet per year. Most of these melt before becoming a hazard, but many hundreds still reach the midlatitude shipping lanes.

Although the southern icebergs breaking from Antarctica are greater in size than their northern counterparts, they are kept from drifting into the midlatitude shipping lanes by strong ocean currents surrounding the Southern Ocean, as described in the next section.

OCEAN CURRENTS

Although Matthew Maury synthesized the first global chart of surface currents, the existence of such currents was long known from the deviation of ships from their expected courses. The more-informed ship captains made use of their knowledge of current flow to shorten their travel time between ports of call. The learned Benjamin Franklin, as Postmaster General, was aware of the importance of the Gulf Stream in speeding shipborne mail and had a current map drawn for the use of the mail packets.

FIG. 14-15a. The thin pack ice covering the Arctic Ocean averages about 10 feet thick compared to voluminous icebergs. The open band of sea water, called a *lead* becomes frozen in winter. (Photo by Arctic Group, Lamont-Doherty Geological Observatory.)

Franklin believed, as did others, that ocean currents were produced by the drag effects of the prevailing winds on the water surface. Maury, among others, proposed that surface currents resulted from differences in the expansion of seawater caused by systematic temperature differences. Currents would then flow from the regions of low-density water where the sea level was high to regions of cooler, higher-density water. It finally turned out that both were partly right. Most of the surface currents can be explained by wind drag on the water and most of the subsurface currents are explained by the effects of density differences; some interaction between wind and density effects does occur.

Surface Currents

The surface currents of the ocean are driven primarily by the prevailing-wind systems described in Chapter 13. The pattern of the surface circulation is also determined by two other forces, once the drifting of the water occurs—namely, the Coriolis force due to the earth's rotation, and friction. Earlier, in Chapter 12, we considered the Coriolis force in some detail in connection with wind motion. Recall that this force increases with latitude from zero at the equator to a maximum value at the poles and that it acts to the right of a moving object in the Northern Hemisphere and to the left in the Southern Hemisphere.

To see how the winds set up the surface ocean circulation, recall that the prevailing *average* winds are dominated by the trades and prevailing westerlies. The former have a strong component from the east in both hemispheres between the equator and 30° and the latter, a strong compo-

FIG. 14-15*b.* Picturesque iceberg in the North Atlantic Ocean. The pack ice in Fig. 14-15 (*a*) is true frozen sea water that averages only 10 feet in thickness but icebergs, which are fragments of huge glaciers, can be hundreds of feet thick. (U. S. Coast Guard photo.)

nent from the west between 30° and 60° as indicated in Fig. 14-16*a*. A uniform, water-covered, nonrotating earth would tend to develop three globe-encircling current belts, one constrained to flow from east to west on either side of the equator and the other to flow from west to east between about 30° and 60° in each hemisphere.

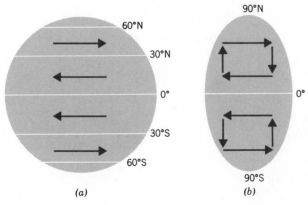

FIG. 14-16. (*a*) the global current pattern generated by the prevailing winds on a uniformly water covered, nonrotating earth. (*b*) Schematic view of the current pattern as modified by rotation and the continental boundaries of the ocean.

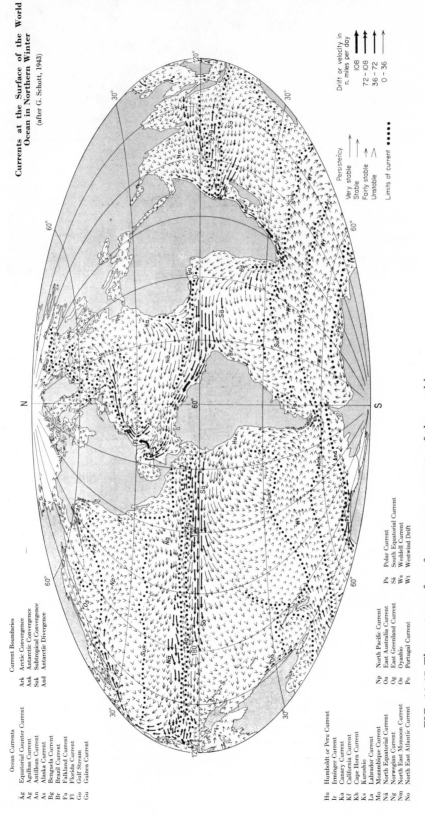

Currents at the Surface of the World Ocean in Northern Winter

(after G. Schott, 1943)

Current Boundaries

Ark	Arctic Convergence
Ank	Antarctic Convergence
Suk	Subtropical Convergence
And	Antarctic Divergence

Ocean Currents

Äg	Equatorial Counter Current
Ag	Agulhas Current
An	Antillean Current
As	Alaska Current
Bg	Benguela Current
Br	Brazil Current
Fa	Falkland Current
Fl	Florida Current
Go	Gulf Stream
Gu	Guinea Current

Hu	Humboldt or Peru Current
Ir	Irminger Current
Ka	Canary Current
Kf	California Current
Kh	Cape Horn Current
Ks	Kuroshio
La	Labrador Current
Mo	Mozambique Current
Nä	North Equatorial Current
Ng	Norwegian Current
Nm	North East Monsoon Current
No	North East Atlantic Current

Np	North Pacific Current
Oa	East Australia Current
Og	East Greenland Current
Os	Oyashio
Po	Portugal Current

Ps	Polar Current
Sä	South Equatorial Current
Ws	Weddell Current
Wt	Westwind Drift

Persistency

Very stable	
Stable	
Fairly stable	
Unstable	

Limits of current

Drift or velocity in n. miles per day

108	
72 – 108	
36 – 72	
0 – 36	

FIG. 14-17. The system of surface ocean currents of the world. (From G. Dietrich, *General Oceanography*, John Wiley & Sons, Inc.)

In the Southern Hemisphere, between the tip of South America and Africa and Antarctica, this uniform flow is nearly achieved as evident in the world current map in Fig. 14-17. But the rest of the real world is not so uniform. North and South America and Eurasia-Africa divide the world ocean into two broad, north-south stretches of sea, the North and South Atlantic and the broader North and South Pacific Oceans.

Each of these oceans can be viewed schematically as north-south belts converging, as they do on the spherical earth, toward the poles, as in Fig. 14-16b. The side boundaries of these water bodies are the continents. These continental borders must tend to set up northward and southward flow patterns at the ocean borders that would form a clockwise rotational pattern or gyre in the Northern Hemisphere and a counterclockwise gyre in the Southern. Inspection of Fig. 14-18 shows that qualitatively each of the oceans shows just such gyres; for example, in the North Atlantic, the current loop consists of the equatorial current, deflected northward to form the Gulf Stream and Westwind Drift, which then turns southward as the Canary Current.

The effects of wind drag and frictional resistance alone would result in current gyres quite symmetrical on opposite sides of the ocean. But in the real ocean, the gyres are quite intense in the western sides of the oceans and relatively weak in the eastern sides, a feature evident in the flow lines in Fig. 14-17. This asymmetry results from the addition of the Coriolis force to the wind and frictional forces. Although wind and frictional forces are fairly constant over the current loop, the Coriolis force varies with the latitude. It increases on the western sides of the oceans as the current flows northward and decreases on the eastern sides as the current flows southward. In order to achieve a proper balance between the variable Coriolis force and the wind and frictional forces, the ocean current must flow much faster on the western than on the eastern sides of the ocean.

A summary of the main current gyres in the oceans of the world is shown in Fig. 14-18, where the western intensification of the currents is evident by the crowding of the flow lines along the western margins of the oceans. So pronounced is the effect that the Gulf Stream reaches a speed of 4 knots and has an average speed of 1.5 to 2 knots. When this speed is averaged over the stream width of 30 to 60 miles and a depth of one mile, the enormous volume of water transported by the Gulf Stream can be determined. This amounts to about 4.5 billion cubic feet per second. This is more than one hundred times larger than all of the water transported by all of the streams of the world.

Currents and Temperature

The interaction or energy exchange between oceans and atmosphere is shown most elegantly by the relationship among winds, current flow, and temperature of the sea and air. As we noted earlier, in Chapter 13,

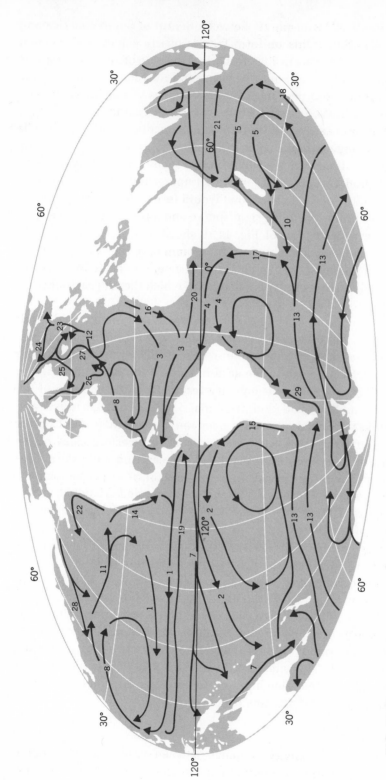

FIG. 14-18. A generalization of the world current chart emphasizing the main patterns and ocean current gyres. (After G. Dietrich *General Oceanography,* John Wiley & Sons Inc.)

the nonuniform heating of the earth and consequently of the atmosphere creates the prevailing-wind systems. These, in turn, set ocean currents in motion. Because surface currents are developed as closed loops or gyres, north- and south-flowing segments of these gyres exist.

Northerly currents carry warm water to higher latitudes (and oppositely, in the Southern Hemisphere) and southerly currents transport cool water to lower latitudes; in this way, the normal tendency of uniformly decreasing temperature with increase in latitude becomes altered. (Unlike wind directions, ocean currents are named according to the direction *toward* which they flow.) Return for a moment to Fig. 12-6 in Chapter 12 and note how the isotherms crossing the oceans are deflected from an east-west orientation. In the North Atlantic Ocean, invasion of huge volumes of warm water by the Gulf Stream causes a strong northward deflection of isotherms extending seaward from the east coast of United States. The British Isles and Scandinavia are relatively warm, inhabited regions compared to the frozen wastes of Labrador and Greenland in the same latitudes across the sea.

The prevailing westerlies transport to Europe heat fed back to the air by the warm sea, heat that the sea picked up at lower latitudes. Air warmed by currents from the south interacts with cold air from the north to produce secondary storm and wind systems that further control the movement of the sea.

Trapped by the relatively fast ocean gyres are large areas of warm, quiet waters. These are known as *sargasso seas* because of the *Sargassum* or sea weed that tends to collect in the calm central waters much as floating leaves collect in the center of a stirred tea cup. These warm regions, shown within the inner closed flow lines of the gyres in Fig. 14-18, play an important part in heating the air and in determining the climate of the continents to windward.

The Deeper Currents

Some very practical problems are related to ocean circulation. The vast increase in population and in industrial processes has led to a vast increase in waste material—as will be discussed further in Chapter 16. The oceans have unfortunately become one of the world's great dumping grounds, leading to the question of how fast the ocean circulation can disperse the waste input.

Relation to surface circulation. Surface circulation is primarily driven by the prevailing world winds. The surface flow, of necessity, must affect the subsurface circulation. For example if two wind-driven currents have a component of flow toward each other a line of *convergence* forms, as

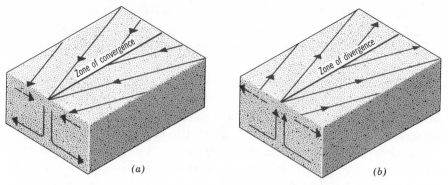

FIG. 14-19. (a) shows convergence, a line along which wind-driven ocean currents have a component toward each other. Descending motion takes place beneath the convergence. Currents have a component of motion away from a zone of divergence in (b) causing upwelling of water from below.

in Fig. 14-19a, a nearly vertical water descent occurs, with subsequent deep horizontal flow away from the region of convergence.

Conversely, if the surface circulation has a component of motion away from a line, as in Fig. 14-19a, a line of divergence is present requiring the ascent of deeper and usually colder water. Divergence is one important cause of *upwelling* of deeper, colder water. Upwelling is of major significance in the biology of the seas. This motion transports nutrient-rich waters to the surface, which would otherwise become depleted by the minute forms of life with which the surface layer often teems.

Density currents. Apart from the surface-current influences, the deep circulation of the oceans is driven primarily by forces developed from differences in density between different water bodies. We have already considered, in connection with the origin of the deep cold water and the freezing of water (pp. 497 and 500), the effect of temperature on water density. Cold water is denser (and heavier) than warmer water of the same salinity and will sink beneath it. The extremely cold *Antarctic bottom water,* formed off Antarctica in the Weddell Sea, is the densest water in the sea. After sinking to the sea bottom, this water continues northward, crosses the equator, and travels as far as 40°N, creeping beneath the waters of the South Atlantic and much of the North Atlantic Oceans. Cold water generated off Greenland is the densest of waters originating in the North Atlantic. This water also sinks and then travels southward as *North Atlantic bottom water* but rises above the colder and denser *Antarctic bottom water* when the two currents meet.

Figure 14-20 is a more detailed three-dimensional diagram of the features in Fig. 14-11 and shows the two deep currents just referred to, together with other currents including the wind-driven surface currents.

In addition to temperature effects, varying salinity also causes density changes that cause vertical water motions, which in turn require horizontal deep current flow. Salinity differences in the open ocean primarily result from the effects of evaporation and precipitation summarized earlier in Fig. 14-13. Hence the driving forces setting up the deep currents pictured in Fig. 14-20 result from density differences generated by a combination of temperature and salinity effects.

OCEAN WAVES

Despite the tremendous importance of ocean currents in regulating world climate and in redistributing heat and salts through the huge volumes of water transported, we really do not see them readily. The dynamic, restless nature of the sea is much more apparent in the waves and swell that travel along the surface of the open ocean and then pound the shores and beaches, often with great destructive effects.

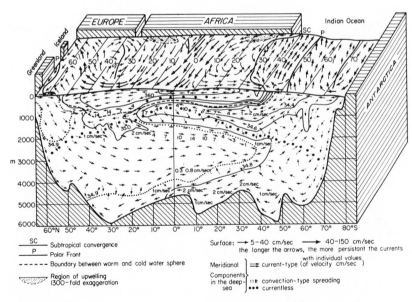

FIG. 14-20. Detailed three-dimensional view of the water circulation in the Atlantic Ocean. The large cold water volume and its high latitude sources is again emphasized. (After G. Wüst.)

Characteristics of Waves

Ocean surface waves are generated by the wind and propagate as undulations of the water surface that travel in the general direction of the wind. All waves can be described in terms of certain characteristic dimensions, partly summarized in Fig. 14-21, where *crest* is the ridge line of the wave, *trough* is the bottom line, *height* is vertical distance from crest to trough, and *length* is the distance between successive crests. As the wave moves, the time it takes for two crests to pass a given point is called the *wave period*.

Wave motion is really the means by which wind energy, transferred to the water, propagates or travels along the surface. If we watch some object that is floating on the water as waves pass, the object merely seems to bob up and down; this indicates that water does not really move horizontally as waves pass across its surface. If we watch the floating object with just a little more care, the object can be seen to move in a small circular orbit as a wave passes, rather than in a simple vertical up-down movement.

Actually, every particle of water describes a nearly uniform circular orbit as a wave passes, as shown, for example, in the striking photograph (Fig. 14-22) of illuminated particles floating in a tank disturbed by waves. From this, we can reconstruct what happens in the water to permit the passage of waves.

In Fig. 14-23, the surface profile of the wave is shown in two positions one quarter of a wavelength apart, the first by the solid line and the second by the broken line. The movement and positions of water particles are shown for three levels. At the extreme left, corresponding to the crest of the wave in the first position (solid line), the water particles are at their highest positions. To the right, in the direction of wave motion, the particles which form the water surface are in positions successively farther from the top. At the wave trough (seventh orbit from the left), the water particles are at the lowest points. As the wave moves to a distance one quarter of a wavelength to the right, each of the water particles moves in the direction of the arrows to the positions connected by the broken lines. By the continuation and repetition of this particle motion, the familiar surface water-wave motion is maintained.

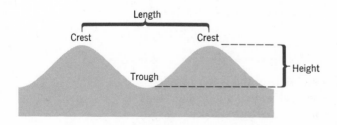

FIG. 14-21. Basic wave terms.

The behavior of particle motion along the vertical is shown by the displacements of the lines which are vertical beneath the wave crest and trough. This effect is nicely shown by the swaying of reeds in shallow water as waves pass.

Wind Waves and Swell

Waves that are growing under the direct influence of the wind are called *wind waves*. Their height when young may be a fraction of an inch and some hours or days later may grow to 30 or more feet; in the same time interval their length can increase from a fraction of an inch to hundreds of feet.

The ultimate height and length that waves achieve depends on three

FIG. 14-22. Laboratory photograph of the orbital motion of water particles beneath surface waves. Explanation in Fig. 14-23. (G. Dietrich, *General Oceanography*, John Wiley & Sons, Inc.)

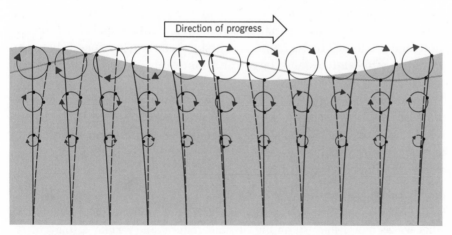

FIG. 14-23. The movement of water particles beneath a single wave. (Explanation in text.)

FIG. 14-24. The sea when rough and wild within a storm. (U. S. Coast Guard Photo.)

factors: (1) the wind speed, (2) the duration of a particular wind, and (3) the *fetch* or straight-line distance over which the wind blows across open water.

As energy is transferred from wind to the water, waves of different lengths arise. Short-wave-length waves grow more rapidly than longer waves; when the ratio of height to length exceeds 1:7, the waves become unstable and break, leaving patches of foaming white water or *whitecaps* along the breaking crests. Whitecaps first appear at wind speeds between 8 and 12 mph and become larger and more numerous as the wind increases.

Much of the energy of the short, breaking waves is transferred to the more stable longer waves, which soon become dominant. Because the wind along the surface does not move as a uniform sheet of air, as in laminer flow (p. 188), but is rather turbulent and gusty in its motion, a sea

FIG. 14-25a. (opposite, above) The sea when calm with low, regular ocean swell in the tropical Pacific Ocean. These were produced as high, irregular wind waves in a distant storm but became uniform and regular as they travelled across the ocean.

FIG. 14-25b. (opposite, below) Aerial view of swell approaching a shore. Note how the waves become sharper and higher as they approach the breaking point. The white zone is the region of white "broken waves."

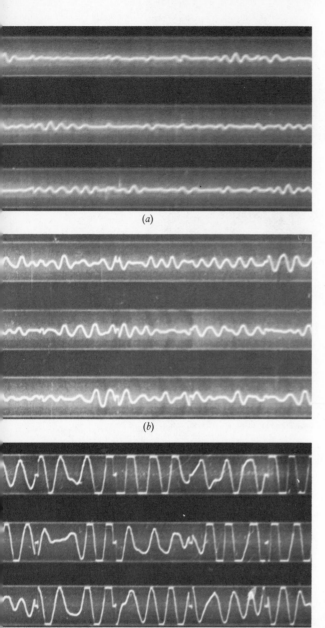

(a)

(b)

(c)

FIG. 14-26. During IGY, an underwater wave meter at Barbados, W. I., sensitive to the pressure changes produced by surface waves, was connected to a wave recorder on shore. The sequence of wave records shows changes in wave size from (a) October 19, 1958 to (b) October 22, 1958 to (c) October 25, 1958 when wave heights were observed between 25 and 30 feet with an average of about 16 feet. Each line shows 5 minutes of wave recording. The Barbados fishing fleet was all but destroyed and sand was deposited in second floor rooms of shore side dwellings. (d) (opposite) We explained the origin of the severe waves in a though c as swell from the intense storm of October 22–24, 1958 in the Atlantic Ocean to the north of Barbados shown in this weather map series. Note the long straight run (fetch) of 50 knot winds directed southward on the west side of the cyclonic storm. These winds caused the Barbados waves.

surface covered with wind waves is quite irregular in the height, length, and horizontal extent of waves. Wind waves have short crests, so that there is considerable horizontal overlap of waves when they are traced horizontally.

Once the wind stops or the wild waves generated within the storm (Fig. 14-24) outrun the area of generation, a pronounced transformation occurs. Crests become lower and more symmetrical and much longer in extent (Fig. 14-25a and b). These regular waves are called *swell* and can travel out across the ocean to break on shores far distant from the storms

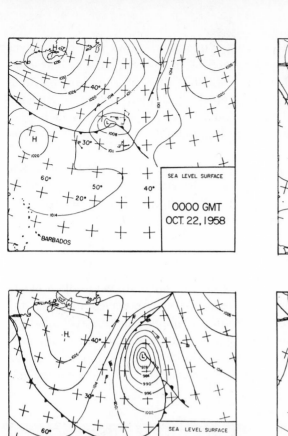

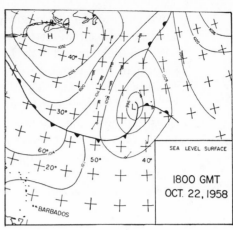

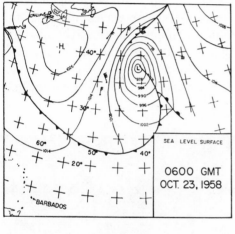

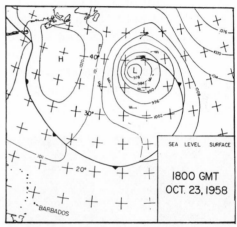

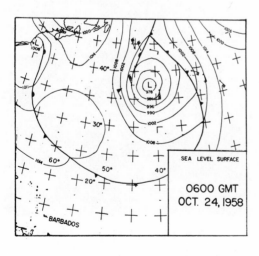

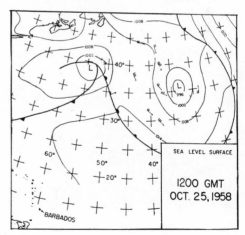

(*d*)

that generated them. Figure 14-26, for example, shows very high waves recorded by an underwater wave meter off the eastern coast of Barbados, W. I., which arrived in perfectly fine weather after traveling from a distant storm in the Atlantic Ocean some 2000 miles away. These unexpected waves inflicted very severe damage on fishing boats and shoreside homes. This IGY experiment in wave recording helped provide a basis for forecasting these dangerous waves.

Many coastal regions are beset by late summer and fall hurricanes. Prior to modern satellite and aviation observations, the first sign of an approaching storm was often the arrival of a strong long-period swell that outran the slower-moving hurricane.

Breakers and Surf

Most of us are familiar with waves not so much from the open ocean but rather from beach and shore experiences. As waves approach the shore, the effect of the shoaling bottom strongly affects wave behavior. Friction with the bottom slows the waves. As they roll in toward the shore, the slightly faster seaward waves overtake the wave closer in, thus decreasing the wave length. As the length decreases, the wave energy gets

FIG. 14-27. Plunging breaker. Where the off-shore bottom has a relatively steep slope, the wave height grows rapidly as it approaches the shore until its energy is released abruptly with the development of a plunging breaker, as seen in this example. (Photo by R. W. Linfield, Chesapeake Bay Institute.)

FIG. 14-28. Spilling breakers. Where the bottom slopes gently breaking begins well off-shore and the wave breaks continuously as it rolls toward the beach.

crowded into a smaller distance so that wave height increases. Because this is a continuous process, heights increase steadily as waves progress toward the beach. Finally, there is more water above than below the sea surface as the wave orbit is transformed from a circle to a vertical ellipse. Also, the orbital motion at the top exceeds the motion below the surface, which is slowed by friction.

Where the slope of the bottom is relatively steep, the crest peaks up rapidly while the lower portion of the wave is retarded so that the crests break in a single crashing breaker—known as a *plunging breaker* (Fig. 14-27). After breaking, the water of the wave crest rushes up the beach as a *wash*. In returning to the sea, the surf forms localized *rip currents,* which, when strong, may pull swimmers out to deep water (such surface rips are erroneously called *undertow,* a nonexistent feature). Earlier, on page 219, we examined the tremendous energy of plunging breakers and the damage wrought by them.

Where the underwater slope is relatively gentle, the process of breaker buildup takes place slowly so that breaking is continuous as the wave rolls toward shore. These *spilling breakers* form picturesque shore scenes as lines of parallel white crests move toward the beach (Fig. 14-28).

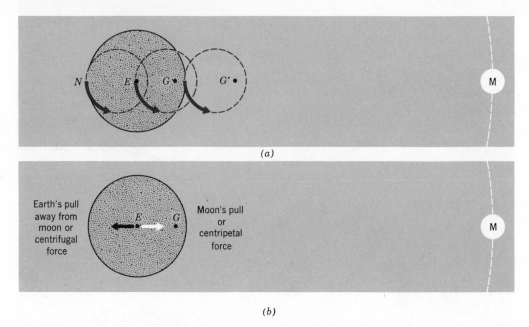

(a)

Earth's pull
away from
moon or
centrifugal
force

Moon's pull
or
centripetal
force

(b)

FIG. 14-29. The origin of the vertical tide-producing force. (See text.)

TIDES

The rise and fall of seawater in some places once a day and in others twice a day, has been known as tide since men first gazed on the sea and its inlets. Tidal variations of water level vary from a fraction of a foot in many places to as much as 50 feet in the Bay of Fundy between Nova Scotia and New Brunswick near the northeast coast of Maine.

Cause of the Tides

Despite the long awareness of tides, their explanation awaited Isaac Newton's monumental theory of gravitation. Newton showed for the first time how the gravitational "pull" of the moon and sun caused the tides. Despite the much smaller size and gravity of the moon, the lunar tide is about twice as strong as that due to the sun. This occurs because the tidal attraction varies as the inverse *cube* of the distances of moon and sun (1/distance^3). Thus, the effect of the sun, 93 million miles away, is much less than that of the moon at a distance of 240 thousand miles.

A very common question about tides is: How can there be two high tides on opposite sides of the earth, one toward the moon and the other away from it? The answer to this question requires some elementary examination of the forces involved.

In Fig. 14-29, the moon at *M* is shown revolving about the earth, whose center is at *E*. The earth in turn tends to revolve about the moon. Because the earth is 80 times more massive than the moon, the point about which the earth revolves is $\frac{1}{80}$ of the way from E to M and actually lies within the earth at G, about three fourths of the way from the center to the surface. The point G is the common center of gravity about which both the moon and the earth revolve. Notice that the earth *revolves* about G; it does not rotate. Rotation is a turning motion about an axis through the center of an object. Whenever an object revolves, every particle within it and on the surface moves in the same path and at the same speed. To see that this is so, push two pencils through a piece of moderately transparent stiff paper. Imagine that one pencil is at E and the other near G of Fig. 14-29. Then by keeping the directions of the two pencils parallel to the initial direction, so as to avoid rotation, trace out a circle with the pencil at E and it will be seen that pencil G moves in the same way and draws a similar circle. Thus, as the earth revolves about G, points toward the moon, E at the center, and N opposite the moon will trace out the same circles. This of course takes a month, during which the moon makes one revolution about the earth.

Now to come back to the forces. The earth is kept revolving about G by the pull of the moon. Since the earth is not pulled any closer to the moon, there must be a force (called centrifugal force) directed opposite to the moon's pull. The two forces must be just balanced at the earth's center. Now the centrifugal force is the same for all points in or on the earth because these points are all moving in the same direction and at

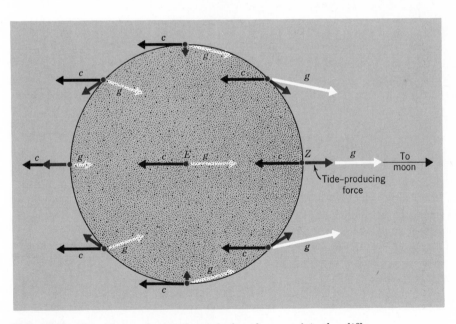

FIG. 14-30. The tide-producing force (colored arrows) is the difference between the force due to the moon's gravitational pull, and the oppositely directed force due to the earth's motion about the common center of gravity c.

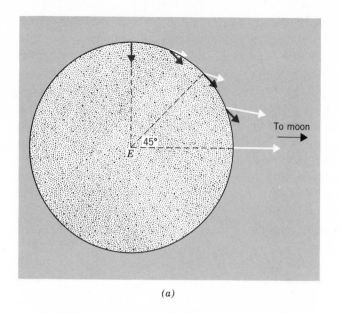

(a)

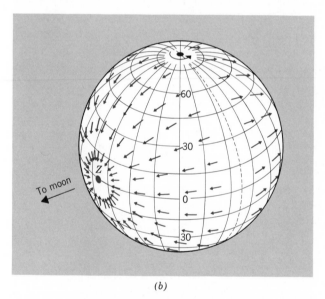

(b)

FIG. 14-31. Because the vertical tide-producing force is so small, the actual accumulation of water to make the high tide results from the horizontal component of the tide producing force over the entire earth as in (a). The distribution and relative size of this force is shown in (b).

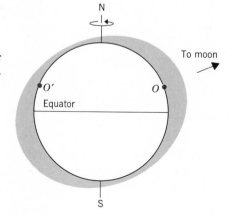

FIG. 14-32. Explanation of the diurnal inequality of the tides.

the same speed. Hence, in Fig. 14-30, the centrifugal force, c, is the same, but the moon's (g) pull varies. On the side of the earth just beneath the moon (Z), it exceeds c by the amount shown by the heavy arrow, which is thus the vertical tide-producing force; at the earth's center the c and g forces are equal and balance each other. On the side opposite the moon, the pull g is less than c, again by the amount T, so that here there is a net vertical tide-producing force directed away from the moon.

Elsewhere from the line of centers of earth and moon, the tide-producing force is also the net of the earth's centrifugal and the moon's centripetal or attractive forces and points as shown by the heavy arrows.

The actual value of the tide-producing forces, shown by the heavy arrows in Fig. 14-30, is found to be very small. For example, the vertical tide force is about one ten-millionth of the earth's gravitational force and would thus be hardly likely to raise the observed tides. Actually, the heaping up of water to make the tidal bulge is a result of the total effect of the small horizontal components of the tide-producing forces as in Fig. 14-31a.

Notice that the horizontal or *tractive* force is greatest at 45° to the line of centers and is zero on this line and at 90° to this line. It is the flow of water in response to the tractive forces that accumulates high water. The distribution of these forces over the globe is shown in Fig. 14-31b. A region of low water or low tide occurs midway between the line of earth-moon centers.

Spring and Neap Tides

Tides produced by the sun form in the same way as that described for the moon, except that the lunar tide is dominant. Twice each month, at times of the new and full moon (p. 27), the tide-producing forces of the sun and moon add together. At these times the spring tide, which is the highest high tide, and of course, the lowest low tide, occurs. At first and third quarter, recall that the moon and sun are at right angles relative to the earth (p. 27). At these times solar high and lunar low tides correspond in position on the earth, as do solar low and lunar high tides. These *neap* tides, midway in the month between the new and full moon, are thus lowest of the high tides and the highest of the low tides.

Tidal Lag and Diurnal Inequality

Friction between the great tide wave and the ocean floor, plus the effects of continental obstacles to the flow of the tides around the earth, cause a lag between the passage of the moon across an observer's meridian and the related high tide. From minutes to many hours may elapse depending on the latitude and local geography.

To understand the reason for the diurnal inequality of the tides, refer to Fig. 14-32. Here the moon is overhead at some latitude north of the equator and the high tides on opposite sides of the earth are not symmetrical about the equator. An observer at 0 will experience the full height of the high tide. But when rotated to 0^1, the high tide experienced will be less. As the moon's declination changes during the month, there will be a change in the types of tide experienced at a particular latitude.

We might also note that the period between true high tides is not 12 hours or half a day. Recall from Chapter 1 that the moon revolves "eastward" as the earth rotates from west to east. The moon thus arrives at the same point in the sky 50 minutes later each day. The "same" tide is thus encountered 24 hours 50 minutes later each day and the intervening high tide is 12 hours 25 minutes later.

LIFE IN THE OCEANS

The sea is teeming with living things. Some are so minute we can only see them with the most powerful microscopes; others, like the great whales, are the largest creatures ever to live on the earth. All forms of life are found in the sea, from the lowest single-celled plants and animals to some of the most advanced of the vertebrates, such as dolphins and whales.

Life in the sea, as on land, also falls into two *kingdoms,* flora and fauna (plants and animals). Unlike on land, marine plant life mostly belong to one of the simplest of the plant groups, the *algae,* and include the common types of seaweed as well as the more important vast assemblage of single-celled individuals.

Another way to classify life in the sea is in terms of living habits. Many are free-swimming forms like fish; others crawl along the bottom, as for example starfish, lobsters, and some forms of clams; some are firmly attached to the bottom, such as sea lilies, coral, and sponges; and a last, huge group of very small, usually microscopic forms, float and drift about with the currents. These drifters and floaters, both plant and animal types that are collectively called *plankton,* play a fundamental role in the total marine *ecology* or interrelationship of the marine communities with each other and their environment.

The Marine Plant World: Phytoplankton

Although we are more familiar with the large varieties of seaweed than with plant plankton, seaweed is relatively unimportant in marine ecology or as a product source. Not long ago it was a major source of iodine until terrestrial mineral sources of this element were found. At present seaweed is still used as a food in the orient. It also finds some use as a fertilizer in regions of very poor soil.

The less visible but more important plant world in the sea includes a variety of minute planktonic forms that, like so many of the familiar land plants, contain the important green chlorophyll. Only the organic substance chlorophyll, in the vast domain of life, can convert raw carbon dioxide and mineral materials into food. The energy to accomplish this conversion is obtained from sunlight. These small floating plant forms are distinguished from planktonic fauna by the name *phytoplankton* (phyto: to bring forth).

The extreme importance of green plants on land and in the sea lies in this food-making ability. Directly or indirectly they constitute the food supply of the animal world; animals may eat plants or eat other animals who eat plants.

Although phytoplankton are found throughout the great expanse of the seas, they are limited to the upper 200 to 300 feet. Without sunlight, which hardly penetrates below this thin layer (p. 488), phytoplankton could not flourish. They are also particularly abundant in regions of upwelling. So numerous are phytoplankton in surface waters that they rapidly deplete the mineral nutrients from the sea unless these are replenished continuously. Upwelling does this by transporting nutrient-rich cool water from the depths.

The common types of phytoplankton, some of which are shown in Fig. 14-33, include (1) *diatoms*, which are elongated but very varied one-celled forms that develop thin siliceous shell-like coverings and range in size from 1/3,000 to 1/30 inches; (2) *peridimians*, which are similar in size to diatoms but have a covering of cellulose rather than silica and have two flagellae or taillike appendages for motion; (3) *coccolithophores*, which are much smaller than the first two forms (less than 1/500 in.) and contain a single flagellum and often a small drop of oil that provides buoyancy; and lastly (4) the *nanoplankton*, or smallest of the group, which are still not too-well known.

A second important planktonic plant group includes the different types of marine bacteria. While phytoplankton synthesize food from raw minerals, bacteria do the opposite. They decompose organic matter into mineral material, principally nitrates and phosphates that once again serve as food for phytoplankton.

Zooplankton

Zooplankton (from zoan, meaning animal) include a huge variety of free-floating animal life. Some varieties are single-celled and others are many-celled individuals that reach several inches in size, in contrast to the more minute plant forms.

Despite the countless number of plant and animal plankton in the thin sunlit upper layer of the ocean, their existence was unknown until 1828, when a British surgeon developed a conical gauze tow net attached to a collecting jar that trapped these small creatures. Continued improvements in tow nets permit them to be opened and closed at any depth so that the forms collected are not contaminated by those from other levels

FIG. 14-33. Some examples of recent phytoplankton photographed by means of electron microscopes at Lamont-Doherty Geological Observatory. (*a*) and (*b*) are diatoms (Courtesy O. Roels); (*c*) is a dynaflagellate type of peridimian, ×1850 (Photo by S. Harrison); (*d*) is a cocolith, ×30,000 (Photo by A. McIntyre).

in the sea (Fig. 14-34). In this way a close correlation has been made between the conditions of temperature and salinity of the surface and deep waters with the species of plankton and deeper forms present.

It has been found in this way that all forms of plankton are quite sensitive to the temperature of the sea and to a lesser extent, the salinity. Some forms of plankton that flourish in the cold surface waters of high

FIG. 14-34. Lowering of a modern plankton tow net. The "sandwich" of several nets can be opened and closed at desired depths. (Lamont-Doherty Geological Observatory.)

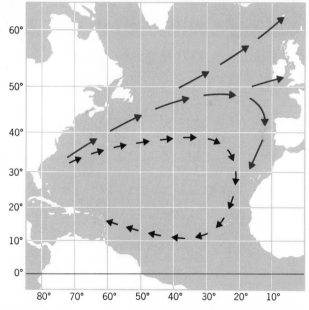

FIG. 14-35. Use of plankton (cocoliths) to determine the position of the North Atlantic current gyre during the last glacial stage (black arrows) when the ocean was colder. Compare with the present gyre (colored arrows) in which the Gulf Stream reaches much further north. (After A. McIntyre in *Science*.)

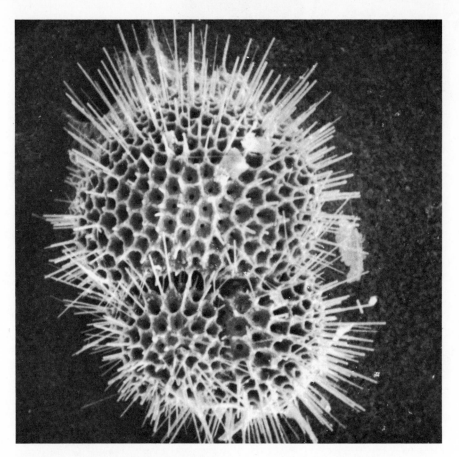

FIG. 14-36. An example of one of the many species of foraminifera present in the seas ×200 (Photo by A. W. H. Be.)

latitudes cannot survive in the warm surface currents of lower latitudes but are found in the deeper cold waters of these latitudes. By plotting the occurrence of planktonic species and types found in the surface and deeper waters, the position and movements of ocean water masses of uniform temperature and salinity can now be delineated. Changes in these marine characteristics show up quickly in the plankton distribution. Figure 14-35 shows how the position of the Gulf Stream is well marked by a particular phytoplankton occurrence.

One group of single-celled microplankton, the foraminifera or "forams" as they are commonly called (Fig. 14-36), are extremely diverse in appearance and occurrence and have provided some of the most valuable information on ocean currents. Then, because their shells are often preserved in the sediments of the deep sea, they provide a means of comparison of current flow in the past with that of the present. More will be said about this in Chapter 15.

Many varieties of zooplankton belong to more advanced forms of

FIG. 14-37. The Ewing piston coring device in the course of deep-sea coring from the deck of the research vessel *Vema*. The weighted cylinder beneath the stabilizing vanes drives the core into the soft sea bottom after free fall. As the coring tube drives into the bottom, a piston rises within the tube causing a suction that causes deeper penetration than is possible by the weight alone. (Lamont-Doherty Geological Observatory photo.)

FIG. 14-38. Close view of the coring device after raising. Note the broken vane atop the tube. (Lamont-Doherty Geological Observatory photo.)

FIG. 14-39. Preliminary examination of the split cores in the laboratory by Prof. Maurice Ewing (upper left) and colleagues. Professor Ewing either invented or developed to a high state of perfection, most of the tools of modern marine geophysics.

invertebrate life but depend for food on the smaller planktonic forms—both plant and animal. Larger forms of vertebrate life such as fish and whales often survive by feeding on the larger zooplankton. The great blue whale swallows 10,000 tons of a particular kind of zooplankton in its life. Other sea life depend on other zooplankton, which in turn depend on phytoplankton, and so on. All of this endless food chain in the sea suggests again the huge total quantity of life that inhabits the sea. It is unfortunately beyond the scope of this chapter to describe the forms and living habits of the millions of species of sea life, large and small, that live in the oceans.

THE SEA BOTTOM

This section is included to remind the reader that much of our modern knowledge of the earth's crust and upper mantle and the history of crustal development is an outgrowth of modern oceanography. Since the middle of the twentieth century, the study of the earth beneath the sea has been a major goal of marine research. Precision depth recorders and subbottom depth recorders (p. 82) have revealed the shape of the sea bottom and the structure of the shallow sediment beneath the bottom surface (Figs. 3-9 and 15-17). The seismic profiler technique (Fig. 3-18) has permitted the penetration of knowledge to thousands of feet into the marine crust. Deep-sea coring procedures (Figures 14-37 to 14-40) have permitted detailed studies of the sediment deposited in the present (Cenozoic) Era and provided tremendous insight into the changing climates of the past.

Most recently the use of the marine magnetometer provided the primary observations leading to the unifying models of sea-floor spreading and plate tectonics that have so revolutionized our knowledge of the earth. In short, much of this book has been concerned with the geological and geophysical results of oceanography.

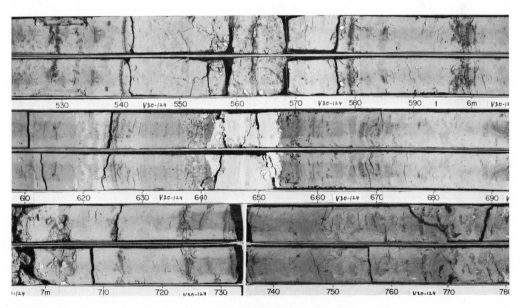

FIG. 14-40. Close view of a portion of a deep sea core No. 124 taken on research vessel *Vema* cruise 20 in the northwest Pacific Ocean. The cores are split and both halves are visible in their trays. The upper right begins at a depth of 5.2 meters below the top of the core. (One meter equals 3.28 feet.) The total core 8.57 meters long spans about 300,000 years. The white zone centered at 6.5 meters is a layer of volcanic ash. Most of the core is a fine-grained sediment (claylike) containing many diatoms. (Lamont-Doherty Geological Observatory photo.)

STUDY QUESTIONS

14.1. How is the science of oceanography related to the history of man?

14.2. What were the contributions of James Cook, Sir John and Sir James Ross, and Matthew Fontaine Maury?

14.3. How is the absorption of sunlight by the oceans related to the low rate of heating and cooling of sea water?

14.4. How do surface currents regulate the surface temperatures of the oceans?

14.5. With the use of isotherms in a simple diagram, show how ocean currents distort the expected uniform east-west temperature distribution.

14.6. What is the principle by which artificial earth satellites can map ocean surface temperatures?

14.7. Draw a diagram showing the typical vertical distribution of temperature in the seas. Label the mixed layer and the thermocline.

14.8. Explain the reason for the mixed layer.

14.9. Why is the bulk of sea water at so low a temperature?

14.10. What is meant by "salinity"? What is the average salinity of the seas?

14.11. Explain why a fresh water lake does not freeze very deeply. Why do icebergs float?

14.12. What is the primary cause of ocean surface currents?

14.13. What is meant by a current gyre? Describe the gyre in the North Atlantic Ocean.

14.14. How do surface currents cause the development of one type of subsurface or deep currents?

14.15. What are density currents? Describe the principal density currents in the North and South Atlantic Oceans.

14.16. Draw a simple, labeled diagram, illustrating the crest, trough, height, and length of a wave.

14.17. If it takes 8 seconds for two wave crests to pass a given point, what is the wave period?

14.18. What conditions determine the size of ocean waves? Distinguish between "waves" and "swell."

14.19. Explain what happens as a wave breaks on a beach.

14.20. Explain the origin of the ocean tides.

14.21. Distinguish between "spring" and "neap" tides.

14.22. What is "phytoplankton" and why is this form of life so important?

14.23. Define "zooplankton" and give some important examples.

chapter 15

The frigid, ice covered Arctic in summertime. To visitors and residents of high northern and southern latitudes, it is quite apparent that we still live in an "ice age." The sky portion of the photo is a composite to show that the sun is continuously above the horizon during the summer season of 24-hour sunlight in high latitudes. (Courtesy of American Museum of Natural History.)

Climates of the Past and Present

It has been said that "Climate is Man." Civilizations flourished where suitable climate prevailed. When climate changed, the sites of human culture also changed. The harsh Sahara was once the cradle of human culture at a time when it was a more humid and well-watered land. Now only rare cave drawings remain to tell us the story of the plant and animal life that once inhabited this barren land.

"Climate" is easier to think of and talk about than to define. It is often defined as the average weather of a region. But weather includes the many elements of temperature, pressure, humidity, clouds, precipitation, visibility, and wind conditions, whereas climate is most commonly thought of, and classified according to, temperature and precipitation. These two elements are certainly the most important climatic determinants. It is true that some regions are characteristically cloudy, others foggy, others very stormy, and so on. From the point of view of climate, these conditions primarily affect temperature and precipitation, with the former being of primary importance.

FACTORS CONTROLLING CLIMATE

Many interrelated factors ultimately determine the climate of a particular region. We can conveniently group these factors into (1) those related to atmospheric composition, especially water vapor, carbon dioxide, cloud particles, and dust; (2) those related directly to sunlight received at the earth's surface—or insolation; and (3) those related to broad permanent or semipermanent weather patterns, in particular as these patterns affect evaporation and precipitation.

Sunshine

In Chapter 11 we already examined in some detail the composition of the atmosphere and referred in particular to the all-important influence of water vapor and secondarily of carbon dioxide in controlling atmospheric temperature.

We also examined in some detail the heat budget of the atmosphere and the mean global temperature distribution. We noted further a number of important factors that bear repetition and elaboration

1. The curvature of the earth causes the intensity of sunlight at the earth's surface to decrease steadily from the equator to the poles. This happens because a sunbeam of a particular width must illuminate an increasingly larger area with increase in latitude, as in Fig. 15-1.

2. With an increase in latitude, an additional loss of surface heating occurs because the amount of atmosphere traversed by sunlight increases by $12\frac{1}{2}$ times from the mean position of the sun over the equator to the tangential position at the poles, as in Fig. 15-2.

3. The inclination of the earth's axis causes a change in the length of the periods of sunlight and darkness (day-night), causing polar regions to experience as much as six months of sunlight in summer and six months of darkness in winter. This is indicated in Fig. 15-3, showing the illuminated and dark hemispheres at the times of the equinoxes and solstices. The actual variation in the lengths of the period of sunlight with latitude is shown in Table 15-1 for the times of the solstices.

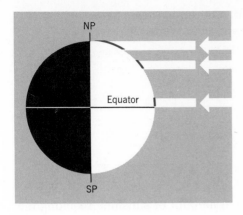

FIG. 15-1. With increase in latitude, solar radiation (expressed by sunbeams of constant width) becomes spread over an increasingly larger surface area thereby causing a poleward decrease in surface temperatures.

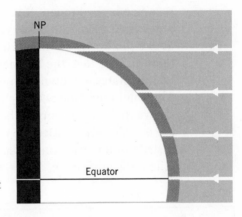

FIG. 15-2. With increase in latitude, sunlight penetrates an increasing thickness of atmosphere thereby decreasing the amount of radiant energy reaching the ground.

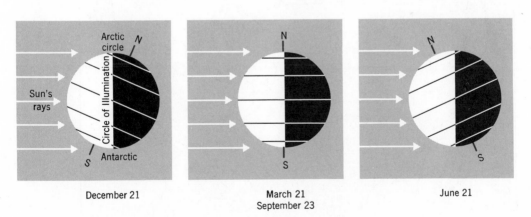

December 21

March 21
September 23

June 21

FIG. 15-3. Circle of illumination separating the illuminated from the dark hemisphere. (See frontispiece satellite photos of the earth.) In winter (December 21 drawing) the Southern Hemisphere has a longer period of daylight and is the warm hemisphere. In summer (June 21 drawing) the Northern Hemisphere enjoys more sunlight.

TABLE 15-1 Variation in Lengths of the Period of Sunlight, in Hours, Between the Longest Day in Summer and the Shortest Day in Winter at Different Latitudes

Latitude	0	10	20	30	40	50	60	90
Summer solstice	12	12.58	13.21	13.93	14.85	16.15	18.50	(6 months)
Winter solstice	12	11.41	10.78	10.06	9.15	7.85	5.50	0

4. We must realize that the actual amount of sunshine reaching the ground also depends on the amount of clouds present in addition to the length of the day. A secondary factor controlling amount of sunshine received is the topography; lowlands or valleys will tend to lose morning and evening sun, the effect varying with the seasonal elevation of the sun above the horizon. Mountaintops will, of course, receive additional sunlight. In Fig. 15-4, the average distribution of sunshine in the contiguous United States is shown as a percentage of the maximum possible sunshine. The combined effects of factors 1 to 4 are summarized in Fig. 15-5, which shows the distribution of the total amount of solar radiation received at the earth's surface. Despite the closed contours of constant radiation, the distribution is essentially parallel to latitude zones, an observation of great climatic significance.

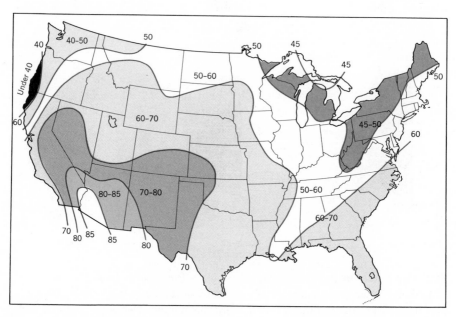

FIG. 15-4. The distribution of sunshine over the United States given in percentage of possible sunshine. Effects of mountains and clouds are included. (Modified from H. Landsberg after Kincer.)

5. The first factors refer to the amount of sunlight reaching the ground. Recall that the sunlight-absorption quality of the earth's surface varies widely both from land to water and over the land itself. Also, the conservative heating and cooling of the oceans has a further controlling influence on the climate of the overlying atmosphere.

FIG. 15-5. The distribution of solar radiation over the world given in units of thousands of calories per square centimeter per year. (After H. E. Lansberg.)

GENERAL PATTERN OF ANNUAL WORLD PRECIPITATION (INCHES)

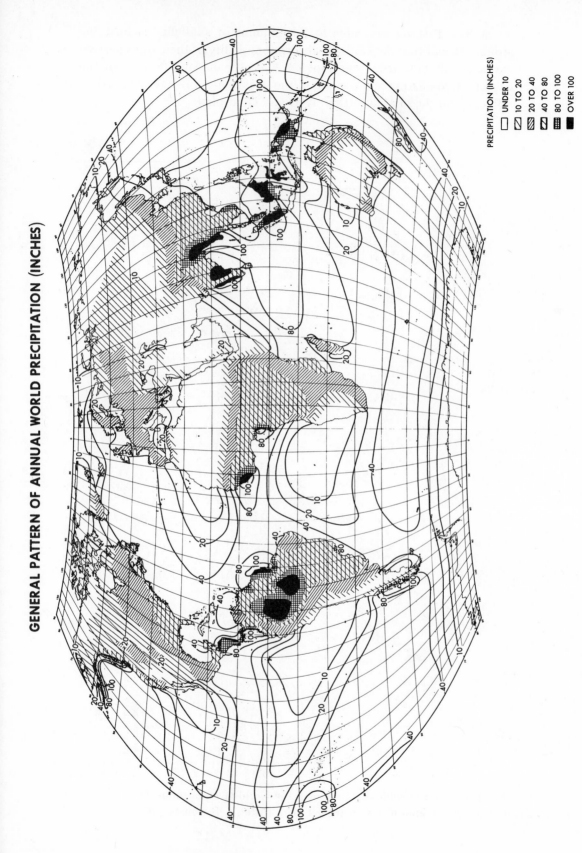

PRECIPITATION (INCHES)

☐ UNDER 10
▨ 10 TO 20
▧ 20 TO 40
▨ 40 TO 80
▦ 80 TO 100
■ OVER 100

FIG. 15-6. World precipitation in inches per year. (Environmental Data Service, NOAA.)

The sunshine factors just reviewed exert the primary controls on global climate. The direct consequence of the radiation factors just examined is the global temperature distribution shown in Figs. 12-16 and 12-17. Further details of the radiational control of climate will be considered later in this chapter.

Permanent and Semipermanent Weather Patterns

In Chapter 13 we examined the mean pressure and wind belts of the world and noted the following.

1. A belt of low pressure (the doldrums) containing very warm and very humid air surrounds the globe in the equatorial region. The characteristic rising air motion in the doldrums releases huge quantities of rainfall.

2. A belt of mean high pressure (the horse latitudes) surrounds the earth in subtropical latitudes, roughly 30° north and south. Descending air in this zone becomes warmed by adiabatic compression with a consequent decrease of relative humidity, producing arid to semiarid surface conditions.

3. Beyond the horse latitudes in both hemispheres lies a broad zone extending to about 60° in which migratory low pressure or cyclonic storm areas move in fairly well-defined paths.

4. Winds blowing off the sea undergo strong uplift when encountering mountain ranges. Resultant cooling produces clouds, often with the release of heavy precipitation on the windward sides of the ranges.

The curves of evaporation and precipitation in Fig. 14-13 show the effects of these climatic influences over the oceans. On lands, their effect is shown in the distribution of precipitation in Fig. 15-6.

THE ELEMENTS OF TEMPERATURE
AND PRECIPITATION

Temperature and precipitation primarily determine the nature of a region's climate. Before a scheme of climate can be constructed, the statistical data on these elements must be known. Also, a number of very practical considerations in the activity of man also require such knowledge.

Temperature

We have seen that the primary control of global temperature is the distribution and absorption of solar radiation. An important secondary factor is the effect of prevailing winds. When off the sea, these winds moderate

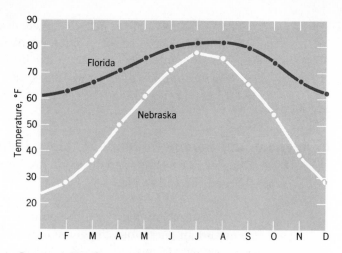

FIG. 15-7. Comparison of annual temperature variation for places with interior continental and coastal marine climates, respectively. Note the smaller annual change in the marine climate curve (Florida) and also its maximum temperature in August compared with July (Nebraska) for most continental regions.

the climate considerably, as in the case of the mellow marine climate of western United States. These climates contrast strongly with that of the continental interior, which undergoes extreme ranges of heat and cold. The annual temperature curves in Fig. 15-7 show clearly the striking difference between two localities that experience a marine and a continental climate respectively.

In describing the temperature of a locality, a number of different values are required, and necessary observational data are available in a series of compilations of the National Weather Service (formerly the U. S. Weather Bureau). For some needs, the hourly variation (or "march") of temperature for particular days is important. For other purposes, only the minimum, maximum, and mean temperatures for a day or a series of days may be required. For general climatic purposes, the daily minimum, maximum, and mean temperatures averaged for an entire month are usually more significant. The change in monthly temperature over a year is called the "annual march of temperature," and the resulting curve is usually an important climatic characteristic of a locality.

Examples of these temperature characteristics are shown in Fig. 15-8, where a shows the daily march of temperature for clear and cloudy days in New York City. The effect of clouds in reducing the maximum temperature by shielding off sunlight is clear. However, the clouds retain surface radiation and keep the air warmer. Figure 15-8b is an example of the march of temperature for the month of January in New York City. Of particular interest here is the pronounced warming spell from January 17 to 22. This anomalous increase in temperature is a characteristic change about January 20 in the northeastern United States, where it is known as the January thaw and is an effect that has come to be known and dreaded by the growing army of experienced skiers in New England. Such characteristic anomalies of temperature that are repeatable are known as climatic *singularities*. Other singularities in which temperatures rise or fall are also well known to climatologists.

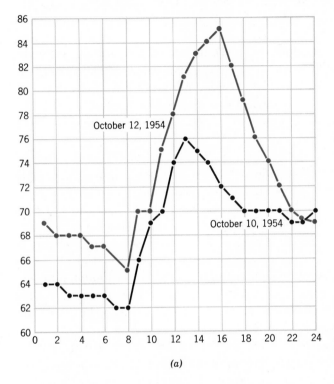

(a)

FIG. 15-8*a*. Effect of clouds in reducing air temperatures by shielding solar radiation. October 12, 1954 was clear and October 10 was cloudy.

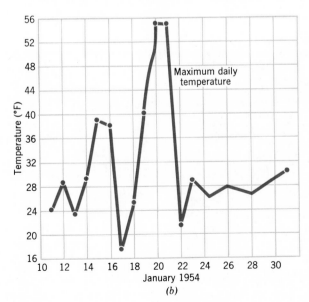

(b)

FIG. 15-8*b*. An example of the "January thaw" or temperature singularity in New York City. The strong warming trend between January 18–22 is a fairly characteristic trend about January 20 in northeastern United States.

The Elements of Temperature and Precipitation 543

We have already examined two cases of the annual march of temperature in Fig. 15-7 where continental and marine climates curves are compared. In addition to temperature curves for specific localities, it is often important to examine temperatures for regions of varying sizes. Fig. 15-9, for example, contains the annual march of temperature for the entire Northern Hemisphere. Temperature data of this kind are important in problems of climate change, as discussed later in this chapter, as well as in studying possible effects of air pollution, as will be examined in the following chapter.

It is always of great interest to see details of temperature extremes. The locations of places in which extremes were recorded for the different continents is shown in Fig. 15-10. Table 15-2 gives the actual data that apply to the numbered locations in this map. Index numbers 1–17 apply to temperature and 18–32 apply to extremes of precipitation as documented in Table 15-3.

Temperature deviations. One of the most useful climatological descriptions of temperature is the deviation of monthly and annual values from the mean value. For example, if the average temperature for a given place for the month of July is 75°F as determined throughout a long interval of time and the mean temperature for a particular July is 78°, the deviation is +3°. The range of such deviations for each locality is of considerable significance in assessing its climate. The variability of temperature about the established mean is much greater in high latitudes than in low ones and is much greater in winter than in summer. This effect depends on the greater weather variability in higher latitudes and in winter as discussed in Chapter 13.

Temperature and cultivation. To the farmer, the horticulturist, or the home gardener, the "safe" time for planting and the time for removing perishable plants is of great concern. Probably the most important tem-

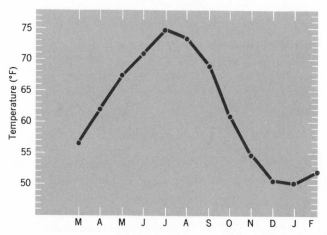

FIG. 15-9. Annual "march of mean temperature" for the entire Northern Hemisphere.

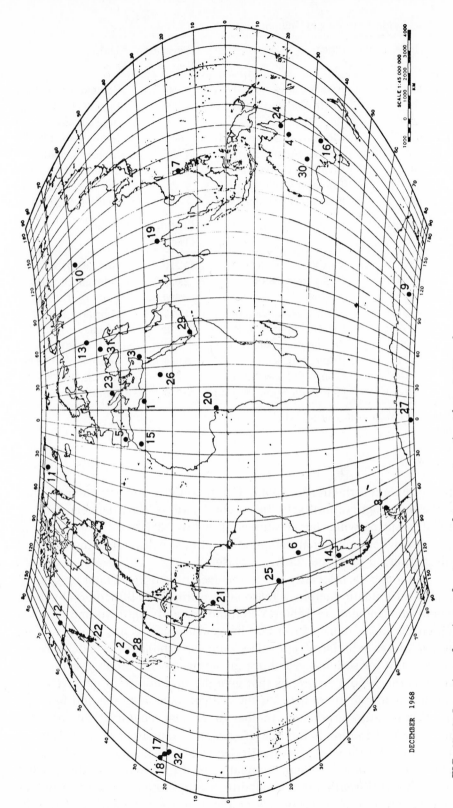

DECEMBER 1968

FIG. 15-10. Location of regions of extremes of temperature (numbers 1–17 and Table 2) and extremes of precipitation (numbers 18-32 and Table 3).

TABLE 15-2   Worldwide Extremes of Temperature

Key Number	Highest °F	Place	Elevation, Feet	Date
1	136	Azizia, Libya	380	September 13, 1922
2	134	Death Valley, Calif.	−178	July 10, 1913
3	129	Tirat Tsvi, Israel	−722	June 21, 1942
4	128	Cloncurry, Queensland	622	January 16, 1889
5	122	Seville, Spain	26	August 4, 1881
6	120	Rivadavia, Argentina	676	December 11, 1905
7	108	Tuguegarao, Philippines	72	April 29, 1912
8	58	Esperanza, Palmer Pen.	26	October 20, 1956

Key Number	Lowest °F	Place	Elevation, Feet	Date
9	−127	Vostok	11,220	August 24, 1960
10	−90	Oymykon, USSR	2,625	February 6, 1933
11	−87	Northice	7,690	January 9, 1954
12	−81	Snag, Yukon, Canada	1,925	February 3, 1947
13	−67	Ust'Shchugor, USSR	279	January [a]
14	−27	Sarmiento, Argentina	879	June 1, 1907
15	−11	Ifrane, Morocco	5,364	February 11, 1935
16	−8	Charlotte Pass, N.S.W.	[c]	July 22, 1947[b]
17	14	Haleakala Summit, Maui	9,750	January 2, 1961

[a] Exact date unknown; lowest in 15-year period.
[b] And earlier date.
[c] Elevation unknown.

perature limit is the freezing point. Certainly the most useful temperature criterion in agricultural activities is the length of time between the last night in spring when temperatures fall below 32°F and the first time in autumn when nocturnal temperature drops below freezing. This is often referred to as the "growing season."

Although each locality has a frost-free period dependent on general and local climatic factors, a reasonable estimate of this interval can be made from the data in Table 15-4 on the basis of continent and latitude. Often the mean dates of last and first occurrence of nocturnal freezing temperatures will be of adequate value to establish the safe growing season. But for serious agriculture involving specific crops, it is possible to construct or obtain graphs that give the probability of occurrence of particular low temperatures for the critical days in spring and fall.

Temperature gradients. Another important characteristic of temperature distribution is the rate of change of temperature with distance in a particular direction—a quantity described in Chapter 12 as the *horizontal temperature gradient.* A climatologically significant gradient is that along the north-south (meridional) direction in United States. Of particular interest is the difference in this gradient between January and July as shown in Fig. 15-11, where isotherms (in °C) indicate the positions of lines of equal temperature. Suppose we follow the temperature change from the northern Gulf of Mexico to the Canadian border along the nearly linear state boundaries starting with that between Texas and Louisiana. The temperature gradient along this line is from 18°C to −14°C (about 64°F to 17°F). This gradient of 47°F across the United States in a north-south direction in January contrasts strongly with a summer gradient over the same distance in July of only 14°F. The closer spacing of the isotherms on the January chart immediately indicates the presence of a much stronger temperature gradient. Note that in these charts, the isotherms are for temperatures reduced to sea level so that the natural variation of temperature with elevation is removed from the presentation.

Another way to look at this temperature effect compares the changes from January to July at the southern and northern borders of the country. The range in the south is from 28°C to 20°C or a range of 14°F. In the north the seasonal range is from −14°C to +20°C or a range of 63°F. Thus there is a major variation in temperature in the northern part of the country from winter to summer. This in turn is a direct effect of the large seasonal variation in insolation in high latitudes versus the small variation in low latitudes.

TABLE 15-3 Worldwide Extremes of Precipation

Key Number	Greatest Amount, Inches	Place	Elevation, Feet	Years of Record
18	460.0	Mt. Waialeale, Kauai, Hawaii	5075	32
19	450.0	Cherrapunji, India	4309	74
20	404.6	Debundscha, Cameroon	30	32
21	353.9	Quibdo, Colombia	240	10–16
22	262.1	Henderson Lake, B. C., Canada	12	14
23	182.8	Crkvica, Yugoslavia	3337	22
24	179.3	Tully, Queensland	b	31

Key Number	Least Amount, Inches	Place	Elevation, Feet	Years of Record
25	0.03	Arica, Chile	95	59
26	<0.1	Wadi Halfa, Sudan	410	39
27	*0.8	South Pole Station	9186	10
28	1.2	Batagues, Mexico	16	14
29	1.8	Aden, Arabia	22	50
30	4.05	Mulka, South Australia	b	34
31	6.4	Astrakhan, USSR	45	25
32	8.93	Puako, Hawaii	5	13

*The value given is the average amount of solid snow accumulating in one year as indicated by snow markers. The liquid content of the snow is undetermined.

bElevation unknown.

TABLE 15-4 Mean Number of Days Between First
Freezing Temperature in Autumn and Last Freezing
Temperature in Spring (Frost Period)

Latitude	Northern Hemisphere Longitude W.—America				
	50°	80°	110°	140°	
70°	240	277	267	270	
60°	175	241	195	100	
50°	140	171	142	—	
40°	—	20	55	—	
Latitude	Asia Longitude E.				
	170°	160°	130°	100°	70°
70°	290	260	254	265	283
60°	200	209	213	210	203
50°	20	120	177	167	161
40°	—	—	64	95	68
Latitude	Europe Longitude E.				
	40°	10°			
70°	232	150			
60°	168	130			
50°	130	43			
40°	—	—			

Source. (From Landsberg).

Precipitation

Precipitation is barely secondary to temperature as a climatic element. The wetness or dryness of a region is most important in controlling whether it is arable and inhabitable. Where rivers, whose sources lie in more humid areas, cross dry lands, some water is available for agriculture and populations. Where no such water is available, only a bare nomadic existence can be obtained from the parched land.

Precipitation is subject to much greater and more irregular variations within both short and long periods of time than is temperature. In New England, for example, a severe drought developed in the middle years of the decade 1960–1970. Vegetation suffered considerably during this time, when precipitation in many areas fell to about half of its normal value.

The geographic variation of precipitation is also very strong. Unlike time variations of precipitation, the geographic variation is quite constant and can be shown in a distribution chart like that in Fig. 15-6, where the

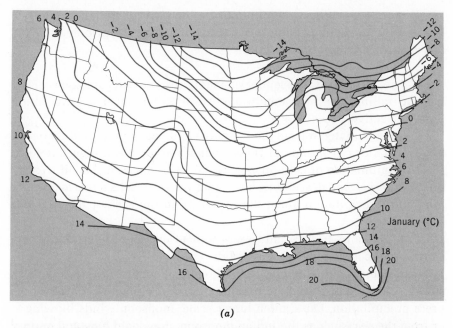

(a)

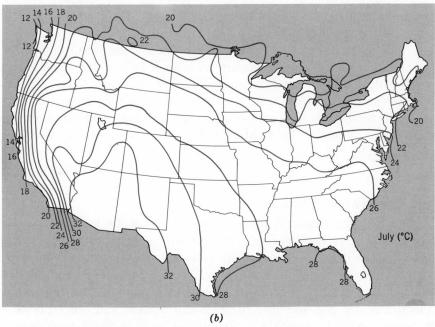

(b)

Fig. 15-11. Temperature gradients across United States from south to north in January (a) and July (b). (After H. E. Landsberg.)

mean global precipitation is shown. Details of the extremes are indicated in Fig. 15-10 and Table 15-3. Some regions experience periodic inundations and in others it practically never rains; precipitation is nearly unknown in places like the Atacama Desert (Index No. 25, Fig. 15-10, Table 15-3).

Factors controlling precipitation. On the basis of the information in this and the preceding chapters, we can summarize the principal factors that control precipitation as follows.

1. The permanent and semipermanent pressure belts are zones of rising and falling air; these conditions cause either moist or arid conditions respectively (see p. 499).

2. Winds blowing in from the sea may yield heavy precipitation on the windward sides of mountain ranges when the air has a high humidity.

3. A combination of cold offshore ocean currents, as exists off western South America, and descending air in high-pressure belts tends to produce minimum precipitation along the coasts, as in the case of the Atacama Desert.

4. The paths of tropical and extratropical cyclones determine where storm-produced rain or snow will fall.

5. Local and regional wind systems greatly affect precipitation. Locally, strong sea breezes encourage precipitation and strong land breezes depress precipitation. On a much larger scale, monsoon winds blowing to or from the sea, such as the Indian monsoon, may yield long and oppressive rainy seasons during the months of onshore winds and the reverse where offshore winds prevail.

6. Inland lakes and large water bodies often cause locally increased precipitation on regions downwind from such bodies. For example, in north-central United States the prevailing northwest winter winds pick up considerable moisture from the Great Lakes, causing frequent heavy snowfalls in areas to the south of Lake Erie and Lake Ontario as the air ascends the slopes of the highlands of central and northern New York and Pennsylvania.

Climate and the Water Balance

Despite the importance of precipitation as a climatic element, its real importance in agriculture lies in the availability of moisture in the soil. Soil moisture, in turn, depends on the ratio of precipitation to evaporation, often expressed as the P/E index or the P/E ratio. For example, the summer months in New England are usually considered the dry season. The soil is dry and watering of grass and flowers and other plants is normally required, compared to the apparently more moist conditions in the spring or autumn. Yet the total amount of rainfall in the warm summer months, due to heavy shower activity, equals or exceeds cold-weather precipitation, when the ground is certainly more moist.

The cause of summer dryness lies not in the lack of rainfall but rather in the high temperature, which causes high evaporation of newly fallen rain as well as soil moisture that may be residual from the preceding cooler season. Water loss by vegetation, as described below, is also important.

To assess the availability of plant and crop moisture, a knowledge of the relationship of precipitation to evaporation is thus required. Formulas for computing the P/E ratio that have been developed from observational data require only a knowledge of the amount of precipitation and the temperature, the factor controlling evaporation; an example of such a formula is one developed by C. W. Thornthwaite:

$$\frac{P}{E} = 11.5\left(\frac{P}{t - 10}\right)^{10/9}$$

where P is the monthly precipitation in inches, E is the evaporation in inches, and t is the average monthly temperature in °F.

Even more reliable estimates of the water balance can be obtained by estimating the possible water loss by direct evaporation from the soil and transpiration from plants themselves. This combined effect is known as *evapotranspiration*, estimates of which are available for the entire country.

SCHEMES OF CLIMATE CLASSIFICATION

In view of the great importance of climate in nearly all aspects of human endeavor and the great variability of the details of climate, many attempts have been made to develop schemes that would readily classify world climate types. From such a classification the expected climate of a particular locality or region could then be estimated. Unfortunately, no single scheme has ever met all of the requirements, but during this century a continual evolution of climate classifications has occurred. Some important and useful classification schemes can be examined here.

Simple Temperature Classifications

The most primitive and generalized of all classification schemes relies solely on the temperature distribution resulting from the global distribution of insolation considered on p. 537. Three broad climatic zones are recognized as the tropical, temperate, and polar zones. The low-latitude tropical zone is characterized by an absence of a cold or winter season owing to the prevailing high angle of sunlight striking the earth. The polar regions of both hemispheres are characterized by a lack of a real warm or summer season. In the middle latitudes, between the two regions of temperature extremes, a combination of both occurs. During the warm or summer season, southerly winds transport tropical air masses into the midlatitudes, bringing typical tropical temperature and humidity. In the winter season, winds from high latitudes bring frigid polar conditions into the temperate zone. Only the mean temperatures are really temperate in this zone.

When all of the factors of continental and marine effects, storm

TABLE 15-5 Köppen Climate Scheme

Climate Groups	Dry Periods	Nature of the Dryness or Cold
	Subdivisions	
A. Tropical rain climate	f, w	
B. Dry climates		S W
C. Warm temperate rainy climates	f, s, w	
D. Cold snow forest climates	f, w	
E. Polar climates		T F

patterns, precipitation and evaporation, mountain effects, and so on, are considered, it becomes clear that these simple broad zones cannot begin to describe the climates that really prevail in them.

Köppen Classification

One of the most widely used climate schemes, often with some modification, is the classification of Köppen of Austria. Although he originally proposed his scheme in 1918, Köppen modified his system several times; the last alteration was in 1936. The Köppen system, which is primarily concerned with the effect of climate on vegetation, is based on the monthly and annual temperature means and the annual precipitation together with its seasonal variation. Köppen climate types are designated by either two- or three-letter symbols, which describe 11 main climate types.

As shown in Table 15-5, five main climate groups, designated by capital letters, are recognized. These five are then modified by the small letters, f, s, and w, which indicate when a dry period, if any, occurs: f shows no dry period; s shows a dry period in summer; and w shows one in winter. In the third column of Table 15-5, the capital letters S, W, and T refer to types of dry climate as steppe, desert, and tundra, respectively, and F indicates perpetual frost.

Climate groups A, C, D, and E are distinguished by temperature and B by precipitation criteria as follows:

A: Coldest month; has a mean higher than 64.4°F (18°C).

C: Coldest month; has a mean between 64.4°F (18°C) and 26.6°F (−3°C).

D: Coldest month's mean is below 26.6°F (−3°C) and the Warmest month is above 50°F (10°C).

E: Warmest month is below 50°F (10°C).

B: Evaporation exceeds precipitation, which is most effective in winter.

These groups, with their modifying climate subdivisions, form the 11 climate types in Table 15-6. The third symbol in the Köppen classification, which may be applied to these 11 types, depends on details of monthly or annual temperature as explained in Table 15-7. In the world map of Fig. 15-12, the distribution of the different climate types is shown by means of these three-letter symbols.

TABLE 15-6 Köppen Climate Types

Climate	Symbol	Type of Climate
(1)	Af	Tropical rainforest
(2)	Aw	Tropical savanna
(3)	BW	Desert
(4)	BS	Steppe
(5)	Cf	Moderate oceanic (west coast) climate
(6)	Cw	Moderate, winter dry, climate
(7)	Cs	Moderate, summer dry, climate
(8)	Df	Continental climate, even precipitation through the year
(9)	Dw	Continental climate, winter dry
(10)	ET	Tundra
(11)	EF	Glacial ice cap

TABLE 15-7 Explanation of the Third (Temperature) Symbol in the Köppen Scheme

Symbol	Condition
a	Mean temperature of warmest month $>22°C$ ($71.6°F$)
b	Mean temperature of warmest month $<22°C$ ($71.6°F$) but at least 4 months with mean temperatures $>10°C$ ($50°F$)
c	Mean temperature of less than 4 months above $10°C$ ($50°F$) but coldest month $>-38°C$ ($-36.4°F$)
d	Coldest month $<-38°C$ ($-36.4°F$)
h	Annual temperature $>18°C$ ($64.4°F$)
k	Annual temperature $<18°C$ ($64.4°F$)

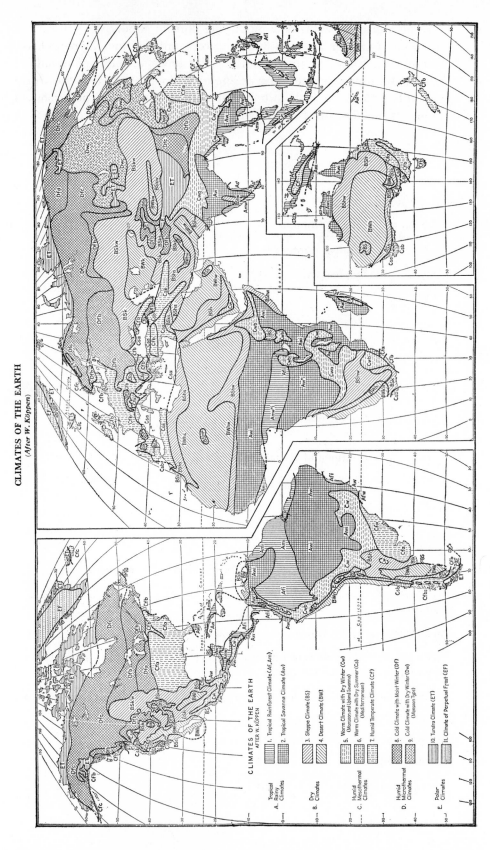

CLIMATES OF THE EARTH
(After W. Köppen)

CLIMATES OF THE EARTH
AFTER W. KÖPPEN

A. Tropical Rainy Climates	1. Tropical Rainforest Climate (Af, Am)
	2. Tropical Savanna Climate (Aw)
B. Dry Climates	3. Steppe Climate (BS)
	4. Desert Climate (BW)
C. Humid Mesothermal Climates	5. Warm Climate with Dry Winter (Cw) (Monsoon and Upland Savanna)
	6. Warm Climate with Dry Summer (Ca) (Mediterranean)
	7. Humid Temperate Climate (Cf)
D. Humid Microthermal Climates	8. Cold Climate with Moist Winter (Df)
	9. Cold Climate with Dry Winter (Dw) (Monsoon Type)
E. Polar Climates	10. Tundra Climate (ET)
	11. Climate of Perpetual Frost (EF)

Climatic atlas of the world. (After W. Köppen.)

Thornthwaite's Humidity Scheme

Thornthwaite developed a climate scheme from the relationship of precipitation zones to types of vegetation. In view of the difficulty in dealing with evaporation in the P/E ratio described on p. 550, Thornthwaite developed a precipitation effectiveness index (P-E index) in which evaporation is replaced by temperature, on which it so closely depends. The P-E index is determined for each month according to the formula

$$P\text{-}E = 115\left(\frac{P}{t - 10}\right)^{10/9}$$

where P is the precipitation and t the temperature. The monthly values are added to obtain the precipitation-effectiveness index and are then used to establish the five main climate types described in Table 15-8.

To the five main groups Thornthwaite then added modifiers, as in Köppen's scheme, that refer to seasonal variation of rainfall, the temperature province, and the thermal-efficiency province. Although these factors can be varied to give 120 different climate types, only 32 types are recognized by Thornthwaite. Even these are nearly three times more than the main Köppen types. Because the Köppen system is simpler and more direct, it has gained wider use.

Thornthwaite subsequently developed two other humidity classification schemes based on the factors of evapotranspiration and a moisture index, respectively. Although he developed a global chart describing climate in terms of the precipitation-effectiveness index, the same was not done for these last two schemes.

A number of other climate systems have been developed with the use of much the same criteria involved in the Köppen and Thornthwaite schemes. All of the systems described so far are based on statistics of temperature and precipitation as related to geography or vegetation. No explanation of the origin of the climate types is involved. So-called dynamic or meteorological schemes have been developed to both describe and explain the climate type in a particular region.

Dynamic Climate Schemes

H. Landsberg developed a genetic climate system in an attempt to give a causal meteorological basis for classifying climates. He employed the two primary meteorologic criteria of primary and secondary circulation features plus the geographic factor of the nature of the surface. Table 15-9, showing Landsberg's system, incorporates the basic meteorological and climatological controls that we have already examined in great detail in preceding chapters. The application of the Landsberg system to particular areas is shown in Table 15-10.

A somewhat analogous climate system is based on air masses and their source regions. In this system climate regions are developed from three basic groups. Group I includes climate types influenced primarily

TABLE 15-8 Thornthwaite's Climate Types

P-E Index	Province Name	Vegetation
(A) $\geqq 128$	Wet	Rainforest
(B) 64–127	Humid	Forest
(C) 32–63	Subhumid	Grassland
(D) 16–31	Semiarid	Steppe
(E) <16	Arid	Desert

TABLE 15-9 Landsberg's Dynamic Climate Classification

Classification	Symbol
Major circulation patterns (primary controls):	
Migrating cyclones	C
Quasi-stationary anticyclones	A
Equatorial convergence	E
Mixtures of the preceding	(CA), (AE)
Secondary or seasonal circulation features:	
Typical monsoons	S
Predominant trade winds	T
Major surface influences:	
Continental	c
Oceanic	o
Mountain	m
Subgroup windward slope	mw
Subgroup lee slope	ml
Glaciated	g
To designate extreme conditions	Subscript e
To indicate mixtures: two type symbols are enclosed in parentheses	

by equatorial and tropical air masses. Group III contains climates of the high latitude polar antarctic and arctic air masses, whereas Group II includes the more variable midlatitude climates influenced by both polar and tropical air masses and is the region of frequent and strong cyclonic storms. The recognition of the climate types still depends on vegetation criteria, as is evident in Fig. 15-13, which shows 13 major types in this system.

Although we have examined only briefly some of the well-known climate systems, it is nevertheless clear that no one system seems to meet all needs and cover all aspects of climate. The system used by a geographer or agriculturist will depend on the particular requirements of the problem.

TABLE 15-10 Examples of Climatic Types in a Dynamic Scheme

Specimen Area	Type Symbol
Northeastern United States	C_c
North Dakota	Cc_e
England	Co
Orkney Islands	Co_e
Cascades	Cm
Poland	C(co)
Coast of S. California	(CA)o
C. Spain	(CA) (co)
N. India	(CA) S
Greenland	(CA) g
Sahara	Ac
Azores	Ao
Great Dividing Range (SE Australia)	Am
Oahu	ATo
Central Amazon Valley	Ec
Gilbert Islands	Eo
Mt. Kenya Area	Em
Guam	(ET)o

The Climagraph

The climagraph is a very useful graphical tool to study and compare climates of a number of places. It is a simple graph of monthly values of temperature (plotted on the abscissa) and precipitation (on the ordinate). Plotted points are connected by lines and are numbered consecutively from 1 to 12, corresponding to the months of the year. The shape and spread of the resulting closed system of points is quite indicative of certain climate types—in particular, continental marine and monsoon conditions. We can deduce that a continental climate, with its large temperature range relative to precipitation, should have a climagraph showing a broad flat curve parallel to the temperature axis (abscissa). Marine climates, with a small range in temperature, should show a much smaller temperature spread, whereas the heavy seasonal precipitation in monsoon regions should show a climagraph with a broad spread parallel to the precipitation axis (ordinate). Examples of climagraphs for continental and marine climates are shown in Fig. 15-14.

Degree Days

Another very convenient tool of the climotologist is the degree day, which has found particular application in heat engineering. The degree day applies only to days whose mean temperature is below 65°F, and is the

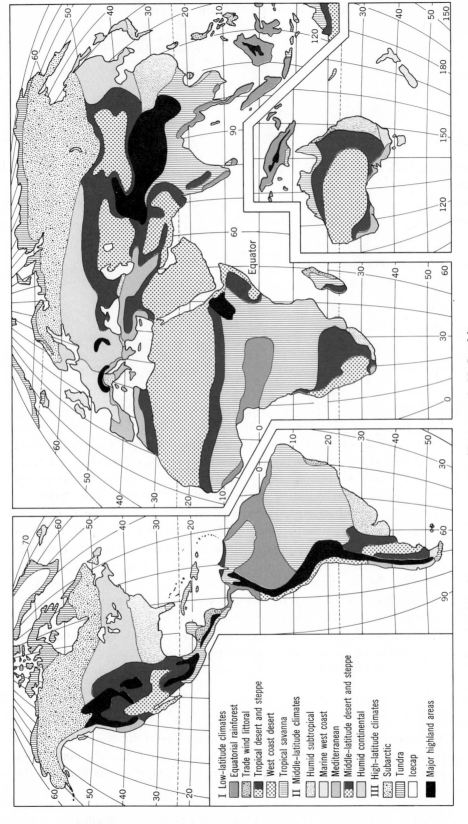

I Low-latitude climates
    Equatorial rainforest
    Trade wind littoral
    Tropical desert and steppe
    West coast desert
    Tropical savanna
II Middle-latitude climates
    Humid subtropical
    Marine west coast
    Mediterranean
    Middle-latitude desert and steppe
    Humid continental
III High-latitude climates
    Subarctic
    Tundra
    Icecap
    Major highland areas

FIG. 15-13. Climatic atlas based on vegetation types. (Courtesy A. N. Strahler, *Introduction to Geography,* John Wiley & Sons, Inc.)

difference between 65° and the mean temperature of a particular day. Degree days are accumulated once the mean temperature first drops below the standard—a value related to the temperature at which artificial building heat is normally used. The number of degree days that have accumulated from any reference day indicates to fuel engineers how much heating, and hence how much fuel, has been used by a customer. Automatic fuel servicing is accomplished by keeping a check on the elapsed degree days.

ICE AGES AND PAST CLIMATES

Of all of the changes in climate known to have occurred during the earth's long history, the ice ages were certainly the most dramatic geologically, and to this day details of their origin remain one of the great unsolved problems of natural science. During several long intervals of time in the

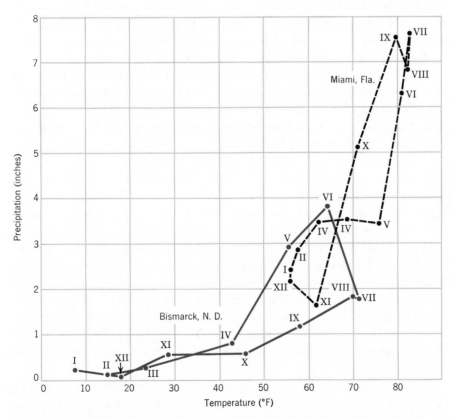

FIG. 15-14. The climagraph in which temperature is plotted against precipitation is a tool for distinguishing continental from marine climates. Continental climates show a broad temperature range and marine climates a broad precipitation range.

geologic past, vast sheets of glacial ice spread across large areas of the continents. At least two great ice ages occurred prior to the beginning of the Cambrian period and two since the beginning of the Cambrian, one in the Paleozoic and one in the Cenozoic Eras.

During the Paleozoic Ice Age, all of the continents of the Southern Hemisphere were affected. Evidence of strong glaciation is now well established for nearly all of Africa from the Sahara Desert to the Cape of Good Hope. The evidence consists of thick deposits of till and broad expanses of glacially grooved bed rock surfaces (see Fig. 6-39). The development of these features was described in the section on glaciation in Chapter 6. Except for parts of India, no evidence of Paleozoic glaciation has been found in the Northern Hemisphere. The significance of this glacial distribution was discussed in Chapter 8 as one of the important evidences of continental drift.

To understand the nature of glacial climates and to explain their origin, it is preferable to concentrate on the most recent ice age. The record of climate is preserved in rocks; hence the details of Precambrian Ice Ages are quite obscured and effaced by many subsequent cycles of erosion. Details of the Paleozoic glaciation have also suffered the geologic ravages of time. Also, the regions affected have been largely uninhabited, and only in the past couple of decades have recent explorations developed a workable body of knowledge of this Southern Hemisphere ice age.

By far, the most complete assemblage of information bearing on ice ages relates to the most recent one. On land, till and erosional features (Chapter 6) are still fairly fresh and widespread. In the sea, the sediments lying beneath the sea contain the most complete and explicit record of Cenozoic climate that can be found.

Climatic Change in the Cenozoic Era

During the early part of the Cenozoic Era, the warm, equable climate inherited from the Mesozoic Era began to change and worsen. According to the fossil record, the Mesozoic was a time of widespread warm climatic conditions. Dinosaurs and other great reptiles abounded over much of the continents and indicate the semitropical nature of the continents at that time. Remains of fossil breadfruit trees of Mesozoic age have been found in the rocks of frigid coastal Greenland just beyond the present ice margin. Such flora occur today only on the islands of the equatorial Pacific Ocean. Imagine the vast change of climate that must have occurred here.

Warm climate persisted into the beginning of the Tertiary period of the following Cenozoic Era. The geologic evidence at hand indicates that the present strong temperature zonation including tropical, temperate, and polar climates was not present in Mesozoic and early Cenozoic times. The trend toward such zonation appeared to begin following early Cenozoic time. This is marked by a disappearance of warm types of plants and animals first in high latitudes and then in lower latitudes.

Beginning of Glaciation and of the Present Climate Zones

Exploration of high-latitude lands has established that glaciation in these regions began at least 10 million years ago, or in the Miocene epoch. In Antarctica typical glacial forms have been blanketed by lava flows whose ages have been determined by potassium-argon dating (see p. 249) to be 10 million years. Hence glaciation must have begun prior to this date. Similar types of evidence indicate a similar age for the beginning of glaciation in Alaska. We thus have strong indications that the frigid polar climate belt was initiated in mid-Cenozoic time.

There is no evidence that tropical temperatures changed significantly from nonglacial to glacial times. The big change seems to have been a strong cooling of the polar latitudes. The midlatitude temperate zone is somewhat of a buffer region between the low- and high-latitude extremes. We can appreciate the change from the past warm globe to the present by considering the nature of the temperature gradient from the equator to the poles. As we travel poleward from the equator, the temperature drops from a mean of just above 80°F to a mean that is below 0°F. This temperature gradient from low to high latitudes contrasts strongly with that in preglacial times, when lands now occupying polar locations were well above freezing.

Pleistocene Glacial-Interglacial Stages

Despite the refrigeration of arctic and antarctic lands, there is no evidence that continental ice sheets extended into middle latitudes until about 750,000 years ago. About this time, a major part of North America and much of Eurasia was invaded by ice sheets as thick as or thicker than the 8000-foot-thick glacial cover on Antarctica today. Prior to the rather recent discovery that the last ice age began at least 10 million years ago, it was thought that all of modern glaciations were confined within the last million years—or to the Pleistocene Epoch, so that the term Pleistocene Ice Age has been applied to the last epoch of glacial climate.

Geologists have known for some time that the great glacial ice sheets advanced and then retreated many times, forming time intervals known as glacial and interglacial stages, respectively, within the Pleistocene epoch. Early studies of North American geology indicated that four major glacial stages occurred, with intervening interglacial stages. The evidence consists of overlapping layers of fresh and weathered glacial till near the margins of former ice sheets. Because later glaciations usually scour and obliterate deposits of former ice sheets, the continental record of earlier glaciations is usually found preserved where tongues of ice extended beyond the local margin of later glaciers. When later tills overlap earlier ones, a clear distinction exists. During the thousands of years of an interglacial stage, the till deposited by the last ice sheet becomes deeply weathered. Subsequent overlapping till deposits thus form a system of tills each representing a glacial stage; the weathered till beneath each fresh

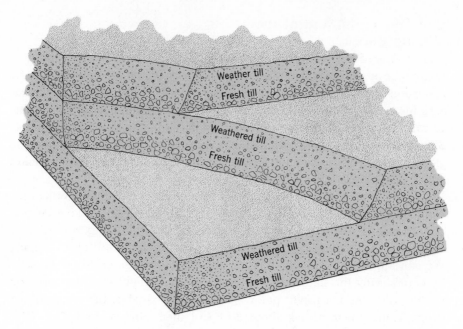

FIG. 15-15. Sketch of successive layers of glacial till showing the
weathered zone in the upper part of the till layer that indicates
interglacial weathering.

deposit corresponds to the interglacial stage—the greater the depth of
weathering, the longer was the interglacial period (see Fig. 15-15).

Because this procedure depends on finding the rather fortuitously
uneroded glacial deposits, there are many uncertainties in establishing
a glacial chronology from a single continent. In Europe, for example, the
results of field work indicate that six glaciations occurred.

By far the most complete record of late Cenozoic climate is found
in the cores of sediments raised from the floor of the deep sea. In much
of the ocean, a rain of sediments has been going on for millions of years
with little disturbance. Different varieties of planktonic life become included in this sedimentation as organisms die. The chemical and biologic
content of the sediments then return to us much information about the
temperature of the surface layer of the seas. Information from a large
number of deep-sea cores indicate that at least eight major changes in
climate occurred in the last 750,000 years.

Some examples of how the deep-sea results were obtained for the
past half-million years is shown in Fig. 15-16, where similar data from
three different criteria by three different groups of investigators were
obtained from three different cores. Data for a core from the Caribbean
Sea (left-hand column) shows the variation in abundance of a warm-water
species of protozoa (one-celled animals) with time as measured in the
particular core. Intervals of very low abundance are interpreted as glacial
and those of high abundance as interglacial. The second core shows the
variation of two isotopes of oxygen (O^{18}/O^{16}) in the calcium-carbonate
shells at depths corresponding to different times in another core from the

Caribbean Sea. Common oxygen is O^{16} but about 1 part in 500 is the heavier isotope, O^{18}. The amount of O^{18} taken up for the carbonate of a shell varies with temperature (decreases as temperature increases) and salinity of the sea, both of which vary with climate. Data for a third core from the tropical Pacific Ocean show the variation of shell calcium carbonate in the core with time.

The diverse data for these cores from different localities are remarkably similar and, for the past 500,000 years, show five of the eight cold intervals referred to above. Many other criteria, mostly biologic, are also used to establish climatic trends that can be interpreted from deep-sea cores.

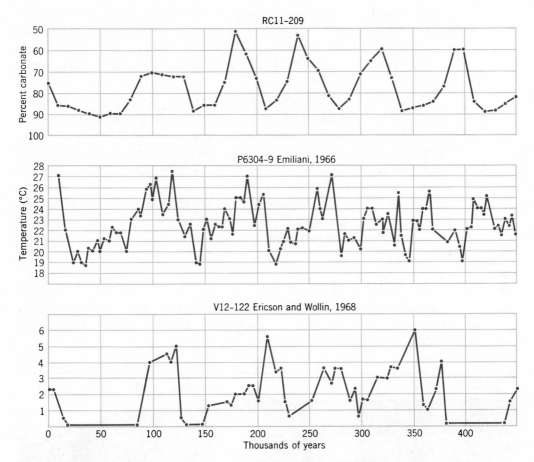

FIG. 15-16. Comparison of temperature curves from three Caribbean cores which indicate the glacial interglacial chronology deduced from the curves. Time is given on the horizontal scale. In the lower core amount of warm and cold water species is the criterion; in the middle, temperature deduced from O^{18}/O^{16} is used; in the upper, percent of carbonate (in microfauna shells) is the criterion. High carbonate concentrations correlate with glacial times. Peaks are interglacial times, troughs, glacial. (After C. Emiliani.)

Chronology of Pleistocene Climate

When deep-sea cores are analyzed, the basic data against which observations are plotted are depths below the core surface. Depths are then translated into time by several means. Ages of carbonate shells can be determined reliably to about 35,000 years by determinations of the $C^{14} - C^{12}$ ratio (see Chapter 7). Other radioactive elements have been used with some success back to several hundred thousand years ago. A common method of time determination is to obtain two or more precise dates in the core by either radiochronology or by correlating certain clear events in one core with the same event elsewhere that has already been dated accurately. Then the rate of sedimentation in the particular core can be worked out and the thickness of sediment can be interpreted as equivalent years.

To most of us, the latter part of the chronology indicated in Fig. 15-16 has the most immediate significance. The evidence accumulated from core analyses, variations of sea level, and continental glaciations on land all indicate that retreat of the last (Wisconsin) glacier began about 18,000 years ago and that all of the Wisconsin-stage ice disappeared about 6000 years ago.

Note well that according to the climatic records in Fig. 15-16, the globe has only recently emerged from a major glacial stage. *There is no reason to assume that we live today in any but another—the most recent—interglacial stage.* Glacial ice will surely advance again. In fact, if you lived on Greenland or Antarctica where a mile to a mile and a half of ice still blankets lands whose total area is much greater than that of the contiguous United States, you would hardly think the Pleistocene Ice Age has ended. Evidence available indicates that the Antarctic and Greenland ice sheets have persisted throughout interglacial as well as glacial stages.

Glaciation and Sea Level

The snow that builds glaciers was evaporated from the oceans initially. Contrary to the present hydrologic cycle (p. 421), this snow did not return

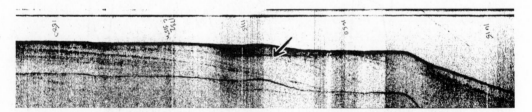

FIG. 15-17. SDR record taken near the edge of the continental shelf off New York City. The submerged terrace at 480 feet below sea level (indicated by arrow) is explained as a wave erosional feature when sea level was much lower in the last (Wisconsin) glacial stage. The lower half of the record is a repeat "echo" of the upper half. (Record by J. Ewing, Lamont-Doherty Geological Observatory.)

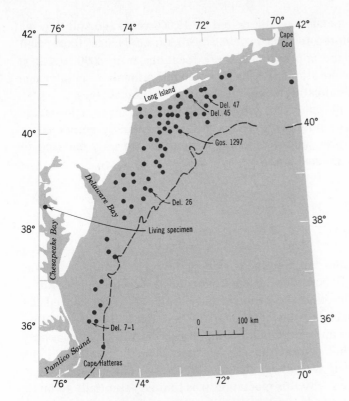

FIG. 15-18. Locations of shallow water forms found in present deeper waters of the continental shelf. (Broken line is present outer boundary of the shelf.) (After I. Marsh and K. Emery.)

to the sea but was trapped within glacial ice. Hence, as ice sheets advanced during glacial stages, the sea level fell. On the basis of calculations of the total increased volume of glacial ice during the Wisconsin stage, the author has estimated that the sea level was lowered from 350 to 450 feet at the maximum of this stage. Such a drop in the ocean level would have exposed large areas of the continental shelves of the world. The prominent surface revealed in the subbottom depth recorder (SDR) record of Fig. 15-17 taken at the outer edge of the shelf southeast of New York City suggests a surface of wave erosion when the sea surface might have stood at this level.

More direct proof of the exposure of the Atlantic continental shelf comes from fossil finds in regions now covered by deep water. For example, oyster shells of the type found only in very shallow-water marine, near-shore environment today, have been found in fairly deep continental shelf water off our east coast (Fig. 15-18)*. Shell ages determined by radiocarbon dating are 8000 to 11,000 years. Other evidence tells us that about

---

*Do not confuse these forms with those shallow water shells transported to the deep sea by turbidity currents as described in p. 143.

half of Wisconsin glacial ice had melted by 12,000 years ago and that sea level was half restored to present level by this time. Hence, these shell positions indicate the approximate shore locations from 8000 to 11,000 years ago. In other locations, fossils of true land animals of greater age have been found, indicating complete exposure of the shelves.

We may thus anticipate that future glaciations will again expose large areas of shallow sea bottoms. Perhaps those occasionally convicted of swindling through the sale of off-shore properties are only engaged in long-range real estate speculation.

CAUSES OF ICE AGES

The origin of ice ages has long been one of the most baffling problems of natural science. To consider the late Cenozoic Ice Age, the one we know best, we must notice that the problem is twofold. (1) What caused the change from nonglacial to glacial climate that resulted in the refrigeration of the high latitudes and the glaciation of Antarctica, Alaska and possibly Greenland? (2) What caused the glacial-interglacial fluctuation of the Pleistocene epoch? If we extend the problem to the Paleozoic Ice Age, we must also explain why this glaciation was limited essentially to lands mostly in the low latitudes of the Southern Hemisphere.

Theories Related To Solar Radiation

In view of the obvious climatic importance of solar radiation, a number of theories have been proposed that depend on variations in the radiation that reaches the earth's surface. According to the volcanic theory, an epoch of unusual volcanic eruptions could have added sufficient volcanic dust to the atmosphere to have reduced its transparency to sunlight, thus cooling the globe. However, there is no evidence for eruptions of the kind that could have produced the slow Cenozoic cooling and that could answer the three problems enumerated above.

Several purely astronomical theories of glacial origin have been proposed during the past century or two. According to one, the passage of the solar system through a cosmic dust cloud reduced the intensity of the sun's radiation at the earth's surface. But in 10 million years the sun has traveled about 45×10^{12} miles, and no suitable cloud is known within this distance. Such a cloud, if it existed, would have to contain broad and intermittently spaced density fluctuations to explain the glacial-interglacial alternations. Another astronomical theory proposes that variations in solar radiation are responsible for the glacial cycles, but this hypothesis does not account for the gross change from nonglacial to glacial conditions that began the Cenozoic Ice Age. Further, there is no present evidence that the sun is a variable star of the necessary kind.

The astronomical theory that has received the widest attention was developed by M. Milankovitch during the second quarter of this century. This Yugoslav scientist calculated the variations in solar radiation arriving at the earth's surface that would result from the combined effects of precession, changes in the obliquity of the ecliptic, and changes in the eccentricity of the earth's orbit (Chapters 1 and 2). However, he did not state quantitatively how large the variations in surface temperature would be. Fluctuations in radiation reaching the earth's surface as calculated by Milankovitch for the past million years are shown in Fig. 15-19. On this curve, troughs represent radiation minima and would cause glacial climate; peaks would cause interglacial warmth. To evaluate the actual temperature effects of these radiation minima, D. Shaw and I applied a mathematical procedure used to compute temperature from the factors involved in the earth's radiational heat balance. Resulting calculations show that an average decrease of only 1.4°C (2.5°F) would occur at 65°N from the effects of the Milankovitch radiation minima. Since temperature changes in historic times equaled or exceeded this amount, the effect would seem to be too small to be a primary cause of glaciation. Also, the Milankovitch effect has been going on continuously during the earth's history, but it can apply only to item (2), the glacial-interglacial alternations.

On the basis of information from deep-sea cores raised from the depths of the Atlantic Ocean and from knowledge about the heat balance of the Arctic Ocean obtained during the International Geophysical Year (1957–1958), Professor Maurice Ewing of Columbia University and I developed a theory of ice ages that attempts to explain (1) the initiation of Cenozoic cooling culminating in the Ice Age and (2) the cause of Pleistocene glacial-interglacial fluctuations. The remainder of this chapter will be devoted to the explanation of this theory.

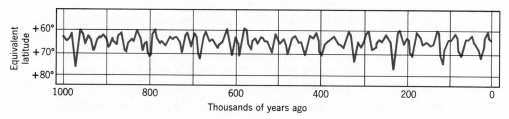

FIG. 15-19. Theoretical variations in solar radiation calculated by M. Milanko-vitch for the last one million years. He expressed the effect of radiation changes as equivalent changes in latitude shown as the present latitude to which 65°N of past time was equivalent. For example, the strong trough about 970,000 years ago indicates that 65°N at this time experienced radiation conditions at 76° today—thus this should have been a cold epoch. According to Milankovitch, troughs are indications of glacial and peaks of interglacial times.

Initiation of Pleistocene Glaciation Through Polar Wandering and Continental Drift

In Chapter 9 we examined the paleomagnetic evidence for polar wandering as well as for continental drift. According to the polar wandering curve in Fig. 8-9a, the present north polar region was occupied by the tropical Pacific Ocean in the early Paleozoic Era. Internal motions slowly displaced surface geography until the central Arctic Ocean reached the position of the North Pole and Antarctica was displaced to the South Pole.

Prior to the Miocene epoch, when open oceans were located at the poles of both hemispheres, extreme high-latitude cooling would not have been possible. We have noticed earlier in this chapter how the oceans are very conservative in their temperature changes. Further, with the poles located in the open oceans, convectional interchange of water between low and high latitudes would also prevent extreme chilling of the polar seas, as cold water would sink and flow equatorward while warm surface waters would flow poleward.

Once the north pole became located in the Arctic Ocean and the south pole in Antarctica, the polar regions were isolated thermally from the rest of the world ocean. The Arctic Ocean, apart from a narrow connection to the north Atlantic, is quite cut off from the rest of the world ocean; the south polar region, being located within a continent, is thoroughly isolated from the southern oceans. In view of the thermal isolation of both polar regions, cooling would have become more and more intense, until high-latitude glaciation occurred in mid-Cenozoic time, as is now well documented.

Since this aspect of the theory was first published in 1956, continental drift has become a well-established crustal process (Chapter 9). Continental drift reinforces the effect of polar wandering in accomplishing the isolation of the Arctic Ocean and the movement of Antarctica to the South Pole.

The combined mechanisms also explain quite readily the restriction of Paleozoic glaciation to lands of the Southern Hemisphere. The South Pole at this time would have been located within the Paleozoic supercontinent of the Southern Hemisphere (Gondwanaland) prior to its splitting and drifting apart. Glaciation of these lands, which were then at high latitudes, is thus just as explainable as the glaciation of Antarctica today. In fact, because most of the Southern Hemisphere continents lie in semitropical to tropical latitudes today, it is scarcely possible to imagine that these regions could have been glaciated in their present locations without much stronger glaciation occurring on the much-higher-latitude lands of the Northern Hemisphere.

Glacial-Interglacial Alternations and the Theory of the Ice-Free Arctic Ocean

In explaining the reason for the glacial-interglacial fluctuations of the Pleistocene epoch, we believe that one of the most important questions to be answered is: Why are the present Arctic lands of Canada and Eurasia, which were so thoroughly glaciated in the glacial stages, no longer glaciated, whereas Antarctica and Greenland still preserve their full glacial cover? How can we explain the selective glaciation and deglaciation of Canada and Eurasia while these other high-latitude regions retained thick ice covers? These lands, particularly northern Canada and Siberia, experience annual temperatures low enough to support glaciers. But the large difference seems to be in precipitation. The total annual precipitation of Arctic Canada and Siberia is only a few inches (see the rainfall map, Fig. 15-6). The prolonged Arctic summer sun would quickly melt a thin snow cover. The reason for this sparse snowfall lies in (1) the generally low moisture content of the cold Arctic air and (2) the lack of an ocean moisture source. Winds from the north sweep over the frozen Arctic Ocean and pick up little moisture. Winds from the west lose their moisture in crossing the mountain ranges of western North America in the case of Canada, and Western Europe in the case of Siberia and eastern Europe. Greenland and Antarctica apparently maintain their ice cover because they are located within relatively warm open oceans that supply them with moisture.

From a study of deep-sea cores we were led to the possibility that the Arctic Ocean might experience alternations between an ice-covered state as at present and an ice-free state. During an ice-free state a strong source of moisture would become available to *initiate* the development of glacial ice sheets in the present high-latitude deserts. Increased snowfall that would occur when the Arctic Ocean ice cover melted would presumably be adequate to persist through the weak summer warmth and would grow progressively thicker each year. Snow reflects from 70 to 80 percent of the sunlight falling onto it, compared to about 10–15 percent for soil and rock. A permanent snow cover would thus make its own new low-temperature climate affecting the surrounding land.

As the permanent snow-ice cover on North America expanded southward, it would begin to draw on the greater precipitation available from southerly sources, as indicated by the precipitation zones in Fig. 15-6. The permanent glacial cover would grow slowly southward each year, always cooling the lands marginal to the ice sheet. Finally, at some latitude, the low-latitude warmth would balance this cooling, and further expansion would cease. In North America this occurred at about the 40th parallel. In Asia, where a great desert occupies the center of the continent, glaciation would be restricted to the Arctic borderland, as did occur in Siberia (see Fig. 6-41).

At the time of maximum glacial extent, resulting cooling caused the surface of the Atlantic Ocean, which supplied the main moisture source for glacial growth in North America and Europe, to experience a strong temperature decrease. The evaporation rate would have been lowered, decreasing the precipitation necessary to maintain the ice sheets, and glacial shrinkage would have taken place. Calculations show that the intense glacial cooling would have caused the Arctic Ocean surface to refreeze by the time a large ice sheet became established. (In this theory the ice-free Arctic Ocean is only necessary to *initiate* glaciation.) Once the ice sheets had retreated to the latitude where Arctic Ocean moisture is necessary for their maintenance, their continued shrinkage would have accelerated.

In the Southern Hemisphere, glacial-interglacial stages have been shown to be essentially synchronous with those of the Northern Hemisphere. Apart from Antarctica, glaciation in South America, New Zealand, and Australia was mostly an extension of high-altitude alpine glaciers that exist today. During glacial times, or at least during the last (Wisconsin) stage, the Alpine glaciers expanded and coalesced along the low-level mountain "aprons." True ice sheets were present only on Antarctica.

Can the Arctic Ocean Remain Ice-Free?

Two fundamental questions regarding the open-Arctic theory of glacial-interglacial stages are (1) was the Arctic Ocean ever ice-free during glacial stages? and (2) if the polar sea ice were to disappear, would the Arctic Ocean remain ice-free?

At present, the answer to the first question has not been determined. However, a large body of evidence relating to the thickness of the Canadian Arctic icesheet suggests most strongly that a local ocean source of moisture must have been present. Only the ice-free Arctic Ocean satisfies this requirement.

The answer to the second question can be attacked by methods of theoretical meteorology. By the use of such methods, proved successful in handling problems of current meteorology, D. Shaw and I attempted an evaluation of the effect of the Arctic heat budget on sea-surface temperatures for present ice-covered conditions and a hypothetical ice-free Arctic Ocean. Results are shown in Fig. 15–20. To test the method, we first calculated the expected present temperature of the Arctic Ocean from the simple heat-balance factors (p. 403). The annual march of temperature shown by the solid curve matches almost exactly the observed monthly temperatures of the polar sea. This result shows that the theory used appears to be workable. We then calculated the annual march of temperature of the Arctic Ocean with the assumption of no ice cover, and made appropriate changes in the terms for reflection of sunlight, increased heat mixing in the ocean, altered heat loss through changed evaporation, and so on. The results, which are indicated by the upper broken curve, show that temperatures would remain above freezing throughout the year if the ice were removed. The lower broken curve shows present sea-surface temperatures of the sea off Jan Mayen Island in the ocean just off the permanent Arctic ice cover. This water is ice-free all year off Jan Mayen so that the close agreement between the calculated ice-free arctic temperatures and the observed Jan Mayen temperatures suggest that the calculations are reliable and that an ice-free polar sea would remain ice-free.

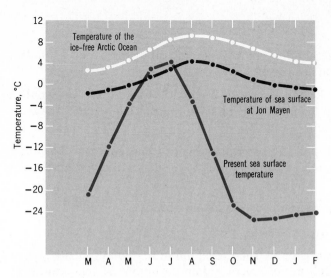

FIG. 15-20. Comparison of curves of the annual march of temperature for the present ice-covered Arctic Ocean, the theoretical ice-free ocean, and conditions at Jan Mayen, an island in the present ice-free northern boundary of the Arctic and North Atlantic Oceans.

Forecasting the Future

We can now try to answer the question of whether we are at present simply in the latest interglacial stage, with other glacials to follow, or whether the world has finally emerged from the Cenozoic Ice Age. According to the Milankovitch theory, fluctuations of radiation of the type shown in Fig. 15-19 must continue, and if this theory is correct, future glacial stages will occur. According to the theory just described, as long as the locations of the north and south poles retain their present thermally isolated locations, the polar latitudes will be quite frigid; and as the Arctic Ocean continues to oscillate between ice-free and ice-covered states, glacial-interglacial climates will continue.

Finally, regardless of which theory one subscribes to, as long as we see no fundamental change in the late Cenozoic climate trend, and the presence of ice on Greenland and Antarctica indicates that no change has occurred, there is no reason not to expect that the fluctuations of the past million years will not continue.

STUDY QUESTIONS

15.1. Explain the factors that control the amount of insolation actually absorbed at the earth's surface.

15.2. Describe very briefly the general climatic conditions that would be expected in the doldrums, horse latitudes, trade wind belts, and belts of prevailing westerlies.

15.3. By means of graphs of temperature against time, compare the daily and annual march of temperature for marine and continental climates.

15.4. What is meant by temperature deviation?

15.5. Define the "growing season."

15.6. Can you explain why the time variation of precipitation at a given place is so much more irregular than variations of temperature?

15.7. Also, why are the geographic variations of precipitation more irregular than that of temperature?

15.8. What are monsoon winds and how do they affect precipitation?

15.9. Explain the P/E ratio and the term "water balance."

15.10. Describe the simple temperature-climate classification.

15.11. Given the Köppen climate symbols for the following places: the Sahara desert, the coast of California, northern Siberia, Nebraska, and Arizona.

15.12. What would be the Thornthwaite climatic province of central Brazil, Arizona, South Dakota, and New York?

15.13. Explain the Landsberg dynamic climate classification scheme.

15.14. How many "degree days" are represented by a day in November in a place where the mean temperature is 55°F?

15.15. What is a fundamental geographic difference between the general regions that were glaciated in the Paleozoic and Pleistocene Ice Ages?

15.16. Describe the main geographic temperature change that occurred in late Cenozoic.

15.17. What is meant by glacial and interglacial stages? Do we live during either of these today?

15.18. What are the evidences that large sheets of ice covered so much of the continents in regions that are ice-free at present?

15.19. Describe the sea level changes that occurred as a consequence of glaciation and deglaciation of the continents.

15.20. Summarize some of the theories of ice ages.

chapter 16

What we are doing to the air we breathe. Industrial effluent along the shore of Lake Erie. (Photo by T. J. Henderson, Atmospherics Inc.)

Man and the Earth

"Never let it be said
And said unto your shame
That all was beauty here
Until you came."
           ANONYMOUS

In the preceding 15 chapters we examined the nature of the earth—our physical environment—in a fairly comprehensive way. We considered the interaction of the many physical forces that have given the earth its present appearance. In the case of the oceans, we examined briefly some important environmental relationships of their plant and animal content. This chapter is more of an essay on the interaction of man with the environment. Each of the areas touched on would require a full chapter or even a book for the more formal exposition given the preceding chapters.

PHYSICAL AND BIOLOGICAL
INTERRELATIONSHIPS

It seems quite safe to say that the earth is unique in the solar system. It is at just the right distance from the sun and has just the right kind of atmosphere to provide the necessary temperature, moisture, and oxygen to maintain the great variety of life that is present now and has been present for hundreds of millions of years. At almost no time in the past three billion years or so is it possible to completely separate the physical and biological processes involved in the evolution of the earth's crust, atmosphere, and oceans from the evolution of life itself. A few examples of how the physical and biologic realms interacted prior to the coming of man should set the background for examining the impact of man on the environment.

Vegetation and Crustal Processes

In Chapter 6 we examined the all-important process of rock weathering and noted that the decay of silicate minerals by a process involving carbonic acid (H_2CO_3) is one of the most important of the processes of rock decomposition and indirectly of rock disintegration. Although much CO_2 from the air can dissolve in rainwater to form carbonic acid, most of this acid as well as other organic (humic) acids accumulate in the soil and groundwater after decay of vegetation. There can be little doubt that the cycles of rock weathering and erosion were speeded up in the course of earth history as vegetation developed and flourished. In turn, the very presence of weathered rock provides the mineral nutrients for the stimulation of plant growth. Weathering and vegetation growth are thus closely related processes. Since weathered rock is eroded more readily than bedrock and often accumulated in geosynclines involved in mountain building, vegetation certainly played its part in such a seemingly remote process as the formation of mountain chains.

Vegetation and the History of the Atmosphere

Conclusions from the study of the oldest rocks make it quite clear that the earth's early atmosphere was devoid of oxygen. The earliest forms of life were thus anaerobic, that is, they obtained the oxygen necessary for their metabolism, not by direct respiration from the atmosphere, but by the breakdown of oxygen-bearing mineral compounds. Small amounts of initial oxygen in the atmosphere probably resulted from the decomposition of water vapor, a process in which hydrogen and oxygen were separated from the H_2O molecule.

With the development of green plants, the process of photosynthesis began to liberate increasingly larger amounts of oxygen as a byproduct of the formation of carbohydrates from CO_2 and H_2O. We owe most of the oxygen in the atmosphere today to the effect of green plants on the environment. The process still goes on, but the rate of natural usage of oxygen just about equals the rate of its production.

Limestones and Marine Fauna

We noted that limestones are one of the three main types of sedimentary rocks. Some limestones have been formed of pure $CaCO_3$ precipitation in the oceans, but much of the total limestone accumulations are all or in part formed of skeletal remains of single- and multicelled marine animals. Many limestones are the products of small zooplankton, particularly foraminifera, that accumulate as limestone ooze on the sea floor. Other limestones have formed from large coral reefs—great limestone masses are composed of the skeletons of small coral, such as those that live in tropical seas today. After burial by later sediments, coral reefs have often developed some of the richest petroleum traps discovered by the oil geologist.

Fossil Fuels

The maintenance of the machinery of modern civilization requires a huge amount of energy. Coal, which is the carbonized end product of the compression and alteration of thick accumulations of plant remains, was the initial main source of the energy needed to drive the industrial revolution. With the discovery of petroleum, oil became the lifeblood of industrial effort. Oil is believed to be the end product of the decay of countless small marine animals following their burial in the sediments of the sea. Today's natural gas is another of the fossil fuels formed in the same manner as petroleum, with which it often occurs.

Unfortunately, as discussed later in this chapter, not only does the combustion of fossil fuels release energy, it also releases huge amounts of combustion waste products that have caused most of the pollution of the once-pure atmosphere.

Ecological Systems

Ecology is the science of the interrelationships of organisms with each other and with their environment. The term ecosystem has been used for a particular environmental unit where the resources available are utilized or recycled by the plant and animal populations of the unit. The physical size of the unit may be large or small and is determined by the areal extent necessary for a compatible population to get adequate resources from the environment. Hence an ecosystem may be a lake, a forest, a mountaintop, or an intermountain valley, and will depend on both geography and climate.

As an example, an environmental unit on land will consist of plants acting as the primary producers of food through the use of sunlight, moisture, carbon dioxide, and soil nutrients. Herbivorous creatures, who consume vegetation, are themselves consumed by carnivores. The wastes and ultimate decay of these primary and secondary consumers complete the cycle by providing plant nutrients following bacterial decay.

In Chapter 14, we examined a similar cycle of marine ecosystems in which phytoplankton are the primary producers; marine bacteria then return nutrients to the sea by their decay of larger organisms that depend upon phytoplankton. As an interesting phase of the ecological cycles, a rather delicate bacterial balance must be maintained. Their overgrowth cause self-destruction. This is prevented to a large extent by the action of protozoa (single-celled animals) that ingest bacteria and keep their numbers under control.

Thus we see that for billions of years a nice environmental balance has developed between elements of the biological and the physical environment. In natural societies, ecosystems tend to flourish until the environment changes. Natural modification of the environment has been frequently expressed as glaciation and great crustal deformation with consequent changes in the ecology of living things.

The Road to Man

By definition we regard man as the culmination of biologic evolution. Biologic changes directly in line of man's development are thus often called "progressive." The direct evolution of man begins with the development and rise of the vertebrates in the Ordovician period (See Table 7-1). Sedimentary rocks of this period contain fish fossils representative of the earliest known vertebrates or backboned animals. A group of fish (the crossopterygians) developed bony fins and internal swimbladders that were connected to their mouths. Such fish could survive droughts by gulping air through their mouths and by using their bony fins to crawl in a primitive fashion from "puddle to puddle." From these fins and swim bladders developed the limbs and lungs of higher vertebrates. First arose the amphibians that were partly but not completely adapted for land living. From a "progressive" amphibian the reptiles arose, followed by a progressive mammallike reptile that in time gave rise to mammals (see Table 7.1). Within the mammals, the primates, the group including man developed. Drawings of the skeletons of fossils from fish to primates are shown in Fig. 16-1, where the evolutionary trend is indicated by similar anatomical features.

The evidence read from the fossil content of rocks indicates that natural environmental upheavals were often accompanied by biological disruptions in which new forms of life were able to develop and flourish. The process of biologic evolution appears to be closely and probably causally related to such revolutions of nature. For example, the Mesozoic Era was a time of widespread low continental elevation and general global warmth. These conditions promoted the dominance of great reptiles, of which the dinosaurs are the most famous examples.

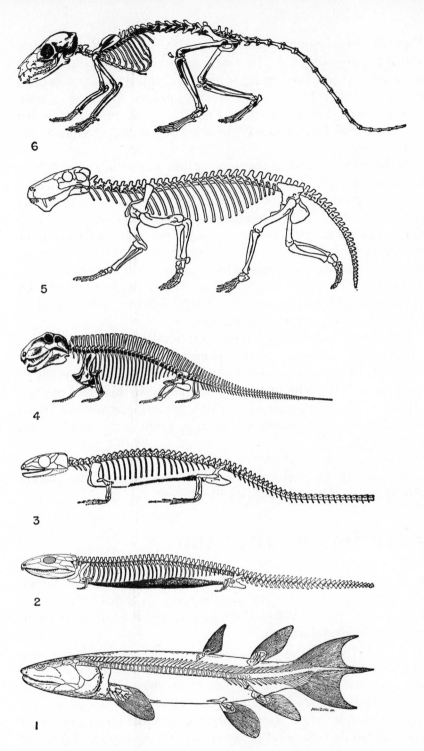

FIG. 16-1. Drawings of fossil forms showing the evolution of vertebrate life from fish to primates. (1) is a "lung-fish," the group from which higher forms of vertebrates evolved; (2) a primitive amphibian; (3) a primitive reptile; (4) a more advanced reptile; (5) an advanced reptile from which mammals evolved; and (6) a tree shrew, a very general type of mammal. (After A. Romer in *Science*.)

Then worldwide mountain-building and continental uplift in general terminated the Mesozoic Era. The vast geographic and climatic changes that occurred so altered the ecology of reptilian habitats that most became extinct. Of some fourteen groups of reptiles present in Mesozoic time, only five survived into the Cenozoic Era. The five include the familiar snakes, turtles and lizards as well as the less-familiar crocodiles and alligators.

With the disappearance of the giant reptiles, a new class of vertebrate life, the mammals, began to achieve dominance. Rapid evolution within the mammalia produced the broad diversification of four-footed animal life so familiar today. Of the many groups of mammals, the *primates* are the most significant on the road to man. This group includes the monkeys, lemurs, apes, and varieties of fossil and recent man. With the exception of man and the baboon, the primates lived and traveled through the trees. Such a form of locomotion was very conducive to the natural selection of great agility, acute stereoscopic vision, and different uses of the fore and hind limbs in the course of evolution.

Within the primates, man is most closely related to the great apes. The evolutionary emergence of true manlike creatures from the more apelike ancestors appears to have been completed by the beginning of the Pleistocene epoch. The first true man, judged by his erect posture, was a form found in Java known as Java Man, or more technically as *Homo erectus*. Examples of the evolution of primate skulls culminating in man is depicted in Fig. 16-2.

The rise of man is also closely related to great environmental changes. *Homo erectus* and later species were found in rocks of the glacial-interglacial stages of middle and late Pleistocene.

THE IMPACT OF MAN ON THE LAND

Gradually, from the caves of tropical and semitropical plains and forests man emerged to flourish and dominate the earth. As global climate ameliorated following the last glacial stage, man spread over all of the continents (except frigid Antarctica).

We have tried to develop the idea that in their natural states man and all of the other organisms live compatibly with one another and in harmony with nature. But with the rise of civilized man, the delicate ecological balance of nature suffered a vast blow. Three major changes occurred: (1) an upset of the natural back pressure that limits population to a size compatible with what the environment can support; (2) a vast increase in the quality and quantity of waste material that now severely threatens the global environment and the related ecosystems; and (3) a large and often irreplaceable loss of natural materials in fueling the basic and the luxury needs of the rapidly expanding population.

In the conversion of raw materials to industrial and domestic use, waste products have virtually saturated much of the environment. It has been estimated, for example, that each American child places 50 times the burden on the environment as does a child in India, because of the great difference in the quality and quantity of consumption.

There are many aspects of man's use of the solid earth or land. These uses involve habitation, waste disposal, agriculture, mineral resources, engineering structures, and recreation. From the standpoint of habitation alone a pressing problem has arisen. In 1650 the earth was peopled by about a half billion individuals. An increase to about seven billion is anticipated by the year 2000—or one individual per habitable square yard. Because populations are not distributed uniformly over the earth but live mostly in urban societies, it is estimated that some 80 percent of the population will reside on only 2 percent of the land.

Use and Disposal of Waste

The other aspects of land use listed above also stem directly from the need to provide for the present and growing society. No satisfactory means has been developed to accommodate the huge waste pile that exists now and that will inevitably be produced in the future. At present (1970), Americans each generate five pounds of garbage and trash per day, double

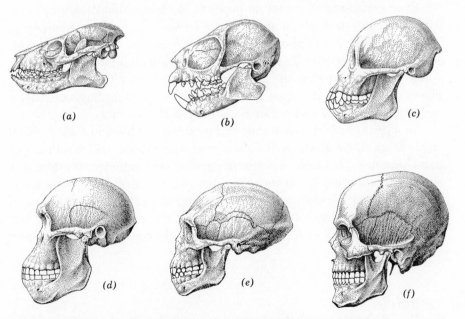

FIG. 16-2. Series of skulls of primates showing the evolution of man within this mammal group. (a) is a fossil lemur; (b) an Eocene age tarsioid; (c) a Miocene ape; (d) is *Australopithicus* an important "link" in the ape-man transition; (e) is *Pithecanthropus* of mid-Pleistocene glacial age, the first of the true men; and (f) modern man. (After A. Romer in *Science*.)

the amount in 1920. An increase to eight pounds is forecast by 1980. For the most part, this garbage tends to be dumped in the most inconspicuous places that can be found. Most of this huge waste can and should be reused. Organic waste must be utilized to replenish the soil. Scrap metal from cans, vehicles, and implements of all kinds can be reprocessed. The degree to which reuse does go on today is relatively minor, especially when we consider figures predicted for 1980 of 48 billion cans, 28 billion bottles, and 7 million junked vehicles per year in the United States.

Agriculture and Engineering

Because of huge wastes in items such as paper, tremendous deforestations have occurred. Despite replanting in the course of lumber operations, huge forests appear to be gone forever. National heritages, such as the great California redwood trees, are under continual threat by exploitational forces. Once lumbered, these magnificent trees cannot be easily replaced. In another direction, the rapid expansion of urban areas has been consuming the natural green environments that once surrounded cities. The preservation of the earth's beauty must be planned for and fought for as cities spring up and older ones expand. Hand in hand with the preservation of our green areas goes the preservation of all the natural wildlife that inhabit them.

Even consideration must be given to the use of land marginal to a wilderness. For example, a great jetport adjacent to a region planned to be forever wild might soon drive most wildlife away. A monumental decision in such a situation was made in the abandonment of plans for a new great airport near the northern border of the Florida Everglades. The combination of drainage and general activity might have destroyed one of the unique wildernesses remaining in this country.

In a further effort to maintain our growing mechanical society, great engineering structures have been required. Huge dams, which on the one hand have provided limited recreational facilities and water for domestic and industrial use have at the same time, forever destroyed large tracts of unsurpassed beauty on the other. In recent years even the collossal Grand Canyon region has been threatened by those who recommend still another dam on the Colorado River to provide water for irrigation and drinking.

Mineral Resources

Finally, we should take note of that area of land use perhaps most closely connected to the geologist, the exploration and use of the earth's store of minerals. The term as used here includes all inorganic material taken from the land, as well as those mineral materials used for fuels, for construction, and for metals.

Coal and oil have been the principal fuels of man. Both are of organic

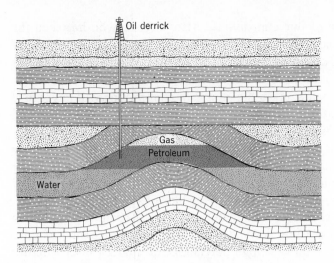

FIG. 16-3. A typical anticlinal oil trap. Petroleum always occurs with salt water. Where a permeable sandstone layer is present the petroleum rises above the water toward the crest of the anticline. If natural gas is present, it too becomes concentrated, but above the petroleum.

origin. Coal has formed from the compaction and decay of huge masses of vegetation that flourished in great abundance at relatively low elevations. Following marine submergence and burial beneath sediments, a decay process removed most of the volatile components, leaving residual carbon and mineral material that remains after combustion. When metamorphosed in a fold mountain belt, the primary bituminous (soft) coal is converted to anthracite, a denser, harder form with still fewer volatiles.

Following its discovery, petroleum was quickly used to feed the fires of the industrial revolution. As steel became the skeleton of our mechanized civilization, petroleum became its life's blood. This fuel has been formed from the decay of huge numbers of small marine animals during certain periods of the geologic past. Rocks of certain ages (for example, the Ordovician, Devonian, Cretaceous, and Tertiary) are locally quite rich in petroleum content. While coal is mined directly as a solid, oil is obtained only by drilling, often to great depths, until the appropriate rock layer is reached.

Even if present in the rocks, the petroleum must be concentrated in amounts that can be pumped up to the surface. The concentration is by natural means only. Oil always occurs with salty, presumably original, marine water. If a proper geological structure is present, the petroleum, being lighter than the water, will rise and become concentrated in the pore spaces of the rock, as illustrated in Fig. 16-3.

But coal and oil are not inexhaustible, despite the occasional discovery of new oil fields, and at the same time their use contributes to the major pollution of our atmosphere. Also, anyone who has seen the visual pollution of scenery now covered by porcupinelike assemblage of oil derricks such as cover the hills around Los Angeles, must deplore this contribution to our culture. But derricks can be removed someday. However, the vast black scars left in the course of coal-mining will never heal in the forseeable future.

With the harnessing of atomic energy for other than military use, optimism about the fuel problem developed. This optimism so far has been short-lived because of the thermal effects on inland waters taken in to cool the atomic reactors and also because of the high toxicity of the radioactive waste material as well as the lack of suitable waste storage.

Other examples of nonmetal materials are glass-sand and building stones. Most of our glass is derived from industrial processing of pure sandstones—rocks with a very high percentage of quartz (SiO_2). The use of glass in buildings and motor vehicles and the most wasteful use in expendable bottles not only depletes the original raw material but also requires the destruction of natural beauty where the sand is quarried. The tremendous edifices of concrete and our "marvelous" superhighway system have already depleted large reserves of sand and gravel, the principal aggregate material of concrete. Some of the most readily available deposits of sand and gravel have been the hills (called *kames*) formed of glacial deposits. Many of these forms, which are of natural beauty and great geologic interest, have vanished because of the insatiable requirements of our "civilized" culture.

Occasionally, some planning is introduced into the destruction of nature. When huge volumes of Manhattan schist were excavated for the New York City subway tunnels, a large quantity of this rock was trimmed and used in the construction of the City College of New York. When the Delaware Aqueduct (the long water tunnel supplying New York City with most of its water) was being constructed, rock excavated from the deep tunnel was crushed to make the aggregate for the concrete lining of the tube. Admittedly, cost saving was a factor in these examples, but they do show that more complete use of our resources may go a long way toward conservation.

To say a word about our metallic resources, recall that most ore deposits (workable ocurrences of metal-bearing minerals) are found in rather rare natural accumulations of minerals. As with coal and oil, it has taken nature a very long time to accumulate metalliferous mineral deposits. Remember that aluminum and iron are the third and fourth most abundant elements in the earth's crust. Perhaps fortunately, they occur in silicate minerals from which extraction is very difficult. Hence, economically useful deposits of these metals are the much rarer but simple oxides. Only recent new finds here "saved" them from depletion in the very near future. But how long can we continue the discard of tin-plated iron cans before the rarer tin oxides vanish? The list goes on and on.

THE IMPACT OF MAN
ON THE ATMOSPHERE

The ravaging of the land, of its natural wealth and beauty, despite its great concern to conservationists and naturalists, is quite secondary to the effect of man on the air and water environment. Both these media are so critically essential to the maintenance of life and its ecological balance in the environment that the accumulating effects of man have become very urgent.

Man's impact on the atmosphere is double-pronged, involving compositional changes and thermal changes. The first is the more obvious and once again is the result of waste discharge in the form of gasses and finely divided solid particles (particulates). Most of the waste material spewed into the air is a product of fuel combustion. Locally, particularly noxious volatiles issue from industrial processing.

The inherent problem in pollution is that prior to man's departure from the natural state, all processes in the biologic world that involved energy consumption were internal to the various organisms that inhabited the earth. Waste material produced in the use of energy sources were a normal part of environmental ecology. It is the by-products of external energy conversion that provide the insult to nature. On a small scale the atmosphere can accommodate pollutional insults, but the ever-increasing quantity of wastes has strained its natural capacity of absorption and dispersal. The present (1970) annual quantity of combustion pollutants is estimated at 200 million tons and is mostly added to the air in relatively crowded urban centers.

The Pollutants

Carbon dioxide (CO_2) is the direct result of the complete combustion of carbon fuels. Although this gas is produced in huge amounts by motor vehicles and all of the industrial and domestic fuel combustion, it is not directly harmful. Its effect is primarily on the heat content of the atmosphere, as explained in the next section. Carbon monoxide (CO) forms from the incomplete combustion of carbon. It is chemically active and very poisonous. A health hazard occurs with a concentration of only 100 parts per million over several hours. Such a condition exists in smoke-filled rooms and poorly ventilated areas with strong motor-vehicle exhaust. Sulfur dioxide (SO_2) and sulfuric acid (H_2SO_4) are produced from combustion in smaller amounts than CO, but they are much more toxic. When sulfuric acid is inhaled during respiration, permanent tissue damage can occur. Hydrogen sulfide (H_2S), a common industrial waste by-product (characterized by a typical rotten-egg smell) is quite lethal in large doses and is known to have been responsible for many deaths and more illnesses. Many processes lend themselves to the formation of several oxides

FIG. 16-4. The edge of a smog layer between the California coast and the mountains to the east. The effect on visibility is obvious. (Photo by T. J. Henderson, Atmospherics, Inc.)

of nitrogen, of which nitric oxide (NO_2) is the most toxic to life. Sunlight decomposes nitrogen oxides into ozone, a very active and poisonous form of oxygen. Ozone concentration is one of the hazard indexes of smog concentration. Certain industrial processes generate hydrogen fluoride (HF), one of the most corrosive of chemicals, which can cause plant damage even in concentrations of one part per billion. Carbon disulfide (CS_2) is a foul-smelling and noxious gas produced in large quantity in paper mills.

In addition to these inorganic vapors, a number of equally toxic volatile compounds of an organic nature are produced. These include ethylene, formaldehyde, and a number of solvents. Ethylene is an important hydrocarbon found in the exhaust fumes of automobiles, buses, and trucks. A few parts per billion damage flowering plants, many of which are important food-bearing types. Formaldehyde, a closely related chemical, is a very irritating component of smog and is generated in stock-yard incinerators. Organic solvents are found in the air in industrial regions and vary from being simply irritating to quite poisonous.

With the advent of nuclear bomb testing and nuclear power plants, an entirely new threat of potentially high toxicity has arisen. No one at present can predict the real hazard from radioactive gasses that are liberated. The dangers are both medical and genetic.

Finally, in addition to the true gaseous pollutants, there is the huge mass of airborne soot, which is finely divided and very active carbon particles. Carbon has the ability of absorbing or sticking to poisonous gas molecules, particularly the heavy hydrocarbons that often form simultaneously in the combustion process that generates the soot. Extremely poisonous vapors that would often be filtered out in the upper respiratory tract are carried deep into the lungs by fine carbon particles. Prolonged and renewed contact is quite hazardous. In large cities present soot-fall rates can be one pound per square foot each year. When averaged over a square mile, this comes to 25 million pounds of soot per year—and when extended to the size of an average city, the annual soot fall is in hundreds to thousands of millions of pounds.

Despite all of man's contribution to the souring of the atmosphere, we must not forget that nature helps complicate the local pollution problem through the effect of atmospheric temperature inversions. The trapping of pollutants by inversions was explained in the case of California smog on p. 420.

Will We Have Enough Oxygen?

So far we considered only the addition of pollutional components to the atmosphere. It has been argued that all of the combustion referred to above would seriously and significantly deplete the oxygen supply of our atmosphere. This seems all the more serious when the effect of insecticides in destroying the blue-green algae of the sea is included. Because atmospheric oxygen is a by-product of photosynthesis, a decrease in marine algae might seriously upset the balance of the production and use of oxygen in the atmosphere and oceans.

The oxygen problem was investigated by W. S. Broecker, who reached the optimistic conclusion that no serious oxygen depletion is in sight. According to Broecker's analysis, the present annual consumption of oxygen by natural processes just about equals the annual rate of production by photo-synthesis. If man were to so increase his industrial activities as to completely burn all known organic fuels, less than 3 percent of the available oxygen would be used. Then, stored in the ocean is a relatively large oxygen content amounting to about $\frac{1}{250}$th that in the atmosphere. Hence, the present store of oxygen in nature seems more than adequate to persist with little change into the foreseeable future.

The Thermal Problem

In Chapters 11 and 15 we examined the factors that determine the air temperature at the earth's surface. The present mean global temperature of 59°F (15°C) is maintained by a nice balance (the earth's heat balance) between the solar radiation received by the earth and the amount stored in the atmosphere and ultimately reradiated. A significant change either in the amount received or the amount stored in the atmosphere (the greenhouse effect) will change the temperature we experience.

As explained previously, water vapor and carbon dioxide are the principal atmospheric absorbers of radiant energy, and of the two, water vapor is far more important. Any process increasing the quantity of these gases will increase atmospheric heat absorption and hence the temperature. In the past hundred years, the amount of carbon dioxide in the atmosphere increased by about 15 percent. The immediate increase of temperature of 1.8°F from 1900 to 1940 was at first blamed on the carbon dioxide produced as a consequence of the industrial revolution. Then from 1940 to 1960, global temperature decreased by 0.6°F.

In part, the possible increased greenhouse effect may be offset by increased atmospheric dust pollution, which may decrease the amount of sunlight reaching the earth's surface and lower atmosphere. An interesting relationship between atmospheric dust and temperature change is shown in Fig. 16-5.

An unresolved but very cogent problem has arisen in connection with the possibility of regularly scheduled supersonic aircraft (SST) operation in the stratosphere. It has been estimated that the hourly combustion of fuel by an SST will release about 83 tons of water vapor and 72 tons of carbon dioxide into the lower stratosphere. It is not clear at present whether these waste products will increase the greenhouse effect and warm the lower atmosphere or develop clouds that will reflect sunlight

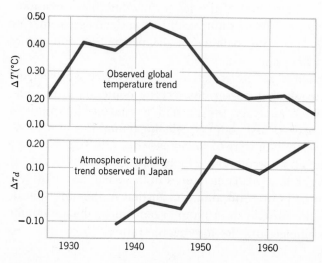

FIG. 16-5. A comparison between global temperature and atmospheric pollutants which implies a temperature trend related to increased pollution. (After P. Furukawa, National Center for atmospheric Research.)

and cool the atmosphere. The entire problem of the thermal effect of pollution is clearly complicated by a tangle of the factors involved—will they cause a net warming or cooling, or balance each other, leaving no significant change?

In addition to the effect of air composition on radiant energy from the sun and earth, there is the effect, more readily estimated, of the direct atmospheric absorption of energy released through the activities of man. In a completely primitive society, an individual puts out only the energy resulting from food utilization. This amounts to about 100 watts per person, put out in the form of heat, or the energy consumed by an average electric light bulb. In societies where energy is added to the air from domestic and industrial technological use, a much greater heat output occurs. The world average, including all societies, amounts to a heat output of more than 1000 watts per person.

For 2 billion people the total heating thus amounts to 2000 billion watts (2×10^{12} watts). In the highly industrialized regions of United States, the energy consumption is of course much greater than the average and equals 10,000 watts per person. If world population increases in the near future to 5 billion people, requiring more industrialized energy consumption, the amount of heat added to the air will reach one hundredth that of the natural radiation balance now existing over the lands.

At present, energy use is increasing at the rate of 4 percent per year. If this continues and we do not run out of usable energy, the heat added to the atmosphere will increase to one third of the natural radiation balance in 200 years. Very serious climatic consequences would occur well before this. Apart from the effects of actual warming of the air, the melting of the Antarctic and Greenland ice caps could raise sea level more than 100 feet.

These are quite firm estimates based on rates of population and power increase and must certainly be considered with the thermal effects of compositional pollution.

THE IMPACT OF MAN ON THE WATER SUPPLY

The water problem, like so many others, is twofold. First, will we have enough water for domestic and industrial progress? Secondly, can we maintain a water supply that will be pure enough for organic consumption?

Before we examine these problems, let us review the nature of water and water supply. If the earth is unique in the solar system in having an oxygen-rich atmosphere as well as being at a distance from the sun that is just about right for a good range of living temperatures, it is also unique in possessing a relatively large supply of water. Water forms the bulk of living cells and tissues; it is a nearly universal solvent and thus provides the vehicle by which nourishment is carried into living cells. It is incidentally of great secondary value to man in providing a medium of recreation and transportation. By its property of absorption of radiant energy, water in vapor form provides most of the greenhouse effect, which maintains the lower atmosphere at a temperature more than 70°F higher than it would otherwise be.

In Chapter 6 we examined the huge weathering and erosional effects of water; in Chapters 11 through 14 we considered its relationship to weather and climate and its behavior in the oceans. In Chapter 4, we looked briefly at the ionic structure of the water molecule and noted the asymmetric nature of the ion distribution. Because of this ionic structure, when water freezes and crystallizes it expands, and in so doing, it increases its volume by about 10 percent. This expansion on freezing, unique in nature, causes ice to float. If water were denser in solid form, ice in polar and cold temperate latitudes would become heavier than water and sink to the bottom of seas and lakes. The insulating property of water would protect the bottom ice during the summer. Ice thickness would increase each winter until much of the earth's water would be solidly frozen! One can imagine the climatic consequences of such an occurrence.

In this section we will examine what man has done to the available water supply, starting with the kinds of sources water is drawn from.

Our Water Supply

Although the seas are our ultimate source of water, they are hardly used, owing to their high salinity. Most of our water is withdrawn at one of several stages in the hydrologic cycle.

1. Freshwater lakes and large streams, where present, have always been a most welcome and easy-to-obtain water source. Unfortunately, expanding populations and water requirements in developed countries have greatly outstripped this supply except in cases of very large lakes as the Great Lakes and similar water bodies.

2. Artificial lakes created by the damming of streams have provided one of the largest sources of urban water. Aqueduct conduits then transmit the water through relatively large distances to the areas of consumption. As early as 312 B.C., Rome recognized the need to reach out for water and began its famous aqueduct system. By 226 A.D., 336 miles of conduits, the longest 62 miles in length, had been constructed. Even earlier, countries in the Mideast desert lands initiated the use of aqueduct tunnels. Possibly the earliest was the tunnel supplying water to Nineveh built in 800 B.C.

In the United States, the Delaware Aqueduct System of New York City transports most of the city's water from large dams on the Delaware and other rivers. The water tunnel, which begins in the Catskill "Mountains," is in places 1000 feet below the surface. In the West, Lake Mead on the Colorado River is one of the largest artificial lakes that provides valuable water for domestic use and irrigation hundreds of miles away in Southern California. In the South, the Tennessee Valley dam system, primarily used for power, is another of the well-known reservoir sources. For each of these mentioned, hundreds of smaller but locally important dams have been constructed with resulting artificial lakes. Recreational fishing and boating have become important by-products of the developments.

3. Throughout much of the world, where lakes or adequate streams are rare or where dam construction is not feasible, simple groundwater wells and artesian-water wells have been the basis of water supply. Simple wells are openings cut below the water table (See Chapter 6). Water-table levels vary seasonally, fluctuating with the evaporation-precipitation ratio. When cut below the dry season (usually summer) water table, groundwater flows from the rock pores and soon fills the opening to the water table level as in Fig. 16-6. When water is withdrawn at a high or continuous rate, especially in commercial use, movement of water through the pore space in the soil cannot match the rate of withdrawal, and the well-water level lowers, giving the condition of *draw-down*. The water-table level near the well forms a *cone of depression* (Fig. 16-6). If the cone of depression falls below the well level, the well will run dry.

Artesian-water wells involve a somewhat more complicated water pattern because the water drawn from the well may have entered the ground at a considerable distance from the point of consumption. As shown in Fig. 16-7, these wells require that a porous, water-bearing rock layer (usually a sandstone) called an aquifer be confined between impervious layers that trap the water in the aquifer. Such a system resembles

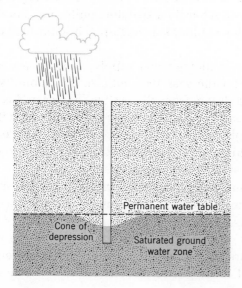

FIG. 16-6. Subsurface geologic relations for a simple well.

the experiment in Fig. 6-18. Rain or snow enters the ground in the highlands and flows through the porous sandstone layer (Fig. 16-7). At the bottom of the well, the pressure is equal to *hdg* (height of water column × density of water × acceleration of gravity) minus the loss due to friction as the water seeps through the pores in the sandstone. If the top of the well is low enough, this pressure might be adequate to maintain a flow at the surface; if frictional loss is too great, pumping is necessary.

Until the huge modern demands on water supply and the complications of waste disposal developed, these sources of water were quite adequate.

Will We Have Enough Water?

The present hydrologic cycle involves the waters in the oceans, atmosphere, and in and on the ground. Only the surface and groundwater phases are of interest to us here. One way of arriving at an idea of the available water is to estimate the water budget of a particular region or the total precipitation minus the amount lost by evaporation and transpiration. The difference is the amount of water that runs off in surface streams or seeps into the ground. When these calculations are made, it is found that some areas are well watered (often places where the demand is least), whereas other regions (often where demand is greatest) experience water deficiencies.

To consider the contiguous United States, only the extreme northwest and the southeast are adequately watered in terms of the demand. Approximately 70 percent of the water that falls on these states is lost by evaporation and transpiration. Of the annual average rainfall over the 48 states of 30 inches, only 9 inches (equivalent to 1300 billion gallons a day) are available for actual withdrawal. Since half of the country is semiarid to arid, the available water is hardly distributed uniformly. For example, about three fourths of this country's average precipitation falls on only one half of the country.

The demands on the available water supply, already very large, are ever-increasing. Figures for the year 1964 indicate that about 340 billion gallons per day were withdrawn from all possible sources in this country. If we divide this by the total population, during the latter part of the 1960s, of some 200 million people, the average water withdrawal is 1700 gallons per person per day. Of this, nearly 51 percent is actually used in industry, 40 percent in irrigation, and only 9 percent as domestic water.

An even larger amount of water is used indirectly in the support of our population. According to John C. Maxwell, the wheat necessary to produce only $2\frac{1}{2}$ pounds of bread transpires about 300 gallons of water to the atmosphere. Also, the alfalfa necessary to produce one pound of steer meat per day requires the much larger volume of 900 gallons. When water for the additional foods plus nonedible crops like cotton are in-

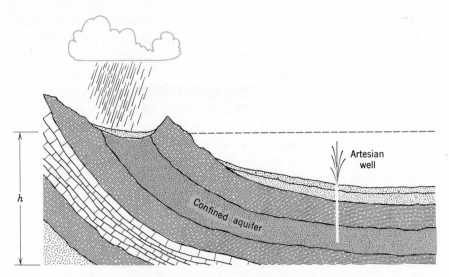

FIG. 16-7. An artesian well. A confined aquifer or water-bearing
permeable layer with access to atmospheric precipitation is required.
The relatively high water pressure at the bottom of the well
causes water to rise naturally within the well. Pumping is usually
required to exhaust the water from the well.

cluded, in addition to that used directly, Maxwell estimates the average
per capita water consumption to be greater than 15,000 gallons per day!
Since these estimates were made (1964), consumption and population have
increased considerably (1970).

We can now make a rough but reasonably accurate estimate of the
water budget for this country. The total daily rainfall over the country,
based on 30 inches per year, is 4300 billion gallons per day. Neglecting
the water lost directly by evaporation and dividing this number by a round
number of 200 million for the population, some 21,500 gallons are available
to support each person. Since more than 15,000 gallons are now used, it
is clear that in the United States, further growth of population and industry will require a change in water habits. If we include in the totals, the
amount lost through evaporation and transpiration, it is evident that only
our reuse of present water has avoided serious deficiencies. Even at
present, at times of drought such as occurred in the northeast in the
mid-1960s, New York City nearly exhausted its available water supply.
Other communities did run out of water and required emergency procedures in the form of actual trucking of water supplies.

Will We Have Pure Water?

Admittedly, there are more aspects of the water-budget problem than we considered above, but in round numbers we are pressing close to the limit. Not only are we using too much, we are also polluting the water beyond belief. Household, industrial, commercial, and agricultural wastes are being introduced into the groundwater supply by industry in recharge wells and by domestic sewage-disposal systems. An even larger amount of waste flows directly into the stream-drainage system of the country. For example, the Great Lakes provide a huge natural water supply for the bordering states as well as for Canada. Being lakes, they represent topographic depressions cut by nature well below the water table. The chief sources of their waters are groundwater and stream inflow. But the groundwater and streams of the northern Midwest also receive the industrial and domestic wastes of large municipalities, and these wastes end up in the Great Lakes.

The same process, of course, takes place along the margins of the country, but the ocean is a much larger "dumping ground" and is also not a source of fresh water. We thus tended to overlook the slower but inexorable pollution of the seas. Freshwater bodies that receive sewage effluent, whether directly or indirectly, soon become seriously polluted. Sewage pollution makes the waters unfit for drinking or recreational use and often adds organisms and chemicals to aquatic life that make them hazardous for human consumption. The great excess of mineral materials, particularly phosphates and nitrates, that are included in most modern sewage provides ideal conditions for eutrophication, in which abundant flowering of aquatic flora occurs, causing a loss of oxygen. Many smaller water bodies have already been choked by such plant growth. The decay of the huge masses depletes the normal oxygen supply of the waters, causing a consequent falloff in both fish and shellfish as the natural ecology is disordered.

The DDT problem was brought into sharp focus when this poisonous insecticide was found in the tissues of fish as far away as the waters around Antarctica. More recently, the bodies of seals found far out on the distant shores of islands in the North Pacific Ocean were found to contain significant quantities of mercury. Closer to home, kitchen tap water in eastern Long Island has been bubbling with detergents present in the well water supply. The list of toxic or simply unpleasant materials in our drinking and recreational waters is long and frightening and needn't be elaborated here in detail.

With the coming of great nuclear power plants, a new form of water damage has occurred—*thermal pollution*. All large air-conditioning and industrial processes utilize water as a coolant, but their effects are rather small. Nuclear plants require the removal of tremendous amounts of heat. Thus, a relatively large percentage of the Hudson and other rivers is drawn in to cool the nuclear generators along the river banks. The water goes in cool and comes out significantly warmer. Local fauna and flora ecologies have been changed noticeably.

CAN WE SAVE NATURE?

There can be little doubt that civilized man has fairly devasted the precious environment that has supported him. Sprawling urban centers in particular have blotted out the beauty of nature and added vast quantities of waste, much of it toxic, to the air, water, and land. There can also be little doubt that man cannot continue in this path much longer without causing self-destruction.

We cannot cure all of the damage, at least not in the immediate future, but we can try to help nature heal the wounds of civilization by (1) minimizing waste of raw materials, (2) decreasing the kind and amount of pollutants, (3) where possible, actually reversing the flow of waste into the environment by recycling and reprocessing, and (4) embarking on constructive and imaginary procedures that improve and utilize existing ecosystems rather than destroy them.

The first three suggestions are closely tied to Federal, state and local reforms. Some reforms enacted or under consideration include (1) controls of the kinds of chemicals in domestic and commercial use that are discharged into the environment as waste; (2) requirements for adequate sanitary sewage treatment prior to discharge into streams, lakes, or seas; (3) requirements for treatment of industrial waste material; (4) reprocessing of organic solid waste for use in replenishing soil minerals; and (5) more adequate control of the uses of the environment to decrease waste (of water, for example).

In addition to controls to save the environment, there is the imaginative use of the environment. As an example, we can look at the seafood farming experiment of O. Roels of Columbia University. Recall that ocean surface waters are quickly depleted of the mineral nutrients necessary for sustaining phytoplankton, which in turn provide food for higher forms of marine life. Because of this, about 90 percent of the fish catch comes from some 10 percent of the oceans—regions where upwelling carries nutrients to surface waters. Roels, a marine biologist, is trying to reinforce nature by artificially pumping deep, nutrient-rich water into surface pools where phytoplankton and forms of seafood life are now growing. If successful, this experiment may represent a major step in increasing the world's food supply.

As regards our dwindling water supply, atmospheric scientists have for two decades been trying to wring some extra water out of the heavens and do what the rainmakers of old pretended to do. Up to the present time only the most modest increases of rainfall have been made through man's efforts, and these increases have always required some special meteorological conditions to begin with.

Rainmaking involves the more complete use of the water already in the environment. By increasing precipitation where more water is needed, a small increase in the rate of water turnover—the hydrologic cycle—is made. This requires tampering with fundamental processes of nature, unlike the simple recycling of solid wastes referred to earlier.

Modern rainmakers have used two different ways of attacking the problem. Recall from Chapter 12 that the condensation process forming clouds requires three conditions: (1) adequate water vapor, (2) cooling of the air to and below the dew point (or temperature at which condensation occurs), and (3) nuclei of condensation. Man can do little about adding water to the air. Hence the twofold attack has involved the other two factors.

Stimulation of rainfall where clouds already exist has been attempted by scientists through "cloud seeding." In this process, small pellets of dry ice (solid carbon dioxide at very low temperatures) are dropped from airplanes into a cloud top. The resultant increased cooling might thus increase condensation, causing cloud droplets to grow to a size permitting them to precipitate to earth. As noted, only very limited but definite increases in precipitation have been triggered by cloud seeding. As an example of our great lack of knowledge, in some experiments, evaporation of large parts of clouds occurred rather than increased condensation.

In other cases, scientists tried to induce precipitation by increasing the numbers of nuclei of condensation present. This is done by burning silver iodide in small furnaces. The resulting smoke contains billions of minute crystals of silver iodide, which are borne by air currents to great heights in the atmosphere. Now silver iodide forms crystals of the same shape as ice crystals. Since water vapor in the air condenses more readily on ice crystals than on other nuclei, the vapor is "tricked" into increased condensation. Far more study and effort is necessary, however, before we can expect really significant returns from this most ambitious plan.

Another example of cooperation with nature lies in the area of power. To gain power, we are polluting the air and water with all of the waste materials and heat referred to earlier. Although windmills have been used by individuals for ages to provide small sources of power, the vast energy of the wind, waves, and tides has never been tapped successfully. But with ever-increasing needs, these sources should be tapped once again.

With all that mankind has learned about nature and with all of the ingenious devices, which as by-products are destroying nature, it is time we turned our full ingenuities to the saving of nature.

STUDY QUESTIONS

16.1. Describe the mutual interaction of rock weathering and plant growth.

16.2. What has been the impact of surface vegetation on the history (and composition) of the atmosphere?

16.3. What is meant by "fossil fuels"?

16.4. Explain the meaning of "ecology" and "ecological systems." Give an example of the physical and biologic factors that may be involved in a particular ecological system.

16.5. Give an example of how a particular ecological system might be upset.

16.6. Summarize briefly the evolution of vertebrates leading to man.

16.7. Give some examples of how man uses the solid earth environment to his advantage and some of the resulting insults caused to the environment as a result.

16.8. What are some of the important pollutants that men's activities have added to the atmosphere.

16.9. What is meant by the "thermal problem"?

16.10. Describe the three main sources of water developed for domestic, commercial, and industrial use.

16.11. Distinguish between a simple well and an artesian well.

16.12. Describe the present water budget as involved in man's total ecology.

16.13. Explain "eutrophication" and the problem resulting from this process.

16.14. Discuss the problem of saving our environment and include some concrete ideas for studying and improving conditions.

appendix A

Properties and Identification of Minerals

PHYSICAL PROPERTIES

Although the definitive nature of a mineral lies in its chemical composition and lattice structure, the identification of a mineral by chemical testing requires laboratory facilities not commonly available. Fortunately most common minerals, as well as many not so common, can be identified quite easily by an examination of their physical properties. After some experience is gained, sight determination of many minerals is usually possible, with no aids being necessary.

Cleavage

The mineral property of splitting along smooth, parallel planes of weakness is known as cleavage. Resulting cleavage planes are flat and shiny, yielding bright reflecting surfaces as they catch the light. Cleavage planes result from characteristic internal atomic structures, or in the case of silicates, from the arrangement of the silicon-oxygen tetrahedra. Variations in the internal structures of different minerals produce variations in (1) the ease with which minerals cleave, (2) the number of cleavage directions, and (3) differences in the angles between cleavage faces. These characteristics are quite distinct and often diagnostic for different minerals. One, two, or three directions of cleavage within mineral crystals are common; four or six, although present, are less common.

Fracture

A break that is not flat and not controlled by a natural plane of weakness is called a fracture. If the mineral is very homogeneous internally, the fracture pattern is usually shell-like in shape and is then called a *conchoidal fracture*. This feature is characteristic of quartz with its very homogeneous, three-dimensional arrangement of the silicon-oxygen tetrahedra.

Hardness

Hardness is the ability to resist being scratched. It is in no way related to the ease with which a mineral breaks. (Brittleness and cleavage determine the ease with which a mineral breaks when dropped or struck with a hammer.) Some minerals are so soft that they can be scratched with a fingernail, whereas others are so hard that they readily scratch glass. Care must be taken to test only the fresh unweathered surfaces, because chemical weathering softens most minerals. Mineralogists have selected ten minerals ranging from the softest to the hardest and have set them up as an arbitrary standard for hardness comparisons. This hardness scale is given in Table A-1. Other convenient tools commonly at hand for testing hardness are fingernail—2.5; steel knife blade—5 to 5.5; and glass—5.5.

In making hardness tests, a fresh smooth surface should be selected for abrasion by the tool. After scratching, the mineral surface should be wiped to be sure the visible trace is a true scratch rather than the powdered tool rubbed off on the harder mineral.

Luster

Luster refers to the quality of light reflection by a mineral. It is defined by a number of descriptive terms whose meanings are probably universal enough so that they have almost a quantitative value. The commonly occurring lusters are vitreous (glassy), waxy (resembling candle wax), pearly, silky, metallic, and earthy.

Streak

Streak is the color shown by a mineral when it is finely powdered. This color can be obtained readily by rubbing the specimen on a hard, flat, but heavily frosted surface. In practice, an unglazed porcelain plate, such as the underside of a house tile, is used and is known as a *streak plate*. The streak is often very diagnostic for minerals, particularly ore minerals, and may differ significantly from the color of the mineral specimen. White minerals always have a white streak. Minerals harder than the streak plate, even if colored, will show no real streak.

Specific Gravity

This is a measure of the density of a mineral and is numerically equivalent to density in the metric system. Again, it expresses the weight compared to that of an equal volume of water, which has a specific gravity of 1. Specific gravity, one of the most diagnostic properties of minerals, un-

TABLE A-1 Mineral Scale of Hardness

Hardness Scale	Mineral Standard
1	Talc
2	Gypsum
3	Calcite
4	Fluorite
5	Apatite
6	Orthoclase
7	Quartz
8	Topaz
9	Corundum
10	Diamond

fortunately requires laboratory apparatus for its determination. With experience, however, one can tell comparative specific gravities by the "heft" of a specimen in one's hand.

Magnetism

A few minerals, particularly certain ores of iron or nickel, are attracted to a magnet. Even if the hand specimen is too weakly magnetic to show this effect, the scrapings may readily respond to a magnet.

Twinning Striations

Certain minerals, particularly the plagioclase feldspars, are composed of innumerable thin, closely joined parallel crystal plates that show as fine parallel striations on those cleavage faces normal to the plates.

DETERMINATIVE MINERALOGY

The procedure of mineral identification, whether for fun, hobby, or serious professional work, is called determinative mineralogy. With only a little experience, a student can learn to apply the physical properties just described to identify unknown minerals. Some practice with known samples is usually necessary in order to acquire a little working knowledge of these properties. Then with the use of tables such as Tables A-2 and A-3, the specimens can be sorted and named. The basis for these two tables is the presence or absence of a colored streak. If a colored streak is present, Table 4-5 is followed. Often, the visual appearance of a mineral is so diagnostic that it can be identified at sight. Occasionally, only one or two properties need be checked. For some a more complete rundown is needed.

TABLE A-2 Mineral Identification: Chart I (Colored Streaks)

Streak	Color	Luster	Hardness	Cleavage		Distinctive Properties	Name and Composition	Geologic and/or Economic Importance
Black	Black	Dull metallic	5.5–6.5	None	*Crystal faces commonly present*	Magnetic High specific gravity	MAGNETITE Fe_3O_4	Iron ore
	Brassy yellow	Metallic	6.0–6.5	None		"Fool's gold" color	PYRITE FeS_2	Most abundant sulfide; source of sulfur
	Lead gray	Brilliant metallic	2.5	3-perfect, cubic		Very high specific gravity	GALENA PbS	Chief lead ore
Red-brown	Red to steel gray	Earthy to metallic	Up to 6.5	None		Varieties: micaceous, fibrous, oolitic, etc.	HEMATITE Fe_2O_3	Iron ore—Great Lakes, Labrador, South America
	Yellow to dark brown	Earthy to submetallic	Up to 5.5	None		Often contains clay-like impurities	LIMONITE $Fe_2O_3 \cdot xH_2O$	Iron ore
Cream to Pale Yellow to Brown	Amber to dark brown	Vitreous to waxy	3.5–4.0	6-perfect,[a] oblique		Heavy mineral with vitreous luster	SPHALERITE ZnS	Chief zinc ore
	Bright green	Vitreous to dull	variable	None		Bright green color; often banded and impure, hence harder than normal; sometimes tufts of needle-shaped crystals	MALACHITE Hydrous copper carbonate	Minor ore of copper
Green	See Table A-3 for details					Light green streak	HORNBLENDE	Important igneous rock-forming mineral

[a]You will rarely be able to recognize all six cleavages even if present. Some of our specimens are crystal aggregates.

TABLE A-3 Mineral Identification: Chart II (White or Colorless Streaks)

Color	Luster	Hardness	Cleavage	Distinctive Properties	Name and Composition	Geologic and/or Economic Importance
Colorless, white, gray, rose, black, etc.	Glassy	7.0	None	Lack of Cleavage Hardness	QUARTZ SiO_2—silicon dioxide	A major rock-forming mineral in all three classes of rocks Glass, optical instruments, sand paper, concrete, etc.
Gray to black	Waxy	7.0	None	Conchoidal fracture Chalky coating	FLINT (a variety of quartz) SiO_2—silicon dioxide	Occurs as concretions in chalk
Usually a deep red	Glassy	6.5 to 7.5	None	Smooth *dull* fracture surfaces Color	GARNET A complex silicate	Sandpaper; jewelry
Olive green	Glassy	Indeterminable with your equipment	None	Commonly granular (like loaf sugar) Olive green color	OLIVINE An iron-magnesium silicate	A major ultrabasic rock-forming mineral
White, tan, salmon, green (rare)	Glassy	6.0	2 good, at right angles	Cleavage Hardness	FELDSPARS ORTHOCLASE $KAlSi_3O_8$ or $K_2O \cdot Al_2O_3 \cdot 6SiO_2$ Potassium aluminum silicate	Chief constituent of granite and related rocks Porcelain, pottery, pottery glass, etc.
White	Glassy	6.0	2 good, almost at right angles	Cleavage; hardness; color Striations on some cleavage faces	ALBITE $NaAlSi_3O_8$ or $Na_2O \cdot Al_2O_3 \cdot 6SiO_2$ Sodium aluminum silicate	Same as for orthoclase
Gray to bluish-gray, dark gray	Glassy	5.0 to 6.0	2 good, almost at right angles	Cleavage; hardness Commonly a massive aggregate Striations on some cleavage faces Bluish-green iridescence on some specimens	LABRADORITE $Na_2O \cdot Al_2O_3 \cdot 6SiO_2$ + $CaO \cdot Al_2O_3 \cdot 2SiO_2$ Sodium calcium aluminum silicate	Major constituent of such igneous rocks as gabbro and basalt, an ornamental stone

Color	Luster	Hardness	Cleavage	Remarks / Tests	Mineral / Composition	Uses
Black, dark green (rare)	Glassy (poor)	5.0 to 6.0	2 good	Cleavages at 60 and 120 degrees . . . Cleavages at 90 degrees . . . Streaks often light green	HORNBLENDE (See also Chart I) / AUGITE / Complex silicates	A constituent of many igneous and some metamorphic rocks
Greenish white	Silky	Indeterminable with your equipment	None	Finely fibrous / Single fibers are white.	CHRYSOTILE (variety of serpentine) / A hydrous magnesium silicate	Asbestos (a heat insulator)
Colorless, white, pink, blue, etc.	Glassy	3.0	3 perfect, at oblique angles (rhombic)	Cleavage; hardness / Effervescence with dilute acid	CALCITE / $CaCO_3$—calcium carbonate	Chief constituent of limestone and marble; fertilizer
Colorless; may be variously tinted.	Glassy	2.5	3 perfect, at right angles (cubic)	Cleavage / Salty taste	HALITE / NaCl—sodium chloride	Common salt, food preservative, chemical industries
Colorless, bronzy to pale green	Pearly	2.0 to 2.5	1 perfect	Cleavage; color / Thin elastic plates or flakes	MUSCOVITE (MICAS) / A complex K, Al silicate	Insulation in electrical appliances and instruments
Black; brown in thin sheets	Pearly	2.0 to 2.5	1 perfect	Cleavage; color / Thin elastic plates or flakes	BIOTITE (MICAS) / A complex K, Al, Fe silicate	Biotite and muscovite are common in igneous and metamorphic rocks
Colorless, white, salmon, gray, etc.	Pearly to glassy	2.0	1 perfect, 2 poor. None if fibrous.	Scratched by fingernail / Flakes not elastic / Some specimens poorly fibrous	GYPSUM / $CaSO_4 \cdot 2H_2O$—hydrous calcium sulphate	Plaster products
White (if pure)	Earthy	Indeterminable Very soft	None	A compact chalk-like powder; color	KAOLIN / A hydrous aluminous silicate	Chief constituent of clay Pottery, bricks, etc.
White to pale green	Pearly	1.0	1 perfect	Scratched by fingernail / Soapy feel / Commonly in flaky aggregates	TALC / A hydrous magnesium silicate	Laboratory sinks and table tops, slate pencils, talcum powder, etc.

appendix B

Properties and Classification of Igneous Rocks

TEXTURES OF IGNEOUS ROCKS

A primary characteristic of igneous rocks is their homogeneous, crystalline nature. Crystals do not simply grow in a side-by-side manner, but are completely interlocked, giving such rock a relatively dense, massive quality. The size and nature of the interlocked grains is referred to as *texture* and is an important feature in the classification scheme usually followed for igneous rocks. Most igneous rocks fall into four broad texture groups.

Phaneritic or Coarse Textures

When magma cools and solidifies in relatively large masses well below the earth's surface, the entire solidification is slow enough for the full and equal crystallization of the components present in the melt. Such rocks only become visible after prolonged erosion of mountain areas has removed thousands of feet of overlying rock. Phaneritic or coarse-textured rocks exhibit mineral grains that are all large enough to be clearly visible to the unaided eye. As a rough, rule-of-thumb estimate, we can say that the individual grains in a rock with granitoid texture, vary around one quarter of an inch in diameter.

Aphanitic or Fine-grained Texture

This texture is applied to rocks that are completely crystalline but whose individual grains are too small to be visible to the unaided eye. They are, however, readily apparent when the rock is viewed microscopically, as with the petrographic microscope. This type of texture, which forms when the magma cools too rapidly for the ions to accumulate into large crystal grains, occurs in surface or near-surface types of igneous rocks.

Glassy Texture

When magma erupts as lava, the very rapid cooling may result in solidification prior to the development of any true crystals. The ions are literally frozen in position within the melt, yielding a rock completely barren of crystal grains. Although in a solid state, such rocks are technically not crystalline solids and fall within a class of materials called glasses—hence the term glassy texture. If a natural glassy rock is remelted in the laboratory and then allowed to cool slowly, it will crystallize into those minerals appropriate to its composition and which it would have formed naturally had it cooled in a deep-seated environment. In the same way, if a coarsely crystalline rock is remelted in the laboratory and then cooled rapidly, it will develop a glassy texture and resemble a natural glass that might have formed as a lava flow.

Pegmatitic Texture

This is the coarsest of all rocks' textures and is associated with rocks in the very last stages of crystallization. Although the pegmatitic rocks commonly encountered in the field exhibit individual crystalline components ranging from a half inch to a few inches in size, it is not unusual to find examples in which individual grains reach a foot or so. More unusually, they are found with grains so large that individual crystals may weigh a great many tons. In the last stages of magmatic crystallization, the melt is rich in the silicate composition of orthoclase feldspar and quartz and occasionally muscovite mica. Intimately associated with this late-stage mixture is a relatively large amount of water, which becomes concentrated in the residue after separation from the magma which has already crystallized. The presence of the water greatly increases the fluidity of the remaining melt as well as lowering the temperature at which final crystallization occurs. Most pegmatitic rocks are composed primarily of the end-stage minerals of quartz and feldspar. Owing to the high aqueous content of the mixture, the highly fluid melt rises into cracks and fissures, forming veins of pegmatitic rock.

Prophyritic Texture

This is a secondary textural term applied to a rock that exhibits groups of minerals with contrasting grain sizes. It results from the magma having undergone two different environments of cooling. After a period of slow cooling, the magma may rise to a higher level and undergo more rapid cooling, resulting in finer grains. The finer-grained material is called the *matrix* or *groundmass*. The larger crystals are called *insets* or *phenocrysts*. Porphyritic textures may occur with any of the primary textures in phaneritic, aphanitic, or glassy rocks.

CLASSIFICATION AND TYPES OF IGNEOUS ROCKS

Igneous rocks are identified particularly in hand specimen on the basis of the texture and mineral composition. The latter is usually reflected in the color of the rock, which becomes important when dealing with aphanitic rocks whose mineral composition is not readily recognized. As with minerals, the actual identification of rocks is a laboratory procedure that requires handling actual rock specimens. We will thus simply review the general scheme of rock classification and dwell on those aspects basic to understanding of the earth's crustal composition and genesis.

Before turning to the actual classification, we can gain an idea of the color criteria by reviewing the chief igneous rock-forming minerals. Quartz, orthoclase and albite are either white or pale shades of salmon pink. The higher the proportion of these minerals, the lighter colored is the rock. The calcium-rich plagioclase feldspars become darker, the higher the calcium content. Dark minerals are hornblende, augite, biotite, and olivine, and the higher the content of this group, the darker is the rock.

A classification scheme for igneous rocks is given in Table B-1. There are almost as many such schemes as there are books on the subject. But the differences are mainly in detail, not in basic organization. It should also be realized that there are not the sharp lines between rock groups implied by the rulings in the table. Rather, there is an almost continuous transition in mineral composition so that the rocks grade into one another. Under mineral composition, only those minerals essential to the standard rock-classification scheme are given. Invariably, rocks contain accessory minerals whose presence or absence does not influence the rock name. Note that the light colored rocks have an abundance of the orthoclase-albite members of the feldspars whose compositions are respectively $KAlSi_3O_8$ and $NaAlSi_3O_8$ and are thus very rich in silicon and aluminum. From this has been developed the term sialic rocks for such silicon-aluminum-rich rocks. The entire group of rocks in this portion of the table is often referred to by the generalized term *sial*. The darker rocks are increasingly rich in magnesium minerals in addition to the silicon always present; hence, the adjective *simatic* is used to modify their descriptions. Collectively, the very dark, magnesium-rich rocks are known as *sima*; the latter are also called *mafic* rocks to describe their high iron (Fe) and magnesium (Mg) composition.

Sialic rocks commonly occur with one or more of the following minerals as accessories: muscovite, biotite, and hornblende. Hornblende and olivine may occur with the rocks of the first column of the simatic groupings.

In Table B-1, the rocks within a vertical grouping have similar chemical compositions and vary only in texture as a consequence of the history of crystallization. Note that three names are emphasized. Granite and granodiorite form perhaps 95 percent of the rocks of the continent. Basalt, a dark fine-grained rock, forms the rock of the ocean basin and may well extend as a more-or-less globe-encircling layer beneath the lighter-weight granitic rock. On continents, basalt is prominent in the form of lava plateaus described in Chapter 3. Basalt is far more common on continents than is granite in the oceans. The rocks classified here are defined in terms of the criteria heading their columns and rows. Thus, granite is a coarse-grained rock containing quartz, orthoclase, and albite, and so on.

Any of the rocks in the table may show true contrasting grain sizes or be phorphritic, in which case the term *porphyry* is added to the name; for example, basalt porphyry. Much of the simatic rock of the ocean basins consists of an olivine basalt porphyry—a basalt composed of olivine phenocrysts.

TABLE B-1 A Classification Scheme for Igneous Rocks

Texture	Mineral Composition				
	Light colored (Sialic)			Dark colored (Simatic)	
	Quartz, orthoclase, and albite	Orthoclase and albite	Albite and Ca feldspars	Labradorite and augite	Olivine
Phaneritic = coarse grained	GRANITE	Syenite	Diorite	Gabbro	Peridotite
	GRANODIORITE				
Aphanitic = fine grained	Rhyolite	Trachyte	Andesite	BASALT	
Glassy	Obsidian Pumice	Obsidian Pumice		Scoria	
Pegmatitic	Granite Pegmatite				

appendix C

Classification of Sedimentary Rocks

TABLE C-1 Summary of sedimentary rock history

Type of Origin	Sediment Source	Lithification Process	Rock
Clastic	Gravel	Cementation	Conglomerate; Breccia
	Sand	Cementation	Sandstone
	Silt	Cementation-compaction	Siltstone
	Clay	Compaction-cementation	Shale
Nonclastic			
Biochemical	Calcarous mud	Compaction-crystallization	Limestone
Chemical	Material in solution	Precipitation	Gypsum
	Material in solution	Precipitation	Rock salt
	Material in solution	Precipitation	Sedimentary hematite

appendix D

Classification of Metamorphic Rocks

TABLE D-1 Classification of the Principal Types of Metamorphic Rocks

Structure		Texture	Principal Foliation-Producing Minerals		Name of Rock	Usually Derived from
Foliated	Slaty Cleavage	Very fine	Mica (Muscovite)		Slate	Shale
		Fine	Mica (Muscovite)		Phyllite	Shale or Slate
	Schistosity	Medium to Coarse	Mica (Muscovite)	The foliation producing mineral constitutes more than 50% of the rock	Mica (Muscovite) Schist	Shale, Slate, Phyllite
			Mica. (Biotite)		Biotite Schist	Iron-bearing igneous or sedimentary rocks
			Hornblende		Hornblende Schist	Mafic igneous rocks or calcareous shale
	Gneissic Banding		Variable—foliation-producing mineral constitutes less than 50% of the rock		Gneiss (e.g., Biotite Gneiss, Granite Gneiss, etc.)	Wide variety of ign. sedimentary or other metamorphic rocks
			Principal Mineral Composition			
Nonfoliated (massive)		Very fine	Mica, Quartz and other contact-metamorphic minerals		Baked Shale (Hornfels)	Shale
		Fine	Serpentine		Serpentine	Ultra-mafic ign. rocks
			Talc, Chlorite		Soapstone	Ultra-mafic ign. rocks
		Fine, Medium or Coarse	Calcite, Dolomite		Marble	Calcareous or Dolomitic Limestone
			Quartz		Quartzite	Quartz sandstone

Index

Numbers with an asterisk are pages containing illustrations.